# Applications of (Big) Data Analysis in A/E/C

# Applications of (Big) Data Analysis in A/E/C

Editors

**Zheng-Yun Zhuang**
**Ying-Wu Yang**
**Ming-Hung Hsu**

MDPI • Basel • Beijing • Wuhan • Barcelona • Belgrade • Manchester • Tokyo • Cluj • Tianjin

*Editors*

Zheng-Yun Zhuang
National Kaohsiung University of Science and Technology
Kaohsiung
Taiwan

Ying-Wu Yang
Zhejiang A&F University
Hangzhou
China

Ming-Hung Hsu
National Penghu University of Science and Technology
Magong
Taiwan

*Editorial Office*
MDPI
St. Alban-Anlage 66
4052 Basel, Switzerland

This is a reprint of articles from the Special Issue published online in the open access journal *Buildings* (ISSN 2075-5309) (available at: https://www.mdpi.com/journal/buildings/special_issues/GV29ERG645).

For citation purposes, cite each article independently as indicated on the article page online and as indicated below:

LastName, A.A.; LastName, B.B.; LastName, C.C. Article Title. *Journal Name* **Year**, *Volume Number*, Page Range.

**ISBN 978-3-0365-8218-4 (Hbk)**
**ISBN 978-3-0365-8219-1 (PDF)**

# Contents

# About the Editors

**Zheng-Yun Zhuang**

Dr. Zheng-Yun Zhuang is a full professor at the Department of Civil Engineering, College of Engineering, National Kaohsiung University of Science and Technology (NKUST). His research interests include OR (operational research) methods/applications, computer science, (big) data analytics, DDDM (data-driven decision making), goal programming, transportation, non-clinical healthcare, supply chain decisions, and property management.

**Ying-Wu Yang**

Dr. Ying-Wu Yang is an associate professor at the College of Landscape and Architecture, Zhejiang A&F University. His research interests include structure health monitoring, structure vibration analysis, and structural detection and strengthening.

**Ming-Hung Hsu**

Dr. Ming-Hung Hsu is a full professor at the Department of Electrical Engineering, National Penghu University of Science and Technology. His research interests include vibration analysis, power systems, electromechanical integration, structural health diagnosis/systems, smart grids, wind power producing, and photovoltaics.

buildings

MDPI

*Editorial*

# Applications of (Big) Data Analysis in A/E/C

**Ming-Hung Hsu [1], Ying-Wu Yang [2] and Zheng-Yun Zhuang [3,*]**

[1] Department of Electrical Engineering, National Penghu University of Science and Technology, Penghu 880011, Taiwan; hsu@gms.npu.edu.tw
[2] College of Landscape and Architecture, Zhejiang A&F University, Hangzhou 311300, China; yangyw06@163.com
[3] Department of Civil Engineering, National Kaohsiung University of Science and Technology (NKUST), Kaohsiung 807415, Taiwan
* Correspondence: wayne@nkust.edu.tw; Tel.: +886-7-3814526

This editorial paper provides an overview of the *Buildings* Special Issue (SI), dedicated to the topic "Applications of (Big) Data Analysis in A/E/C" (where A/E/C stands for architecture, engineering, and construction) and the academic papers it includes. For more details about this SI, please refer to https://www.mdpi.com/journal/buildings/special_issues/GV29ERG645 (accessed on 4 May 2023).

Some of this paper's contents are presented in tabular format for clear organization. Table 1 lists the title and keywords of each article published in the SI. This paper also offers a series of digests based mainly on their abstracts to maintain each study's original contributions, as claimed by the author(s) at publication. Stemming from these reviews, the paper summarises each study's application domains within A/E/C, outlines the phase or role each study plays in the theoretical flow of (big) data analysis, and then concludes with some key observations. This allows for identifying 'hot zones' of research, in which follow-up research and extended studies should be performed, and 'cold zones', awaiting the utilisation and novel application of data and theories/models. Note that in the subsequent text, if not otherwise specified, the discussion of the papers follows the order in which they were published.

Zhuang and Kuo [1] propose and apply a systematic data analysis methodology to analyse experimental data from high-performance concrete (HPC) samples with different admixtures for use as offshore fan foundation grouting materials. Compared to other relevant research, including experimental studies, physics and chemistry of materials studies, and cementitious material portfolio determination studies, this data-driven analysis deeply explores the experimental variables associated with the test data. The authors employ several methods, including correlation analysis, cosine similarity analysis, simple linear regression (SLR) modelling, heat maps, and heat-based tabularised visualisations, to offer a comprehensive and in-depth perspective. The proposed methodology in [1] is easily implementable. The authors validated the results using a pairwise comparison approach (PCA).

The contributions of this work include insights for coherent groups of variables, techniques for double and triple checking, the establishment of a 'knowledge base' consisting of 504 SLR predictive models with their effectiveness (significance) and prediction accuracy (data-model fitness) used in practical applications, an alternative visualisation of the results, three data transforms that can be omitted in future analyses, and three valuable theory-linking perspectives (e.g., for the relationships between destructive and non-destructive tests with respect to the variable categories). The implication that some variables are interchangeable will make future experiments less labour-intensive and time-consuming for pre-project HPC material testing.

**Citation:** Hsu, M.-H.; Yang, Y.-W.; Zhuang, Z.-Y. Applications of (Big) Data Analysis in A/E/C. *Buildings* **2023**, *13*, 1442. https://doi.org/10.3390/buildings13061442

Received: 17 May 2023
Accepted: 26 May 2023
Published: 31 May 2023

**Table 1.** Title and keywords for each paper published in this SI.

| Ref.# | Title | Keywords [1] |
|---|---|---|
| [1] | Unravelling the Relations between and Predictive Powers of Different Testing Variables in High Performance Concrete Experiments: The Data-Driven Analytical Methods | high performance concrete (HPC); data-driven analysis; experimental parameters; pre-project material testing; pairwise comparison analysis; correlation analysis; cosine similarity; predictive regression modelling; heat map; variable transform |
| [2] | An Intelligent Detection Logic for Fan-Blade Damage to Wind Turbines Based on Mounted-Accelerometer Data | energy generation; green energy; wind turbines; fan-blade damage; mechanical condition; accelerometer; monitoring system; fault diagnosis; vibration data; maintenance by prediction; big data analytics |
| [3] | Reliability Analysis of RC Slab-Column Joints under Punching Shear Load Using a Machine Learning-Based Surrogate Model | reliability analysis; RC slab-column structure; machine learning; Monte Carlo simulation; SHapley Additive exPlanation |
| [4] | Real-Time Monitoring for Monolithic Movement of a Heritage Curtilage Using Wireless Sensor Networks | wireless sensor network; historic building; real-time monitoring |
| [5] | Polyethylene (PE)Waste Minimization Study of Cement Mortar with Adding PE Content under Different W/B Ratios | recycling of waste materials; cement mortar; water-to-binder ratio (W/B ratio); waste PE; durability |
| [6] | Real-Time Monitoring for Effects of Vibration and Temperature of Construction Site on Steel Assembly Bracing of Foundation Pit | steel assembly bracing; foundation pit; environmental temperature; vibration analysis; real-time monitoring |
| [7] | Automated Semantic Segmentation of Indoor Point Clouds from Close-Range Images with Three-Dimensional Deep Learning | building information model; 3D point cloud; semantic segmentation; deep learning |
| [8] | Determinants of Data Quality Dimensions for Assessing Highway Infrastructure Data Using Semiotic Framework | highway data quality assessment; data quality dimensions; semiotic framework; decision-making |
| [9] | Applying Deep Learning and Single Shot Detection in Construction Site Image Recognition | construction image; artificial intelligence; deep learning; object detection; single shot multi-box detector (SSD) |

[1] Keywords may be reordered for simplicity of presentation.

Hsu and Zhuang [2] assist industry by establishing a real-time condition-monitoring and fault-detection system with rules for recognising a wind turbine's abnormal operation, mainly caused by different types of fan-blade damage. This system can ensure ideal wind turbine operation by monitoring the health status of the blades, detecting sudden anomalies, and performing maintenance almost in real time. This enables 'maintenance by prediction' actions for unplanned maintenance as a supplement to the 'predictive maintenance' tasks for regular planned maintenance, which is especially significant for wind farms operating in harsh marine or shore environments that are subject to frequent natural disasters (e.g., earthquakes and typhoons). Turbines might fail to endure these because the manufacturers have built them according to the standards developed for areas less prone to natural disasters.

The system's rules are established utilising concepts and methods from data analytics, digital signal processing (DSP), and statistics to analyse data from the accelerometer mounted on the platform of the wind turbine's structure, measuring the vibration signals in three dimensions. The patterns for cases involving fan blade damage are found to establish the rules. By detecting and reporting anomalies effectively, repairs and maintenance can be carried out on faulty wind turbines.

Shen, Shen, and Liang [3] found that reinforced concrete (RC) slab-column structures are prone to punching shear failure, despite their architectural flexibility and easy construction. Punching shear failure is a typical brittle failure, making it challenging to assess slab-column structure functionality and failure probability. Therefore, predicting punching shear resistance and the corresponding reliability analysis are critical issues in designing RC slab-column structures. The authors used a database containing 610 experimental data for machine learning (ML) modelling to enhance the computational efficiency of the reliability analysis of RC slab-column joints. According to the nonlinear mapping between seven selected input variables and the punching shear resistance of slab-column joints, the study established four ML models, namely the artificial neural network (ANN), decision tree (DT), random forest (RF), and extreme gradient boosting (XGBoost) models.

Based on three performance measures, the authors selected XGBoost as the best prediction model; its root mean square error (RMSE), mean absolute error (MAE), and coefficient of determination ($R^2$) were 32.43, 19.51, and 0.99, respectively. Such advantages are reflected in a comparison with five empirical models introduced. The study visualised the prediction process of XGBoost using SHapley Additive exPlanation (SHAP); the importance sorting and feature dependency plots of the input variables explain the prediction process globally. Furthermore, the paper adopts Monte Carlo simulation with an ML-based surrogate model (ML-MCS) to calibrate the reliability of slab-column joints in a real engineering example. The authors obtained 1,000,000 samples through random sampling and calculated the reliability index ($\beta$) of this practical building using Monte Carlo simulation. As a result, they achieved the targeted reliability index under the design provisions. They also conducted a sensitivity analysis of stochastic variables and deeply examined the impact on structural reliability.

Shen, Yang, Yang, Yang, Zhu, and Wang [4] argue that since monolithic movement is a promising technology for relocating historical buildings, corresponding real-time monitoring is of great interest due to the buildings' age and poor structural integrity. However, as related research and practical applications remain limited, the paper proposes a wireless sensor network (WSN)-based strategy as a non-invasive approach to monitoring heritage curtilage during monolithic movement. The collected data show that the inclination of the curtilage is almost negligible. With the aid of finite element simulation, the study found that the crack displacement curves changed from −0.02 to 0.07 mm depending on the direction of movement; however, this value is not enough to cause structural cracks. The deformation of the steel underpinning beam, used to reinforce masonry walls and wooden pillars, is related to the stiffness in different directions. In addition, the strain variations of the steel chassis, which bears the vertical loads from the wooden pillars and masonry walls, are less than 0.04%. This indicates that they are kept within the durable range during monolithic movement. The authors thus prove that the WSN-based approach can potentially be used for the real-time monitoring of the monolithic movement of historic buildings.

Lin, Hung, Wang, and Wen [5] researched the durability of cement mortar prepared using different W/B ratios and percentages of waste PE content. The study was based on the logic that waste can be effectively used in concrete, and the characteristics of concrete can be maintained or enhanced, so the economy of waste management can be significantly increased, thus reducing pollution. They mixed the cement mortar with 0%, 1%, 2%, 3%, and 4% of waste PE and 20% of ground-granulated blast-furnace slag (GGBFS) in W/B ratios of 0.4, 0.5, and 0.6. The results show that slump and flow decrease with increasing waste PE content and increase with increasing W/B ratio; therefore, the setting time becomes shorter as the waste PE content rises. Regarding hardened (mechanical) properties, the specimen strength slightly decreased with increasing waste PE content. Still, the specimens performed better at a later age due to the pozzolanic reaction of slag, which can be verified using a scanning electron microscope.

Yang, Zeng, Liu, Yang, and Li [6] propose a real-time monitoring system—including vibration acceleration sensors, temperature sensors, and static and dynamic strain sensors—to monitor the safety status of a steel (assembly) bracing system in a practical project. It uses 5G wireless networking technology to transmit monitoring data to a cloud server for early warning of abnormal changes and development trends. The authors used real-time monitoring data obtained from a construction site as the inputs for the finite element model. They compared the corresponding results of a numerical simulation with the results from real-time monitoring. The paper concluded that (1) environmental temperature causes significant stress, which can be higher than the initial prestress of the steel bracing system; (2) the stress caused by vertical vibration, mainly from construction vehicles, is not remarkable, but the vertical frequency-weighted acceleration of support vibration is relatively large, which can affect on-site engineering technicians' sense of safety; and (3) the combination of environmental temperature and vertical vibration does not affect the safety of the steel bracing system.

Hsieh and Ruan [7] point out that generating a 3D model from 3D point clouds involves classification, outline extraction, and boundary regularisation for semantic segmentation. In addition, the number of 3D point clouds generated using close-range images is smaller, and they tend to be unevenly distributed. This is not conducive to automated modelling processing. However, the creation of building information models requires acquiring real building conditions. Thus, the authors propose an efficient solution for the semantic segmentation of indoor point clouds from close-range images. They further propose a 3D deep learning framework that achieves better results. The study used a dynamic graph convolutional neural network (DGCNN) 3D deep learning method to learn point cloud semantic features. Moreover, the authors developed more efficient operations to build a module for extracting point cloud features to resolve the problem of inadequate beam and column classifications. They first applied DGCNN to learn and classify the indoor point cloud into five categories: columns, beams, walls, floors, and ceilings. Then, they utilised the proposed semantic segmentation and modelling method to obtain the geometric parameters of each object to be integrated into building information modelling (BIM) software.

According to the experimental results, the overall accuracy rates of the three experimental sections of Area_1 in the Stanford 3D semantic dataset test results were 86.9%, 97.4%, and 92.5%. The segmentation accuracy of corridor 2F in a building was 94.2%. In comparing the length with the actual on-site measurement, they found the root mean square error to be $\pm 0.03$ m. Thus, the method is capable of automatic semantic segmentation from 3D point clouds with indoor close-range images.

Krishna, Ruikar, and Jha [8] identified the critical data quality dimensions affecting highway projects' decision-making process to address the data quality issues posed by the rapid accumulation of highway infrastructure data and its widespread reuse in decision-making. The authors propose addressing these issues by examining data quality, using various approaches to enhance data quality, and making decisions based on data quality information. Firstly, they conducted a state-of-the-art review of data quality frameworks applied in multiple fields to identify suitable frameworks for highway infrastructure data. Next, they identified data quality dimensions of the semiotic framework from the literature and conducted interviews with highway infrastructure stakeholders to finalize the data quality dimension. Then, they used a questionnaire survey to identify the critical data quality dimensions for decision-making. They also identified the importance of each critical dimension at each decision-making level in the highway infrastructure project. This 'semiotic data quality framework' provides a theoretical foundation for developing data quality dimensions to assess subjective data quality. However, further research is required to find effective ways to evaluate current satisfaction with data quality at various decision-making levels.

Lung and Wang [9] observed that although most construction site workers take photos of construction activities, the site manager relies mainly on manual labour to assess construction progress, quality control, and field management to facilitate job site coordination and productivity management. Moreover, it often takes a great deal of time to process the many photos taken, so in most cases, the image data are processed passively and used only for reference. However, using computerised tools, these photos could serve as aids for project management, including construction history records, quality, and schedule management. Thus, the authors propose an image recognition system for construction activities by incorporating image recognition through deep learning, using the powerful image extraction ability of a convolutional neural network (CNN) to automatically extract contours, edge lines, and local features via filters and feed feature data to the network for training in a fully connected way. The system is effective in image recognition, which helps identify subtle differences. The authors adjusted the parameters and structure of the neural network for use with a CNN. They selected objects like construction workers, machines, and materials for a case study. A CNN can be used to extract individual features for training, which improves recognizability and helps project managers make decisions regarding construction safety, job site configuration, progress control, and quality management, thus enhancing the efficiency of construction management.

Table 2 below summarises the application domains of the papers reviewed above. Table 3 then summarises the phases of (big) data analytics touched on by the SI papers.

**Table 2.** Application domains of the SI papers.

| Application Domain | SI Papers | Count (Multiple Answers) |
|---|---|---|
| Structures and Structural Eng. | [2–4,6] | 4 |
| Materials or Geotechnical Eng. | [1,5] | 2 |
| Construction Design, Measure, Information Systems and BIM | [3,6–8] | 4 |
| Transport/Energy Planning and Related Building Management | [1,2,4,8,9] | 5 |

**Table 3.** Phases of big data analytics touched on by the SI papers.

| Phase Involved or Resolved | SI Papers | Count (Multiple Answers) |
|---|---|---|
| Data Collection and Curation | [4–6,8,9] | 5 |
| Data Pre-processing | [1,2,4,6,7] | 5 |
| Data Analysis | [1,3–6,8] | 6 |
| Prediction | [1–3,7,9] | 5 |
| Forecasting | N/A | 0 |
| Decision-making and Support | [2,3,5,8,9] | 5 |

Tables 2 and 3 show that the SI papers involve highly diversified application domains and phases. However, several observations can be made:

1. Big data theories and techniques have been applied less to the materials or geotechnical (earth) engineering domains than to the other three domains, thus pointing to opportunities for future research;
2. More and more studies (including 6/9 reviewed here, as shown in Table 2) involve two interdisciplinary application domains;
3. As seen in Table 3, an increasing number of studies applying the theories or methods of big data analysis involve two or even three phases within the methodological flow of big data. This trend is noteworthy;
4. Rarely do studies address the forecasting function (0/9 in this review). This also indicates an opportunity for its first-time application in A/E/C studies;
5. By contrast, many papers contribute to the data collection and curation phase (5/9). Still, for the application of (big) data analysis in A/E/C, it is possible that many datasets still need to be collected and/or curated (e.g., using proper IoT or sensor devices). However, much room remains for subsequent analytical or knowledge exploration studies following this initial phase.

As a concluding remark, we would like to highlight the two areas for future studies in the above list. Researchers may consider focusing on the current research 'hot spot' and conducting research based on the existing solid ground; alternatively, they may wish to propose new ways to apply the theories or methods to new cases (to fill in vacant areas, i.e., the 'cold spot'). Either way, researchers may find the papers in this SI helpful.

**Author Contributions:** Conceptualization, Z.-Y.Z., Y.-W.Y. and M.-H.H.; Writing—original draft preparation, M.-H.H.; Writing—review and editing, Z.-Y.Z. All authors have read and agreed to the published version of the manuscript.

**Funding:** This editorial essay was funded by the National Science and Technology Council (Taiwan, ROC), grant number 111-2410-H-992-011, the Natural Science Foundation of Zhejiang Province (China), grant number LGG20E080002, LQ21A020009, the Science Foundation of Zhejiang Sci-Tech University (China), Grant number 19052460-Y, and the Department's Quality Improvement and Leap Projects of National Kaohsiung University of Science and Technology (NKUST) (Taiwan, ROC), grant number 112TSD00-1.

**Data Availability Statement:** This editorial essay involves no data outside the quoted SI, nor any data that are not publicly citable.

**Conflicts of Interest:** The authors declare no conflict of interest.

## References

1. Zhuang, Z.Y.; Kuo, W.T. Unravelling the Relations between and Predictive Powers of Different Testing Variables in High Performance Concrete Experiments: The Data-Driven Analytical Methods. *Buildings* **2022**, *12*, 1545. [CrossRef]
2. Hsu, M.H.; Zhuang, Z.Y. An intelligent detection logic for fan-blade damage to wind turbines based on mounted-accelerometer data. *Buildings* **2022**, *12*, 1588. [CrossRef]
3. Shen, L.; Shen, Y.; Liang, S. Reliability Analysis of RC Slab-Column Joints under Punching Shear Load Using a Machine Learning-Based Surrogate Model. *Buildings* **2022**, *12*, 1750. [CrossRef]
4. Shen, L.; Yang, B.; Yang, Y.; Yang, X.; Zhu, W.; Wang, Q. Real-Time Monitoring for Monolithic Movement of a Heritage Curtilage Using Wireless Sensor Networks. *Buildings* **2022**, *12*, 1785. [CrossRef]
5. Lin, K.T.; Hung, C.C.; Wang, H.Y.; Wen, F.L. Polyethylene (PE) Waste Minimization Study of Cement Mortar with Adding PE Content under Different W/B Ratios. *Buildings* **2022**, *12*, 2117. [CrossRef]
6. Yang, Y.; Zeng, H.; Liu, X.; Yang, B.; Li, Y. Real-Time Monitoring for Effects of Vibration and Temperature of Construction Site on Steel Assembly Bracing of Foundation Pit. *Buildings* **2023**, *13*, 450. [CrossRef]
7. Hsieh, C.S.; Ruan, X.J. Automated Semantic Segmentation of Indoor Point Clouds from Close-Range Images with Three-Dimensional Deep Learning. *Buildings* **2023**, *13*, 468. [CrossRef]
8. Krishna, C.M.; Ruikar, K.; Jha, K.N. Determinants of Data Quality Dimensions for Assessing Highway Infrastructure Data Using Semiotic Framework. *Buildings* **2023**, *13*, 944. [CrossRef]
9. Lung, L.W.; Wang, Y.R. Applying Deep Learning and Single Shot Detection in Construction Site Image Recognition. *Buildings* **2023**, *13*, 1074. [CrossRef]

buildings

MDPI

Article

# Unravelling the Relations between and Predictive Powers of Different Testing Variables in High Performance Concrete Experiments: The Data-Driven Analytical Methods

**Zheng-Yun Zhuang and Wen-Ten Kuo ***

Department of Civil Engineering, National Kaohsiung University of Science and Technology, Kaohsiung 807, Taiwan
* Correspondence: wtkuo@nkust.edu.tw; Tel.: +886-7-3814526 (ext. 15201)

**Abstract:** This study proposes and applies a systematic data analysis methodology to analyse experimental data for high-performance concrete (HPC) samples with different admixtures for offshore fan foundation grouting materials uses. In contrast with other relevant research, including experimental studies, the materials physics and chemistry studies, or cementitious material portfolio determination studies, this data-driven analysis provides a deep exploration of the experimental variables associated with the test data. To offer complete and in-depth perspectives, several methods are employed for the data analyses, including correlation analysis, cosine similarity analysis, simple linear regression (SLR) modelling, and heat map and heat-based tabularised visualisations; the outcome is a proposed methodology that is easily implementable. The results from these methods are validated using a pairwise comparison approach (PCA) to avoid unnecessary interference between data variables. There are several potential contributions from this work, including insights for cohered groups of variables, techniques for double check and 'third check', an established 'knowledge base' consisting of 504 SLR predictive models with their effectiveness (significance) and prediction accuracy (data-model fitness) used in practical applications, an alternative visualisations of the results, three data transforms which can be omitted in a future analysis, and three valuable theory-linking perspectives (e.g., for the relationships between destructive and non-destructive tests with respect to the variable categories). The implication that some variables are interchangeable will make future experiments less labour intensive and time consuming for pre-project HPC material testing.

**Keywords:** high performance concrete (HPC); experimental parameters; data-driven analysis; pre-project material testing; pairwise comparison analysis; correlation analysis; cosine similarity; predictive regression modelling; heat map; variable transform

**Citation:** Zhuang, Z.-Y.; Kuo, W.-T. Unravelling the Relations between and Predictive Powers of Different Testing Variables in High Performance Concrete Experiments: The Data-Driven Analytical Methods. *Buildings* **2022**, *12*, 1545. https://doi.org/10.3390/buildings12101545

Academic Editors: Ahmed Senouci and Jan Fořt

Received: 13 August 2022
Accepted: 20 September 2022
Published: 27 September 2022

**Publisher's Note:** MDPI stays neutral with regard to jurisdictional claims in published maps and institutional affiliations.



## 1. Introduction

There are many types of experiments for testing a concrete material sample. As an example, in a study [1] involving the selection of high-performance concrete (HPC) admixtures for offshore wind farm construction, potential experiments are classified into the following three categories:

(Cat1) fresh mechanical properties,
(Cat2) hardened mechanical properties,
(Cat3) the durability measures.

In the study, slump flow and the time required to flow through a V-shaped funnel were included in (Cat 1). Compressive strength (CS), ultrasound pulse velocity (USPV), and electrical resistivity on surface (ERoS) were included in (Cat 2), while anti-sulphate capability (ASC) and rapid chloride permeability (RCP) were included in (Cat 3). These categories allowed the researchers to determine the superior admixtures that included cement, fly ash, silica fume, super plasticiser, and water for grouting before building (non-floating) wind turbines with foundations constructed in the sea. One analysis, mentioned only in the paper's Appendix, was of particular interest because it concluded that the relationship between CS and RCP could be identified (and established) as follows:

$$\text{RCP} = 7966.72 + (-97.76)\text{CS}. \quad (1)$$

This equation was formulated and validated through an extensive series of exploratory data analyses in that study; further details are provided in Appendix A of this study. Figure 1 provides visualisations of the final 'effective process' that resulted in Equation (1) using K-means (i.e., a non-supervised machine learning approach) and simple linear regression (SLR) modelling.

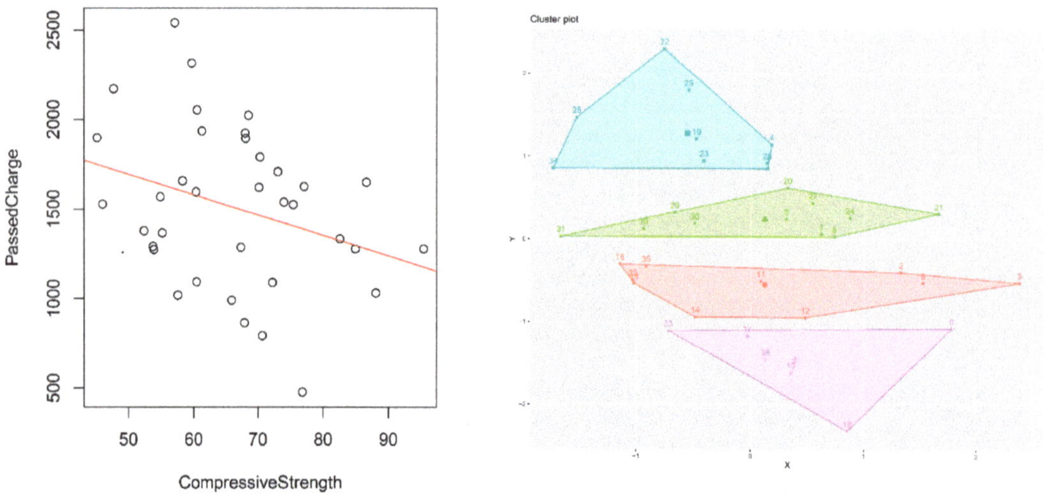


(**a**) (**b**)

**Figure 1.** Relationship between CS and RCP: (**a**) a rough trend was observed between CS and RCP; (**b**) visualising the result of clustering in the previous study. (Data source: Re-plot and New Plot).

However, a practical benefit of the above process has not been described in the literature: such an outcome might reduce the effort to perform sample tests because with Equation (1), one of the two experimental results (e.g., RCP) can be anticipated (e.g., by CS), thereby reducing time and effort. Nevertheless, the practical benefits of this encouraging result are still limited because it was only validated for *one pair* of variables. Therefore, this study seeks to answer the question: Among *numerous pairs* of parameters tested for the HPC samples, does *any other pair* exist in which one parameter can be used to predict another?

In this study, a full set of data related to the experiments is sourced, and a data-driven analysis is performed following a pairwise comparison approach (PCA). To a large extent, this study identifies *all pairs* of concrete sample parameters in the available datasets sourced from an HPC laboratory, providing in-depth and cross-categorical views of the relationships between each variable pair and offering scientifically grounded insights about the experimental values that can be used to anticipate others.

A systematic data analysis methodology is designed and proposed utilising numerous methods, including correlation coefficient, cosine similarity, SLR modelling, and dimensional alternation (domain transform) of the variables (before estimating the parameters of the SLR model to ensure the linearity of each established model), in addition to other supplemental methods (e.g., the heatmap visualisation technique, and viewing the results from different perspectives). Using this methodology, the analysis reveals essential information about all Cat2 and Cat3 variables, the relationship between each pair of test variables, the variable's ability to predict other variables, and the accuracy with which the variables can predict one another. The set of information obtained from the analysis (i.e., the 'knowledge base') can be used to benefit researchers and practitioners.

Since every pair of variables fitted with a model with sufficient predictive power (in terms of data-model fitness and model significance or effectiveness) can be summarised from the knowledge base, the results of some tests can be anticipated by using the results of other tests (if the law allows not testing every item). This could be a significant benefit to material testers whose time is valuable, and could also indirectly help reduce the cost and complexity of construction projects.

In this paper, other discussions are presented for the insights gained, particularly for the identified pairs of variables for which the results are positive (i.e., effective information). Implications are thus drawn from several aspects of the analysis, such as selecting a proper dimensional alternation method to convert the independent variable data for the established SLR model, performing both 'double check' and 'third check' by using the cosine similarity index and the relevant statistical descriptors of SLR models for the main results in the total correlation matrix, confirming theories in the existing literature or standards (between the test variables), and revealing the truth between the different destructive and non-destructive tests. Extensive discussions are also given for the utilisation of the developed knowledge base and future applications of the proposed methodological framework, as well as the time-saving effects on making material tests in each case (if the schedule for the construction project is tight). These also associate the results with theories and practices.

Section 2 reviews the literature related to the subject (i.e., the primary research question) and the methodologies to conduct this study, as well as the main data analysis methods applied. Section 3 presents the results, and Section 4 discusses the primary positive outcomes in terms of the application of methods, the practical insights gained, and the theoretical aspects of the study. Finally, Section 5 concludes this paper.

## 2. Literature Review and Methods

### *2.1. Background of the Problem*

#### 2.1.1. Renewable Energy (RE) Planning and Wind Farm Construction

For both developing and developed economies, stable and adequate supplies of energy are mandatory. However, as world's fossil fuel-based energy resources are limited, renewable energy (RE) has emerged as a viable solution to replace conventional electric power generation [2,3]. In addition, utilising RE is also an 'environmentally sound' solution for a country to meet its sustainable development goals (SDGs) [4], this is also addressed by ESG (environmental, social, governance) targets that are established for business institutions [5].

Despite disagreements over the definitions of 'RE', 'green energy', 'sustainable energy', and 'clean energy', a classification of RE that is commonly accepted refers to the type of RE resource, including solar, wind, hydropower, geothermal, biomass, marine, and others [6–8]. This is due to the fact that the technological aspects to exploit these types of RE resource are usually totally different, thereby leading to salient 'watersheds' that can be told between them.

A stream of research adopting this view of classification is related to the 'portfolios', either in terms of selecting the optimal investment portfolio (by a company's decision makers) [9] or determining a country's optimal RE portfolio (by energy planners or operators within its long-term energy policy [10]. In research related to portfolios, uncertain or risky issues are also addressed [11,12]. However, in contrast to these operational management topics, more studies focus on the technologies to exploit various aspects of RE.

As an example, the technologies to exploit wind resources usually involve aspects of EE (electrical engineering), fluid dynamics/mechanics, and construction engineering (CE). Moreover, the first step in building a wind farm is typically to construct the foundations before erecting the wind turbines. For onshore or inland turbines, it is possible for the local weather conditions to be directly utilised as engineering parameters (e.g., steel structures, concrete materials, etc.). However, for offshore turbines, regardless of their working mode (i.e., fixed or floating, depending on the water depth), underwater foundations are always required, which necessitates the well-proportioned HPC material used for grouting [13].

Constructing new wind farms is necessary for a country with unique conditions to exploit its wind resources. For offshore wind farms, in the Taiwan Strait, numerous projects have taken advantage of the very shallow waters of the 'Taiwan Bank' [14] and the high wind availability [15]; similar projects have been developed on the edges of the Mediterranean Sea [16,17]. For on-shore or in-land wind farms, projects have been initiated near the Sahara Desert, e.g., budget has been allocated for the construction of wind farms in Algeria [18].

The motivation of this study is to explore and better understand the parameters (or the parametric performance) of the HPC materials (the numerical values for which can be obtained experimentally) used for grouting the structural bases of offshore wind turbines. This may benefit the selection process before the real admixture of the HPC is determined and applied as grouting, and it may also generate a novel knowledge set to be used for anticipating the potential outcomes of experiments.

#### 2.1.2. The Special Weather/Sea Conditions in Taiwan and the Effects for Wind Turbines

As discussed in several studies, the soil of Taiwan is rich in sulphates and crystalline salts due to its geological features which result from frequent crustal deformation [19,20]. Osmosis of sulphate that occurs in concrete has caused serious problems for inland structures; the gypsum reaction is one such reaction, which can inflate the concrete and peel the surface off from the structure [1]. Therefore, anti-sulphate capability is typically a critical durability parameter in this context.

In addition, the high temperature and high humidity of Taiwan may accelerate these effects and decrease the life of structures. Moreover, the frequent earthquakes in the Circum-Pacific Belt and the typhoons that originated in the West Pacific region may also cause unexpected damage to structures. The damage is 'unexpected' because such events usually are inherently random, as is the scale of the event, e.g., the 921 Taiwan Earthquake of 1999, the Typhoon Morakot, etc. [21,22]. These types of natural disasters should be considered to be different from the damage caused by other types of unexpected events, such as fires [23].

These same conditions also apply to the Taiwan Strait, within which the Taiwan Bank is a very large but shallow continental shelf as a potential site for constructing offshore wind farms. Despite a high availability wind field above the sea all year, the land under water is susceptible to acid attacks (such as sulphates, chlorides, etc.), experiences the same extreme climate conditions discussed above (high temperature and high humidity), suffers from frequent natural disasters (earthquakes and typhoons) [24], and is scoured by the changing ocean currents throughout the year [25]. Therefore, quality HPC material is always required for wind turbines for its good workability during grouting, its beneficial mechanical properties during construction, and its long-term durability.

Thus, the parameters of an HPC sample (i.e., the experimental values from individual tests for all HPC samples which become 'variables' during data analysis) should be further studied and clearly understood. In addition, a suitable method is needed to determine the optimal HPC materials for construction based on the parametric data of the samples being tested during the experiments.

The parameters included in this study are organised and shown in Figure 2. It should be noted that, unlike in previous studies, electrical resistivity on surface (ERoS) is considered to be a durability property (parameter) of HPC samples in this research. This is believed to be a more reasonable consideration based on recent studies [26–28].

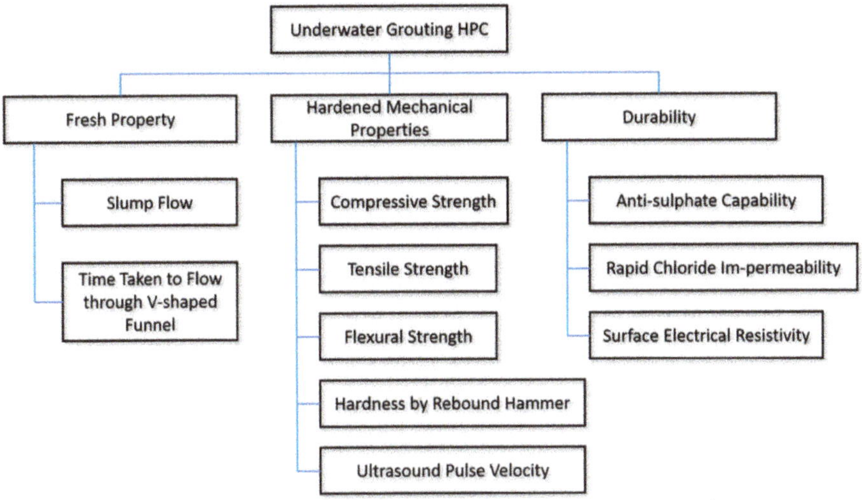


**Figure 2.** HPC Sample Parameters (Organised).

In Figure 2, the two fresh-property parameters are excluded from this study. These two tests are usually related to the workability of an HPC material during the engineering process, and their data formats are usually incompatible because the 'data sampling periods' of these two experiments are different than those in the other two categories. In contrast, the other eight variables have a quite large overlap in terms of their data sampling periods, so their data formats are compatible for the analysis (see Section 3.1).

In this study, the barrier between the hardened mechanical properties and durability properties is also removed intentionally. By doing so, not only can the pairs of parameters within a certain category be explored, but so can the pairs of parameters across two categories, so that relations between them and the intensities of the relations can be identified and evaluated. As will be shown later in Section 2.3, these relations include correlations, similarities, and predictive associations in terms of the ability to use one parameter to anticipate another (and how 'safely' it can be used).

### 2.2. About the Parameters

This section provides a review and discussion of the eight testing parameters of HPC analysed in this study (see Figure 2), regardless of the barrier between the hardened mechanical and durability parametric categories (see Section 2.1.2).

#### 2.2.1. Compressive Strength (CS)

The CS of a concrete material (sample) is defined as its ability to withstand stress without failure. In concrete design and quality control, CS is the most generally specified property. The standard test method(s) for CS is (are) defined by ASTM C39 [29].

#### 2.2.2. Tensile Strength (TS)

In this study, the splitting tensile strength (TS) of cylindrical specimens is accepted as the main tensile TS measurement for a concrete (and HPC) material (sample). Its test(s) can be performed in accordance with ASTM C496 [30].

#### 2.2.3. Flexural Strength (FS)

The primary property for a concrete material (sample) to demonstrate is CS, but its tensile behaviours are also important. When reinforced concrete (RC) is subjected to tensile force, cracks and expansion will occur, depending largely on the TS of the concrete material. Direct tension tests of concrete are seldom performed, mainly because the device that holds the specimen may introduce secondary stresses that introduce error in the measurement. Another test for estimating the TS of a concrete material (sample) is the ASTM C78 (third point) flexural loading test, the result of which is referred to as flexural strength (FS) in this study [31].

#### 2.2.4. Hardness (by Rebound Hammer) (HbRH)

The rebound hammer method utilises non-destructive testing techniques to measure the concrete strength. This test can be performed in accordance with ASTM C805 [32].

#### 2.2.5. Ultrasound Pulse Velocity (USPV)

USPV is another common non-destructive index. The structural compactness in the specimen can be estimated according to the ultrasonic transferring rate (i.e., the speed of ultrasound waves going through the sample). The more complete the hydration reaction inside the concrete, the smaller the pores will be; thus, the detected USPV increases. ASTM C597 details the method(s) to test USPV [33].

#### 2.2.6. Anti-Sulphate Capability (ASC)

Sulphate penetrates the surface of the concrete by infiltration, and a chemical reaction is produced when it combines with calcium hydroxide in the cement; when this occurs, the concrete expands, which further peels off the surface or destroys it. The experiment to measure the anti-sulphate capability of the HPC material samples can be carried out according to ASTM C1012; this method evaluates the resistance of the HPC to sulphate attack by measuring the rate of weight loss in an immersed specimen [34].

#### 2.2.7. Rapid Chloride (Im-)Permeability (RCP)

Another indicator of the durability of concrete is its ability to resist chloride ion penetration. This can be performed by testing the concrete material samples using a rapid chloride ion rapid test (RCPT) with an applied electrical voltage. This test method can be performed in accordance with ASTM C1202 [35].

#### 2.2.8. Electrical Resistivity on Surface (ERoS)

To evaluate concrete compactness and the durability of cement composites in an HPC sample, the ERoS can be measured. The higher the electrical resistance, the more compact the substrate, which increases durability. This study considers ERoS to be a durability variable (see Section 2.1.2), which can be tested using ASTM C876 [36].

### 2.3. *The Methods: A Brief Review*

#### 2.3.1. Correlation Analysis

This study uses the Pearson correlation coefficient (*P-Co-Co*) method as the basis of the analysis to identify the correlation between each pair of experimental parameters (variables). The computation of a *P-Co-Co* is typically defined as the following:

$$PCoCo(X,Y) = \frac{n\sum_{i}^{n}(X_iY_i) - \left(\sum_{i}^{n}X_i\right)\left(\sum_{i}^{n}Y_i\right)}{\sqrt{n\sum_{i}^{n}(X_i^2) - \left(\sum_{i}^{n}X_i\right)^2}\sqrt{n\sum_{i}^{n}(Y_i^2) - \left(\sum_{i}^{n}Y_i\right)^2}} \quad (2)$$

where $X$ and $Y$ are variables between which the *P-Co-Co* is to be calculated; $n$ is the data size (length) of each variable; and $i$ is the identifier for a specific data entry.

There are two primary reasons for using the *P-Co-Co* method:

(1) All of the variables are continuous, so the two variables when searching for correlations must also be continuous; therefore, the *P-Co-Co* meets the analytical purpose.
(2) Developed more than 100 years ago, *P-Co-Co* is the most common (and standard) method to compute and examine the correlation between two continuous variables.

#### 2.3.2. Cosine Similarity Analysis

The cosine similarity (*Cos-Sim*) analysis diverges from the traditional thinking in statistics, treating experimental parameters as 'vectors', rather than 'variables'; in so doing the similarity between two series of parametric values from the tests can be justified by the *Cos-Sim* index in the dimensional space. The standard computational process for this index can be written as:

$$CosSim(X,Y) = \frac{\sum_{i}^{n} X_i Y_i}{\sqrt{\sum_{i}^{n} X_i^2}\sqrt{\sum_{i}^{n} Y_i^2}} \tag{3}$$

where $X$ and $Y$ are variables being treated as vectors between which the index is to be calculated (so $X_i$ or $Y_i$ is the $i$-th element of vector $X$ or vector $Y$); the other symbols are as defined previously.

The reasons to apply *Cos-Sim* in this context are as follows:

(1) *Cos-Sim* has been proven to be an effective supplemental measure to *P-Co-Co* in real applications [37,38];
(2) *Cos-Sim* also fits the data variables, i.e., the variables considered in this study are all continuous and positive, so the vectors are all meaningful in the multi-dimensional data space.

#### 2.3.3. Simple Linear Regression

In this study, the predictive power and the relevant statistical descriptions between a pair of variables are justified by building several SLR models. Every SLR model chooses a data transform (method) to convert the independent variable (i.e., the RHS variable) prior to parameter estimation, so the model remains linear. A typical SLR model is defined as follows:

$$Y_i = \alpha + \beta X_i + \varepsilon_i \tag{4}$$

where $\alpha$ and $\beta$ are the intercept and the regression coefficient associated with the only independent variable $X$ in the model, respectively, which are to be estimated; given the $i$-th data tuple in the dataset, $(X_i, Y_i)$, $\varepsilon_i$ is the residual of this data tuple with respect to $Y = \alpha + \beta X$; other symbols are as defined previously.

The SLR model $Y = \alpha + \beta X$ can then be plotted as the best line fit (i.e., the 'AB-line') that includes all data points $(X_i, Y_i), \forall i \in \{1, 2, \ldots, n\}$ in the data space. This model also satisfies the requirement to identify the causal relationship and predictive power between two variables 'pair-wisely', i.e., the PCA used in the overall analysis (which also applies to the correlation analysis and the cosine similarity analysis).

In our analysis, every time two variables are paired, one variable becomes the dependent variable $Y$ and the other one becomes the independent variable $X$. This then yields a total number of $C_2^8 \times 2! = 56$ 'base models' of (X, Y), and in each case, the two paired variables are interchangeable (to appear on the RHS or LHS of the model, i.e., 2!).

#### 2.3.4. Data Variable Transforms

As discussed above, each of the 56 'basic models' justifies the predictive relation and power between a certain pair of variables and the relevant statistical descriptions for the SLR model. However, the models are established based on the assumption that no variable is transformed, i.e., both variables on the RHS and LHS in the model use their data values in the *source domain*.

In this study, we transform the values of the independent variable on the RHS of each of the 56 models in nine ways in prior and build nine models by re-estimating the parameters of the SLR model that fit the independent variable in the *transform domain*. For comparison purpose, the models also include the initial case of 'no transform' (i.e., the base model), and are written as:

(1) RHS Variable Transform: $X' = X^1$ (no transform; the basic model)
(2) RHS Variable Transform: $X' = X^2$ (square)
(3) RHS Variable Transform: $X' = X^3$ (3 power)
(4) RHS Variable Transform: $X' = e^X$ (e power X)
(5) RHS Variable Transform: $X' = 2^X$ (2 power X)
(6) RHS Variable Transform: $X' = \sqrt{X}$ (square root of X)
(7) RHS Variable Transform: $X' = \sqrt[3]{X}$ (triple root of X)
(8) RHS Variable Transform: $X' = \log(X)$ (log of X, base 10)
(9) RHS Variable Transform: $X' = \lg(X)$ (log of X, base 2)

With all these transforms performed prior to establishing the model, the model can be simplified as Equation (5) uniformly, and the linearity for all other SLR models derived from a basic model also holds:

$$Y = \alpha + \beta X' \tag{5}$$

For a model with (X, Y) as the RHS and LHS variables, the 56 cases already included their counterpart, i.e., (Y, X) as the RHS and LHS variables, respectively (see Section 2.3.3), meaning that all cases would be considered. As such, since there are nine models derived from each 'basic model' totally, there would actually be 56 × 9 = 504 SLR models in this work using the same set of data.

2.3.5. Indicators Used to Justify the SLR Modelling Results

Other than P-Co-Co and Cos-Sim which are in themselves 'indicators' for paired data variables, in this study, the following measures are used to tell the quality of an established SLR model. These include:

(1) The estimated value of parameter $\alpha$, $\alpha^*$: this is the *intercept* of the predictive SLR model that fits the data, which is one of the two parameters defining the model in Equation (5);
(2) The estimated value of $\beta$, $\beta^*$: this is the *slope* of the predictive SLR model, which is another parameter that defines the model in Equation (5);
(3) The $p$ value of the entire SLR model: this value indicates whether the model is *significant* or not; the significance of an SLR model usually dictates whether the model is *reliable* and the extent to which it can be trusted;
(4) The $p$ value for $\alpha^*$: this $p$ value indicates whether the estimated parameter $\alpha^*$ in the SLR model is *significant* or not;
(5) The $p$ value for $\beta^*$: this $p$ value indicates whether the estimated parameter $\beta^*$ in the SLR model is *significant* or not;
(6) The R-square value, $R^2$: this value usually connotes the *data-model fitness*, i.e., how well does the obtained SLR model fit the given data of $(X_i, Y_i)$, wherein the parameters in this case are typically estimated using the OLS (ordinal least square) method; and
(7) The R-square value, $(R^2)^*$: this is another R-square value that is adjusted based on $R^2$ and the sample size; we primarily observe the traditional $R^2$ value in this study, since the two values are usually very similar.

To visualise the summarised and tabularised computational results, they are also plotted as heat maps to provide a clear view. A heat map is a visualisation tool that is frequently used in the field of data-driven decision-making (DDDM) since Toussaint Loua provided its first application in 1873.

## 3. Results

### *3.1. Original Datasets*

The experimental datasets for the eight hardened mechanical or durability parameters, i.e., the eight variables of the HPC samples to be analysed, are summarised in Table 1. The datasets are presented following the order provided in Section 2.2.

**Table 1.** Source data from experimental tests: (a) for CS and TS; (b) for FS and HbRH; (c) for USPV and ASC; (d) for RCP and ERoS.

**(a)**

| **Parameter of HPC Sample** | | **CS of Sample at Days (MPa)** | | | | | **TS of Sample at Days (MPa)** | | | | |
|---|---|---|---|---|---|---|---|---|---|---|---|
| **W/B Ratio** | **% of FA and SF** | **1** | **7** | **28** | **56** | **91** | **1** | **7** | **28** | **56** | **91** |
| 0.28 | FA 0% SF 0% | 42 | 65.3 | 73.92 | 82.54 | 95.43 | 7.32 | 12.74 | 16.15 | 17.24 | 22.05 |
| | FA 0% SF 10% | 31.7 | 56.3 | 68.53 | 70.16 | 88.04 | 5.39 | 9.82 | 12.86 | 15.16 | 21.04 |
| | FA 10% SF 0% | 28.2 | 60 | 68.1 | 75.33 | 84.9 | 4.28 | 10.29 | 13.13 | 15.33 | 19.57 |
| | FA 10% SF 10% | 27.1 | 46.4 | 54.88 | 67.37 | 72.19 | 4.41 | 8.8 | 10.69 | 14.89 | 18.81 |
| | FA 20% SF 0% | 27.9 | 42.5 | 53.88 | 60.42 | 70.66 | 4.17 | 7.17 | 8.8 | 12.29 | 14.95 |
| | FA 20% SF 10% | 25.8 | 40.5 | 52.37 | 65.95 | 76.74 | 3.71 | 6.94 | 8.2 | 12.31 | 17.89 |
| 0.30 | FA 0% SF 0% | 35.7 | 50.6 | 60.5 | 70.3 | 86.59 | 5.12 | 10.31 | 14.17 | 16.2 | 20.46 |
| | FA 0% SF 10% | 23.9 | 48 | 57.07 | 61.28 | 77 | 4.88 | 8.75 | 11.7 | 15.16 | 20.01 |
| | FA 10% SF 0% | 17.7 | 49.2 | 59.67 | 68.03 | 73.01 | 3.08 | 9.5 | 12.14 | 15.27 | 18.83 |
| | FA 10% SF 10% | 16.1 | 38.7 | 47.71 | 58.25 | 60.35 | 3.31 | 7.21 | 8.77 | 14.87 | 18.24 |
| | FA 20% SF 0% | 17 | 35.6 | 46.02 | 53.74 | 57.53 | 2.53 | 6.29 | 7.81 | 12.99 | 15.41 |
| | FA 20% SF 10% | 14.6 | 34.5 | 45.17 | 55.17 | 67.92 | 2.23 | 5.9 | 7.22 | 13.61 | 17.75 |

**(b)**

| **Parameter of HPC Sample** | | **FS of Sample at Days (MPa)** | | | | | **HbRH of Sample at Days (MPa)** | | | | |
|---|---|---|---|---|---|---|---|---|---|---|---|
| **W/B Ratio** | **% of FA and SF** | **1** | **7** | **28** | **56** | **91** | **1** | **7** | **28** | **56** | **91** |
| 0.28 | FA 0% SF 0% | 6.29 | 11.66 | 13.49 | 13.74 | 14.21 | 37.61 | 36.58 | 33.09 | 31.26 | 27.92 |
| | FA 0% SF 10% | 3.79 | 9.79 | 10.5 | 12.18 | 12.48 | 45.41 | 42.57 | 43.68 | 40.36 | 33.37 |
| | FA 10% SF 0% | 6.54 | 10.62 | 10.77 | 12.31 | 13.39 | 53.47 | 50.51 | 44.97 | 43.24 | 37.17 |
| | FA 10% SF 10% | 4.8 | 8.79 | 9.95 | 10.84 | 11.71 | 58.47 | 55.56 | 50.42 | 52.23 | 45.45 |
| | FA 20% SF 0% | 5.64 | 8.37 | 10.05 | 10.56 | 10.82 | 63.48 | 65.68 | 59.89 | 60.78 | 57.55 |
| | FA 20% SF 10% | 3 | 7.56 | 8.93 | 9.35 | 10.18 | 37.61 | 36.58 | 33.09 | 31.26 | 27.92 |
| 0.30 | FA 0% SF 0% | 6.15 | 10.61 | 12.41 | 12.81 | 13.39 | 29.78 | 29.01 | 25.95 | 19.5 | 23.94 |
| | FA 0% SF 10% | 3.38 | 8.82 | 10.02 | 11.83 | 12.41 | 44.37 | 38.81 | 40.92 | 36.83 | 30.73 |
| | FA 10% SF 0% | 6.24 | 9.77 | 10.19 | 11.99 | 13.26 | 52.45 | 49.86 | 44.24 | 43.01 | 34.15 |
| | FA 10% SF 10% | 4.45 | 8.52 | 9.4 | 10.46 | 10.95 | 58 | 54.49 | 48.24 | 48.84 | 41.53 |
| | FA 20% SF 0% | 5.34 | 7.7 | 9.37 | 10.18 | 10.63 | 63.35 | 67.26 | 58.07 | 59.07 | 54.52 |
| | FA 20% SF 10% | 2.89 | 7.47 | 8.33 | 9.16 | 10.01 | 29.78 | 29.01 | 25.95 | 19.5 | 23.94 |

**(c)**

| **Parameter of HPC Sample** | | **USPV of Sample at Days (m/s)** | | | | | **Weight Loss at Days for ASC (%)** | | |
|---|---|---|---|---|---|---|---|---|---|
| **W/B Ratio** | **% of FA and SF** | **1** | **7** | **28** | **56** | **91** | **28** | **56** | **91** |
| 0.28 | FA 0% SF 0% | 4263.7 | 4794.7 | 4923.3 | 5278 | 5653 | 5.07 | 3.35 | 0.57 |
| | FA 0% SF 10% | 3620.3 | 4135.3 | 4415.7 | 4781 | 4981.3 | 4.92 | 3.08 | 0.51 |
| | FA 10% SF 0% | 3816 | 4336.3 | 4545.7 | 4739.3 | 4859.7 | 4.81 | 2.47 | 0.27 |
| | FA 10% SF 10% | 3300.7 | 4380.3 | 4685 | 4773 | 4908.7 | 2.23 | 1.32 | 0.21 |
| | FA 20% SF 0% | 3389.3 | 3930.3 | 4313.7 | 5454.3 | 5823 | 2.21 | 1.28 | 0.18 |
| | FA 20% SF 10% | 3648.7 | 4167.3 | 4634.7 | 4910.7 | 5748.3 | 1.25 | 0.79 | 0.05 |
| 0.30 | FA 0% SF 0% | 4104 | 4598 | 4698 | 5236 | 5877.7 | 6.18 | 4.18 | 2.4 |
| | FA 0% SF 10% | 3442 | 3949 | 4359 | 5078.3 | 5287 | 5.76 | 4.15 | 2.03 |
| | FA 10% SF 0% | 3777.3 | 4384 | 4578.3 | 4815.3 | 5386.7 | 5.21 | 3.22 | 1.04 |
| | FA 10% SF 10% | 3308.7 | 4407.3 | 4612.7 | 4819.7 | 5338 | 3.31 | 2.3 | 0.71 |
| | FA 20% SF 0% | 3128 | 3786.3 | 4231.3 | 5342.3 | 6048 | 3.81 | 1.87 | 0.61 |
| | FA 20% SF 10% | 3423.3 | 3905.3 | 4593.7 | 5111.3 | 5993 | 2.41 | 1.94 | 0.3 |

**Table 1.** *Cont.*

| **(d)** | | | | | | | | | |
|---|---|---|---|---|---|---|---|---|---|
| **Parameter of HPC Sample** | | **Charge Passed for RCP (Coulombs)** | | | **ERoS Tested at Days (kΩ-cm)** | | | | |
| **W/B Ratio** | **% of FA and SF** | **28** | **56** | **91** | **1** | **7** | **28** | **56** | **91** |
| 0.28 | FA 0% SF 0% | 1539.7 | 1334 | 1278.36 | 21.43 | 25 | 37.89 | 77.11 | 93.11 |
| | FA 0% SF 10% | 2022 | 1622.1 | 1031.66 | 15.96 | 24.22 | 40.22 | 78.78 | 93.33 |
| | FA 10% SF 0% | 1893.7 | 1525.6 | 1277.91 | 14.03 | 25.44 | 40.67 | 77.78 | 93.44 |
| | FA 10% SF 10% | 1569.9 | 1286.3 | 1090.04 | 15.27 | 26.33 | 43 | 74.67 | 97.45 |
| | FA 20% SF 0% | 1274.3 | 1093.2 | 793.1 | 16.48 | 20.44 | 42.33 | 82 | 96.78 |
| | FA 20% SF 10% | 1378.3 | 990.4 | 476.593 | 22.02 | 28.89 | 47.33 | 76.33 | 93.33 |
| 0.30 | FA 0% SF 0% | 2051.9 | 1790.5 | 1650.65 | 25.67 | 46.92 | 72.44 | 82.78 | 98.53 |
| | FA 0% SF 10% | 2541.3 | 1934.9 | 1626.58 | 12.63 | 44.67 | 74.55 | 83.67 | 94.45 |
| | FA 10% SF 0% | 2315.2 | 1922.7 | 1709.15 | 11.96 | 46.22 | 71 | 82 | 90.67 |
| | FA 10% SF 10% | 2170.6 | 1658.2 | 1596.96 | 10.1 | 47.11 | 79.89 | 85.45 | 97.67 |
| | FA 20% SF 0% | 1528.6 | 1291.3 | 1019.57 | 16.6 | 41 | 77.22 | 79.22 | 98.78 |
| | FA 20% SF 10% | 1897.7 | 1366.8 | 864.455 | 27.76 | 58.72 | 81.45 | 94.89 | 98.33 |

### *3.2. Correlation Analysis*

Table 2 shows the P-Co-Cos between each pair of variables in a correlation matrix, while no variable is transformed. Figure 3 visualises the results using a heat map.

**Table 2.** Correlation coefficients between each pair of variables (no transform).

| | **CS** | **TS** | **FS** | **USPV** | **HbRH** | **ERoS** | **ASC** | **RCP** |
|---|---|---|---|---|---|---|---|---|
| **CS** | 1 | 0.875333 | 0.824243 | 0.43716 | −0.39808 | 0.29562 | −0.2662 | −0.31178 |
| **TS** | 0.875333 | 1 | 0.784765 | 0.622967 | −0.40234 | 0.562644 | −0.37139 | −0.32165 |
| **FS** | 0.824243 | 0.784765 | 1 | 0.310713 | −0.30525 | 0.205377 | 0.036492 | −0.00205 |
| **USPV** | 0.43716 | 0.622967 | 0.310713 | 1 | −0.23428 | 0.645174 | −0.56893 | −0.58283 |
| **HbRH** | −0.39808 | −0.40234 | −0.30525 | −0.23428 | 1 | −0.23448 | −0.04958 | −0.05542 |
| **ERoS** | 0.29562 | 0.562644 | 0.205377 | 0.645174 | −0.23448 | 1 | −0.54312 | −0.33091 |
| **ASC** | −0.2662 | −0.37139 | 0.036492 | −0.56893 | −0.04958 | −0.54312 | 1 | 0.817532 |
| **RCP** | −0.31178 | −0.32165 | −0.00205 | −0.58283 | −0.05542 | −0.33091 | 0.817532 | 1 |

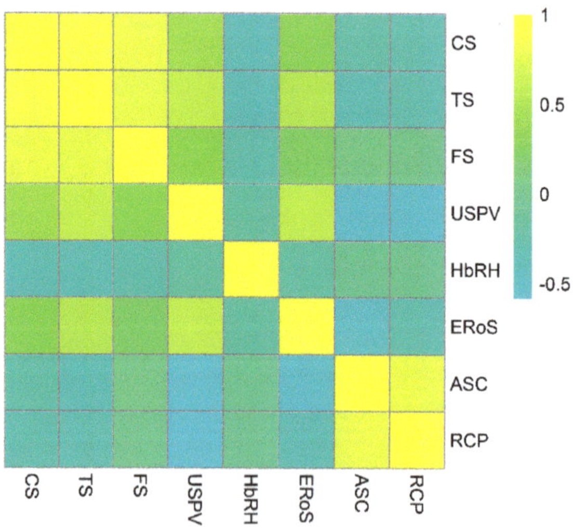


**Figure 3.** Visualisation of the correlation matrix (no transform).

To ensure an objective basis using the adopted measures (see Section 2.3) and taking the data variables pair-wisely, every data variable should have an equal length and be tested on the same set of testing days for the same set of HPC samples. Fortunately, the collected datasets have followed these conditions. Inside the red boxes shown in Table 1, experimental data are provided for all data variables on day 28, day 56, and day 91, over all HPC samples. The result is eight variables with an equal data length of $n = 36$, correspondingly; these form the basis of subsequent data analyses.

Each sub table in Table 3 shows a correlation matrix produced when variable $X$ in each pair (of variables) is transformed using one of the methods discussed in Section 2.3.4. In these sub tables, the cells are shown in different colours to visualise the results directly (which is analogous to using a separate heat map: dark green for 1, gradient green (lighter and lighter) for (less) positive values, white for 0, gradient red (heavier and heavier) for (more) negative values, full red for −1). In addition, note that the diagonal elements are white boxes with '—' entries to indicate they do not contain any meaningful information.

**Table 3.** Correlation coefficients between each pair of variables, with different RHS variable transforms: (a) $X' = X^2$; (b) $X' = X^3$; (c) $X' = e^X$; (d) $X' = 2^X$; (e) $X' = \sqrt{X}$; (f) $X' = \sqrt[3]{X}$; (g) $X' = \log(X)$; (h) $X' = \lg(X)$; (h) RHS Var. Transform: $X' = \lg(X)$.

| (a) | CS | TS | FS | USPV | HbRH | ERoS | ASC | RCP |
|---|---|---|---|---|---|---|---|---|
| CS | — | 0.858867 | 0.816058 | 0.427017 | −0.40513 | 0.30664 | −0.27887 | −0.30783 |
| TS | 0.880561 | — | 0.776248 | 0.605032 | −0.42425 | 0.573121 | −0.40888 | −0.32631 |
| FS | 0.824674 | 0.77404 | — | 0.308186 | −0.33502 | 0.203073 | 0.032173 | −0.00081 |
| USPV | 0.423489 | 0.606038 | 0.292183 | — | −0.22041 | 0.638643 | −0.56441 | −0.5831 |
| HbRH | −0.42554 | −0.42297 | −0.33213 | −0.18482 | — | −0.20776 | −0.07007 | −0.10351 |
| ERoS | 0.331792 | 0.607924 | 0.221374 | 0.691903 | −0.24499 | — | −0.58517 | −0.3763 |
| ASC | −0.19251 | −0.27938 | 0.063966 | −0.502 | −0.09105 | −0.48745 | — | 0.748105 |
| RCP | −0.31659 | −0.32915 | −0.05971 | −0.55487 | −0.05955 | −0.29833 | 0.817113 | — |
| **(b)** | **CS** | **TS** | **FS** | **USPV** | **HbRH** | **ERoS** | **ASC** | **RCP** |
| CS | — | 0.835226 | 0.79966 | 0.414718 | −0.40593 | 0.313178 | −0.28814 | −0.30018 |
| TS | 0.871874 | — | 0.754935 | 0.575965 | −0.43141 | 0.566401 | −0.43047 | −0.32086 |
| FS | 0.822683 | 0.761367 | — | 0.30597 | −0.36006 | 0.199425 | 0.027196 | −0.00102 |
| USPV | 0.409015 | 0.588051 | 0.273135 | — | −0.2067 | 0.630624 | −0.55874 | −0.5819 |
| HbRH | −0.43656 | −0.43129 | −0.34513 | −0.13617 | — | −0.1822 | −0.09105 | −0.15006 |
| ERoS | 0.358293 | 0.639143 | 0.229501 | 0.725676 | −0.25038 | — | −0.61744 | −0.41401 |
| ASC | −0.15917 | −0.22425 | 0.069324 | −0.44791 | −0.11265 | −0.43577 | — | 0.686281 |
| RCP | −0.31363 | −0.32865 | −0.106 | −0.52188 | −0.04927 | −0.26239 | 0.790176 | — |
| **(c)** | **CS** | **TS** | **FS** | **USPV** | **HbRH** | **ERoS** | **ASC** | **RCP** |
| CS | — | 0.321631 | 0.336406 | 0.219982 | −0.17004 | 0.133833 | −0.17608 | −0.08968 |
| TS | 0.592665 | — | 0.459743 | 0.280597 | −0.25998 | 0.254154 | −0.2681 | −0.1505 |
| FS | 0.737376 | 0.6098 | — | 0.304023 | −0.40063 | 0.144114 | −0.04536 | −0.04884 |
| USPV | 0.358691 | 0.526558 | 0.211007 | — | −0.16596 | 0.597529 | −0.53474 | −0.56972 |
| HbRH | −0.1884 | −0.23478 | −0.14302 | 0.057846 | — | −0.09112 | −0.11434 | −0.20829 |
| ERoS | 0.084827 | 0.292588 | 0.036046 | 0.62018 | −0.03992 | — | −0.31469 | −0.26921 |
| ASC | −0.14749 | −0.17135 | 0.0691 | −0.3713 | −0.14213 | −0.32648 | — | 0.589916 |
| RCP | −0.31593 | −0.33037 | −0.08779 | −0.53969 | −0.04701 | −0.27681 | 0.800249 | — |

**Table 3.** *Cont.*

| (d) | CS | TS | FS | USPV | HbRH | ERoS | ASC | RCP |
|---|---|---|---|---|---|---|---|---|
| CS | — | 0.323844 | 0.337948 | 0.220534 | −0.17107 | 0.135049 | −0.17721 | −0.09065 |
| TS | 0.6749 | — | 0.52721 | 0.330775 | −0.3096 | 0.331323 | −0.3241 | −0.18401 |
| FS | 0.778735 | 0.667112 | — | 0.304588 | −0.41544 | 0.163534 | −0.01711 | −0.0279 |
| USPV | 0.384434 | 0.557977 | 0.24204 | — | −0.1861 | 0.615276 | −0.54784 | −0.57721 |
| HbRH | −0.23342 | −0.27959 | −0.17728 | 0.037031 | — | −0.11131 | −0.11806 | −0.23117 |
| ERoS | 0.10796 | 0.333232 | 0.039413 | 0.64 | −0.04944 | — | −0.36326 | −0.30311 |
| ASC | −0.1655 | −0.21562 | 0.066715 | −0.4307 | −0.12331 | −0.40243 | — | 0.665464 |
| RCP | −0.31737 | −0.33044 | −0.06312 | −0.55599 | −0.05269 | −0.29562 | 0.813711 | — |
| **(e)** | **CS** | **TS** | **FS** | **USPV** | **HbRH** | **ERoS** | **ASC** | **RCP** |
| CS | — | 0.880394 | 0.824872 | 0.441192 | −0.392094 | 0.288458 | −0.258755 | −0.312252 |
| TS | 0.865763 | — | 0.782128 | 0.625573 | −0.384953 | 0.549375 | −0.346861 | −0.314787 |
| FS | 0.822981 | 0.789199 | — | 0.312032 | −0.288658 | 0.205933 | 0.038388 | −0.003242 |
| USPV | 0.443633 | 0.630941 | 0.319695 | — | −0.241226 | 0.647813 | −0.570721 | −0.582105 |
| HbRH | −0.377150 | −0.386542 | −0.286130 | −0.256779 | — | −0.246965 | −0.040729 | −0.032930 |
| ERoS | 0.274790 | 0.535778 | 0.194523 | 0.618160 | −0.227516 | — | −0.520020 | −0.306866 |
| ASC | −0.317586 | −0.422217 | 0.014438 | −0.593223 | −0.015192 | −0.559757 | — | 0.840345 |
| RCP | −0.305280 | −0.313657 | 0.029273 | −0.590432 | −0.044729 | −0.342215 | 0.801181 | — |
| **(f)** | **CS** | **TS** | **FS** | **USPV** | **HbRH** | **ERoS** | **ASC** | **RCP** |
| CS | — | 0.881580 | 0.824551 | 0.442364 | −0.389739 | 0.285838 | −0.256134 | −0.312183 |
| TS | 0.861465 | — | 0.780105 | 0.625348 | −0.378213 | 0.543715 | −0.337961 | −0.311825 |
| FS | 0.822392 | 0.790519 | — | 0.312473 | −0.282887 | 0.206023 | 0.038979 | −0.003723 |
| USPV | 0.445731 | 0.633517 | 0.322639 | — | −0.243537 | 0.648594 | −0.571243 | −0.581773 |
| HbRH | −0.369097 | −0.380444 | −0.278953 | −0.263651 | — | −0.250834 | −0.038112 | −0.025904 |
| ERoS | 0.267583 | 0.526391 | 0.190537 | 0.608824 | −0.224994 | — | −0.512241 | −0.298850 |
| ASC | −0.333218 | −0.434350 | 0.008473 | −0.595380 | −0.001036 | −0.558613 | — | 0.841635 |
| RCP | −0.302340 | −0.310183 | 0.039716 | −0.591281 | −0.039640 | −0.344671 | 0.792474 | — |
| **(g)** | **CS** | **TS** | **FS** | **USPV** | **HbRH** | **ERoS** | **ASC** | **RCP** |
| CS | — | 0.883184 | 0.823110 | 0.444438 | −0.384503 | 0.280268 | −0.250707 | −0.311714 |
| TS | 0.851240 | — | 0.774336 | 0.623212 | −0.363411 | 0.530608 | −0.319290 | −0.304959 |
| FS | 0.820950 | 0.792906 | — | 0.313347 | −0.271000 | 0.206050 | 0.040100 | −0.004813 |
| USPV | 0.449834 | 0.638541 | 0.328445 | — | −0.248149 | 0.650004 | −0.572171 | −0.580974 |
| HbRH | −0.351451 | −0.367043 | −0.263505 | −0.276192 | — | −0.257982 | −0.033478 | −0.012764 |
| ERoS | 0.252926 | 0.507179 | 0.182086 | 0.589881 | −0.219712 | — | −0.496794 | −0.283008 |
| ASC | −0.352850 | −0.440057 | 0.004440 | −0.582619 | 0.031369 | −0.536518 | — | 0.825594 |
| RCP | −0.295152 | −0.301861 | 0.060098 | −0.589648 | −0.027139 | −0.347025 | 0.769558 | — |
| **(h)** | **CS** | **TS** | **FS** | **USPV** | **HbRH** | **ERoS** | **ASC** | **RCP** |
| CS | — | 0.883184 | 0.823110 | 0.444438 | −0.384503 | 0.280268 | −0.250707 | −0.311714 |
| TS | 0.851240 | — | 0.774336 | 0.623212 | −0.363411 | 0.530608 | −0.319290 | −0.304959 |
| FS | 0.820950 | 0.792906 | — | 0.313347 | −0.271000 | 0.206050 | 0.040100 | −0.004813 |
| USPV | 0.449834 | 0.638541 | 0.328445 | — | −0.248149 | 0.650004 | −0.572171 | −0.580974 |
| HbRH | −0.351451 | −0.367043 | −0.263505 | −0.276192 | — | −0.257982 | −0.033478 | −0.012764 |
| ERoS | 0.252926 | 0.507179 | 0.182086 | 0.589881 | −0.219712 | — | −0.496794 | −0.283008 |
| ASC | −0.352850 | −0.440057 | 0.004440 | −0.582619 | 0.031369 | −0.536518 | — | 0.825594 |
| RCP | −0.295152 | −0.301861 | 0.060098 | −0.589648 | −0.027139 | −0.347025 | 0.769558 | — |

### 3.3. Cosine Similarity Analysis

Table 4 shows the Cos-Sim between each pair of variables in a matrix, with no variable transformations. Figure 4 visualises the same results using a heat map.

**Table 4.** The Cos-Sim indices between each pair of variables (no transformation).

| | CS | TS | FS | USPV | HbRH | ERoS | ASC | RCP |
|---|---|---|---|---|---|---|---|---|
| **CS** | 1 | 0.991704 | 0.994792 | 0.986874 | 0.923028 | 0.970084 | 0.766091 | 0.929148 |
| **TS** | 0.991704 | 1 | 0.985801 | 0.978604 | 0.899398 | 0.974761 | 0.726442 | 0.906988 |
| **FS** | 0.994792 | 0.985801 | 1 | 0.990266 | 0.938099 | 0.971088 | 0.802627 | 0.951239 |
| **USPV** | 0.986874 | 0.978604 | 0.990266 | 1 | 0.947857 | 0.983320 | 0.771200 | 0.940099 |
| **HbRH** | 0.923028 | 0.899398 | 0.938099 | 0.947857 | 1 | 0.918275 | 0.765450 | 0.916059 |
| **ERoS** | 0.970084 | 0.974761 | 0.971088 | 0.983320 | 0.918275 | 1 | 0.712888 | 0.913845 |
| **ASC** | 0.766091 | 0.726442 | 0.802627 | 0.771200 | 0.765450 | 0.712888 | 1 | 0.909864 |
| **RCP** | 0.929148 | 0.906988 | 0.951239 | 0.940099 | 0.916059 | 0.913845 | 0.909864 | 1 |

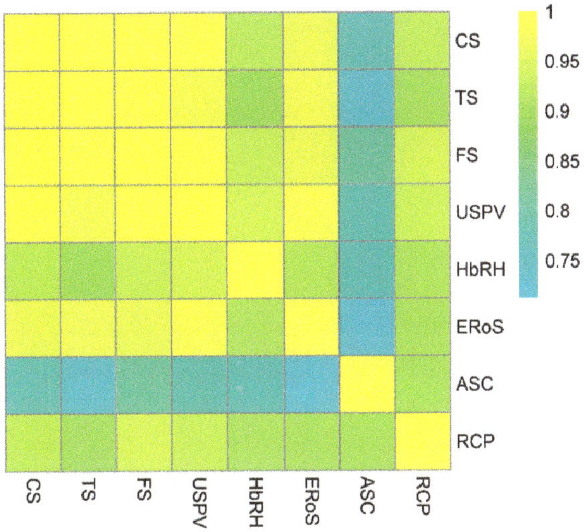


**Figure 4.** The correlation matrix visualised (no transform).

Some studies have successfully treated variables as vectors and used the Cos-Sim between two vectors (see Section 2.3) to confirm the correlation between two variables (but not vice versa), and have observed that a higher P-Co-Co (between −1 and 1) most often indicates a higher Cos-Sim (between 0 and 1). Since in Figure 4 most variable-pairs with higher Cos-Sim indices are observed to have higher P-Co-Cos in Figure 3 (relative to other pairs), the findings from the previous studies are confirmed in terms of P-Co-Co.

Since this outcome can provide justification for performing subsequent analyses, results were not obtained in this study for the other eight transforms. Relevant tables and figures are omitted, as those outcomes are expected to be analogous.

### 3.4. Regression Analysis

This section summarises the results for estimating and establishing the 504 SLR models (see Section 2.3.4). The details are summarised in the web page at the following URL: http://www.DDDM.nkust.edu.tw/download/HPCStudy2_TheUltimateDataExperiments.html (accessed on 3 April 2022) (and in Appendix B), while some initial entries in Table A1 are listed in Table 5 for clarification. A guideline for reading these tables is provided below.

**Table 5.** Results from estimating and developing the models (some initial entries).

| M# | Y | X | *P-Co-Co* | $\alpha^*$ | $\beta^*$ | $p(\alpha^*)$ | $p(\beta^*)$ | $p(M)$ | $R^2$ | $(R^2)^*$ |
|---|---|---|---|---|---|---|---|---|---|---|
| 1 | TS | CS | 0.875333 | −3.667713 | 0.278209 | 0.0466 | 0.0000 | 0.0000 | 0.766208 | 0.759331 |
| 2 | TS | CS | 0.858867 | 5.766322 | 0.001985 | 0.0000 | 0.0000 | 0.0000 | 0.737652 | 0.729936 |
| 3 | TS | CS | 0.835226 | 8.989167 | 0.000018 | 0.0000 | 0.0000 | 0.0000 | 0.697602 | 0.688708 |
| 4 | TS | CS | 0.321631 | 14.568831 | 0.000000 | 0.0000 | 0.0558 | 0.0558 | 0.103447 | 0.077077 |
| 5 | TS | CS | 0.323844 | 14.565644 | 0.000000 | 0.0000 | 0.0540 | 0.0540 | 0.104875 | 0.078548 |
| 6 | TS | CS | 0.880394 | −22.324474 | 4.575331 | 0.0000 | 0.0000 | 0.0000 | 0.775094 | 0.768479 |
| 7 | TS | CS | 0.881580 | −40.931240 | 13.814583 | 0.0000 | 0.0000 | 0.0000 | 0.777183 | 0.770630 |
| 8 | TS | CS | 0.883184 | −62.850941 | 42.785409 | 0.0000 | 0.0000 | 0.0000 | 0.780015 | 0.773545 |
| 9 | TS | CS | 0.883184 | −62.850941 | 12.879692 | 0.0000 | 0.0000 | 0.0000 | 0.780015 | 0.773545 |
| 10 | FS | CS | 0.824243 | 4.243931 | 0.104528 | 0.0000 | 0.0000 | 0.0000 | 0.679376 | 0.669946 |
| 11 | FS | CS | 0.816058 | 7.757808 | 0.000752 | 0.0000 | 0.0000 | 0.0000 | 0.665950 | 0.656125 |
| 12 | FS | CS | 0.799660 | 8.962874 | 0.000007 | 0.0000 | 0.0000 | 0.0000 | 0.639457 | 0.628852 |
| 13 | FS | CS | 0.336406 | 11.087036 | 0.000000 | 0.0000 | 0.0448 | 0.0448 | 0.113169 | 0.087086 |
| 14 | FS | CS | 0.337948 | 11.085907 | 0.000000 | 0.0000 | 0.0438 | 0.0438 | 0.114209 | 0.088156 |
| 15 | FS | CS | 0.824872 | −2.696166 | 1.710449 | 0.1088 | 0.0000 | 0.0000 | 0.680414 | 0.671015 |
| 16 | FS | CS | 0.824551 | −9.616035 | 5.155510 | 0.0004 | 0.0000 | 0.0000 | 0.679884 | 0.670469 |
| 17 | FS | CS | 0.823110 | −17.693156 | 15.910370 | 0.0000 | 0.0000 | 0.0000 | 0.677510 | 0.668025 |
| 18 | FS | CS | 0.823110 | −17.693156 | 4.789498 | 0.0000 | 0.0000 | 0.0000 | 0.677510 | 0.668025 |
| 19 | USPV | CS | 0.437160 | 3.857498 | 0.017583 | 0.0000 | 0.0077 | 0.0077 | 0.191108 | 0.167318 |
| 20 | USPV | CS | 0.427017 | 4.456269 | 0.000125 | 0.0000 | 0.0094 | 0.0094 | 0.182344 | 0.158295 |
| 21 | USPV | CS | 0.414718 | 4.659517 | 0.000001 | 0.0000 | 0.0119 | 0.0119 | 0.171991 | 0.147638 |
| 22 | USPV | CS | 0.219982 | 5.005173 | 0.000000 | 0.0000 | 0.1973 | 0.1973 | 0.048392 | 0.020403 |
| 23 | USPV | CS | 0.220534 | 5.004976 | 0.000000 | 0.0000 | 0.1962 | 0.1962 | 0.048635 | 0.020654 |
| 24 | USPV | CS | 0.441192 | 2.670380 | 0.290147 | 0.0027 | 0.0071 | 0.0071 | 0.194651 | 0.170964 |
| 25 | USPV | CS | 0.442364 | 1.485802 | 0.877204 | 0.2362 | 0.0069 | 0.0069 | 0.195686 | 0.172030 |
| 26 | USPV | CS | 0.444438 | 0.079830 | 2.724581 | 0.9630 | 0.0066 | 0.0066 | 0.197525 | 0.173923 |
| 27 | USPV | CS | 0.444438 | 0.079830 | 0.820181 | 0.9630 | 0.0066 | 0.0066 | 0.197525 | 0.173923 |
| 28 | HBRH | CS | −0.398077 | 65.693972 | −0.390663 | 0.0000 | 0.0162 | 0.0162 | 0.158465 | 0.133714 |
| 29 | HBRH | CS | −0.405125 | 52.917551 | −0.002891 | 0.0000 | 0.0142 | 0.0142 | 0.164126 | 0.139542 |
| 30 | HBRH | CS | −0.405925 | 48.479311 | −0.000027 | 0.0000 | 0.0140 | 0.0140 | 0.164775 | 0.140210 |
| 31 | HBRH | CS | −0.170036 | 40.133608 | 0.000000 | 0.0000 | 0.3215 | 0.3215 | 0.028912 | 0.000351 |
| 32 | HBRH | CS | −0.171068 | 40.138531 | 0.000000 | 0.0000 | 0.3185 | 0.3185 | 0.029264 | 0.000713 |
| 33 | HBRH | CS | −0.392094 | 90.813888 | −6.291765 | 0.0001 | 0.0180 | 0.0180 | 0.153738 | 0.128848 |
| 34 | HBRH | CS | −0.389739 | 115.838473 | −18.857628 | 0.0007 | 0.0188 | 0.0188 | 0.151897 | 0.126953 |
| 35 | HBRH | CS | −0.384503 | 144.146418 | −57.514972 | 0.0020 | 0.0206 | 0.0206 | 0.147843 | 0.122779 |
| 36 | HBRH | CS | −0.384503 | 144.146418 | −17.313732 | 0.0020 | 0.0206 | 0.0206 | 0.147843 | 0.122779 |
| ⋮ | ⋮ | ⋮ | ⋮ | ⋮ | ⋮ | ⋮ | ⋮ | ⋮ | ⋮ | ⋮ |

In Table A1, 'X' determines the eight 'main phases' defined by the eight variables used as the independent variable. In the table, the different background colours denote these phases as blocks. In each main phase fixing the independent variable, there are seven subphases defined by other variables used as the dependent variable '$Y$'. Since each subphase involves nine transforms (refer to the transform descriptions in Section 2.3.4), each block in Table A1 contains 7 × 9 = 63 SLR models, giving 63 × 8 = 504 models in total.

For each model, M# represents the unique model number assigned, with the P-Co-Co between the independent and dependent variables ($X$ and $Y$) given. This is followed by the estimated model parameters: $\alpha^*$ (the estimated value of parameter $\alpha$) and $\beta^*$ (The estimated value of $\beta$); the $p$ values for $\alpha^*$, $\beta^*$ the entire SLR model, $p(\alpha^*)$ and $p(\beta^*)$; and the R square values, $R^2$ and $(R^2)^*$.

As an example, the model with M# = 10 in Table 5 is the model established in the main phase '$X$ = CS' and in the subphase '$Y$ = FS' to identify the relationship between CS and FS of the HPC samples. Since it is the first model of the '$X$ = CS, $Y$ = FS' subphase, transform method (1): $X' = X^1$ is used, which means no transform is performed for the data of variable $X$ in this model (see Section 2.3.4). From the table, the estimated parameters for this model are: $\alpha^*$ = 4.243931 (meaning the regression line intercepts with the $Y$ axis at Y = 4.243931) and $\beta^*$ = 0.104528 (this is the regression line's slope, meaning one unit of increase in CS leads to an increase of 0.104528 in FS). Therefore, the established model can be written as:

$$FS = 4.243931 + 0.104528 \times CS \tag{6}$$

For the same model, $p(\alpha^*) = p(\beta^*) = p(\text{M}) = 0$. Despite the fact that in this table the $p$ values are truncated to 4 digits past the decimal so that '$p = 0$' may represent a very small value of $p$, it could still be inferred that both parameters estimated by the model are very significant, and the model itself is quite significant (i.e., it is very reliable and can be trusted to a large extent).

Moreover, for this model, $R^2$ = 0.679376 and $(R^2)^*$ = 0.669946, meaning that the data-model fitness is acceptable, since over 2/3 of the variability is explained by the established model. However, this is only slightly above the acceptability threshold of 0.6 to claim data-model fitness which was established due to this study's scientific foundation (i.e., natural science than social science investigation, so using 0.6 than 0.4 or even 0.2 is more reasonable).

The full forms of all 504 models are provided in Table A2, which can also be accessed on the web page. For example, Equation (6) for the model with M# = 10 above is the same full form as displayed in Table A2.

### 3.5. Additional Information

Additional information about the SLR models is retrieved from Table A1 and rendered in terms of the transform method used to convert the data value of the independent variable, X. For the analytical targets, only the $p$ and $R^2$ values for all models are considered to conserve space. The reason for considering these values is that the $p$ value of any regression model represents its significance, and $R^2$ indicates the data-model fitness (or the model's explanation power for the variations in the data), both of which are essential for qualifying SLR models.

The results for the $p$ values of the entire model are listed in Table 6. Each sub table contains all pairs of variables (X, Y); $X$ and $Y$ are the RHS (independent) and LHS (dependent) variables in Equation (5), respectively. In the sub tables, the rows are identified by $X$ and the columns are identified by $Y$. The significant results, indicating the SLR model is reliable and the model's predictive power can be trusted, are shown in red font. The significance of the $p$ values is determined using the threshold: $p < 0.10$, which is typically the most relaxed condition that is acceptable by statisticians. In addition, '$p = 0$' may mean a very small value of $p$.

**Table 6.** SLR models' $p$ values for each pair of variables, by different RHS variable transform methods: (a) $X' = X^2$; (b) $X' = X^3$; (c) $X' = e^X$; (d) $X' = 2^X$; (e) $X' = \sqrt{X}$; (f) $X' = \sqrt[3]{X}$; (g) $X' = \log(X)$; (h) $X' = \lg(X)$.

| (a) | CS | TS | FS | USPV | HbRH | ERoS | ASC | RCP |
|---|---|---|---|---|---|---|---|---|
| **CS** | 0.000000 | 0.000000 | 0.000000 | 0.009394 | 0.014243 | 0.068910 | 0.099559 | 0.067789 |
| **TS** | 0.000000 | 0.000000 | 0.000000 | 0.000093 | 0.009916 | 0.000259 | 0.013286 | 0.052104 |
| **FS** | 0.000000 | 0.000000 | 0.000000 | 0.067451 | 0.045792 | 0.234888 | 0.852231 | 0.996245 |
| **USPV** | 0.010063 | 0.000090 | 0.083775 | 0.000000 | 0.196446 | 0.000028 | 0.000336 | 0.000190 |
| **HbRH** | 0.009670 | 0.010165 | 0.047811 | 0.280548 | 0.000000 | 0.224030 | 0.684669 | 0.548015 |
| **ERoS** | 0.048053 | 0.000084 | 0.194446 | 0.000003 | 0.149833 | 0.000000 | 0.000178 | 0.023698 |
| **ASC** | 0.260660 | 0.098907 | 0.710912 | 0.001811 | 0.597407 | 0.002567 | 0.000000 | 0.000000 |
| **RCP** | 0.059945 | 0.049972 | 0.729410 | 0.000445 | 0.730079 | 0.077182 | 0.000000 | 0.000000 |

**Table 6.** *Cont.*

| (b) | CS | TS | FS | USPV | HbRH | ERoS | ASC | RCP |
|---|---|---|---|---|---|---|---|---|
| CS | 0.000000 | 0.000000 | 0.000000 | 0.011907 | 0.014034 | 0.062907 | 0.088345 | 0.075272 |
| TS | 0.000000 | 0.000000 | 0.000000 | 0.000237 | 0.008612 | 0.000317 | 0.008775 | 0.056392 |
| FS | 0.000000 | 0.000000 | 0.000000 | 0.069549 | 0.031002 | 0.243582 | 0.874895 | 0.995302 |
| USPV | 0.013253 | 0.000162 | 0.107008 | 0.000000 | 0.226450 | 0.000037 | 0.000397 | 0.000197 |
| HbRH | 0.007769 | 0.008633 | 0.039253 | 0.428430 | 0.000000 | 0.287537 | 0.597416 | 0.382366 |
| ERoS | 0.031897 | 0.000027 | 0.178153 | 0.000001 | 0.140798 | 0.000000 | 0.000060 | 0.012067 |
| ASC | 0.353814 | 0.188561 | 0.687871 | 0.006154 | 0.513013 | 0.007894 | 0.000000 | 0.000004 |
| RCP | 0.062508 | 0.050343 | 0.538354 | 0.001097 | 0.775369 | 0.122102 | 0.000000 | 0.000000 |

| (c) | CS | TS | FS | USPV | HbRH | ERoS | ASC | RCP |
|---|---|---|---|---|---|---|---|---|
| CS | 0.000000 | 0.055766 | 0.044846 | 0.197340 | 0.321462 | 0.436468 | 0.304312 | 0.602958 |
| TS | 0.000140 | 0.000000 | 0.004788 | 0.097387 | 0.125694 | 0.134707 | 0.113898 | 0.380960 |
| FS | 0.000000 | 0.000079 | 0.000000 | 0.071435 | 0.015465 | 0.401719 | 0.792810 | 0.777263 |
| USPV | 0.031694 | 0.000970 | 0.216709 | 0.000000 | 0.333375 | 0.000119 | 0.000780 | 0.000287 |
| HbRH | 0.271163 | 0.168105 | 0.405339 | 0.737545 | 0.000000 | 0.597118 | 0.506679 | 0.222821 |
| ERoS | 0.622799 | 0.083328 | 0.834675 | 0.000055 | 0.817184 | 0.000000 | 0.061580 | 0.112349 |
| ASC | 0.390676 | 0.317673 | 0.688828 | 0.025775 | 0.408285 | 0.051972 | 0.000000 | 0.000153 |
| RCP | 0.060509 | 0.049079 | 0.610654 | 0.000681 | 0.785420 | 0.102184 | 0.000000 | 0.000000 |

| (d) | CS | TS | FS | USPV | HbRH | ERoS | ASC | RCP |
|---|---|---|---|---|---|---|---|---|
| CS | 0.000000 | 0.054008 | 0.043813 | 0.196189 | 0.318493 | 0.432274 | 0.301173 | 0.599052 |
| TS | 0.000006 | 0.000000 | 0.000954 | 0.048784 | 0.066138 | 0.048389 | 0.053810 | 0.282676 |
| FS | 0.000000 | 0.000009 | 0.000000 | 0.070884 | 0.011745 | 0.340585 | 0.921102 | 0.871662 |
| USPV | 0.020615 | 0.000406 | 0.154958 | 0.000000 | 0.277152 | 0.000065 | 0.000543 | 0.000228 |
| HbRH | 0.170646 | 0.098650 | 0.300968 | 0.830220 | 0.000000 | 0.518074 | 0.492845 | 0.174923 |
| ERoS | 0.530833 | 0.047033 | 0.819475 | 0.000026 | 0.774611 | 0.000000 | 0.029431 | 0.072328 |
| ASC | 0.334737 | 0.206599 | 0.699058 | 0.008735 | 0.473680 | 0.014964 | 0.000000 | 0.000009 |
| RCP | 0.059279 | 0.049027 | 0.714567 | 0.000430 | 0.760238 | 0.080032 | 0.000000 | 0.000000 |

| (e) | CS | TS | FS | USPV | HbRH | ERoS | ASC | RCP |
|---|---|---|---|---|---|---|---|---|
| CS | 0.000000 | 0.000000 | 0.000000 | 0.007071 | 0.018024 | 0.087981 | 0.127552 | 0.063732 |
| TS | 0.000000 | 0.000000 | 0.000000 | 0.000045 | 0.020430 | 0.000520 | 0.038214 | 0.061495 |
| FS | 0.000000 | 0.000000 | 0.000000 | 0.063928 | 0.087751 | 0.228218 | 0.824097 | 0.985030 |
| USPV | 0.006726 | 0.000037 | 0.057342 | 0.000000 | 0.156394 | 0.000019 | 0.000278 | 0.000196 |
| HbRH | 0.023359 | 0.019873 | 0.090691 | 0.130591 | 0.000000 | 0.146475 | 0.813553 | 0.848792 |
| ERoS | 0.104815 | 0.000758 | 0.255601 | 0.000059 | 0.182040 | 0.000000 | 0.001151 | 0.068695 |
| ASC | 0.059099 | 0.010315 | 0.933397 | 0.000137 | 0.929925 | 0.000386 | 0.000000 | 0.000000 |
| RCP | 0.070213 | 0.062484 | 0.865421 | 0.000150 | 0.795613 | 0.041055 | 0.000000 | 0.000000 |

| (f) | CS | TS | FS | USPV | HbRH | ERoS | ASC | RCP |
|---|---|---|---|---|---|---|---|---|
| CS | 0.000000 | 0.000000 | 0.000000 | 0.006903 | 0.018790 | 0.091034 | 0.131594 | 0.063793 |
| TS | 0.000000 | 0.000000 | 0.000000 | 0.000045 | 0.022941 | 0.000610 | 0.043804 | 0.064114 |
| FS | 0.000000 | 0.000000 | 0.000000 | 0.063534 | 0.094571 | 0.228012 | 0.821431 | 0.982807 |
| USPV | 0.006440 | 0.000034 | 0.054960 | 0.000000 | 0.152343 | 0.000019 | 0.000274 | 0.000198 |
| HbRH | 0.026739 | 0.022083 | 0.099447 | 0.120253 | 0.000000 | 0.140053 | 0.825341 | 0.880793 |
| ERoS | 0.114620 | 0.000975 | 0.265660 | 0.000081 | 0.187065 | 0.000000 | 0.001404 | 0.076639 |
| ASC | 0.047043 | 0.008122 | 0.960882 | 0.000128 | 0.995214 | 0.000399 | 0.000000 | 0.000000 |
| RCP | 0.073097 | 0.065603 | 0.818108 | 0.000146 | 0.818453 | 0.039532 | 0.000000 | 0.000000 |

| (g) | CS | TS | FS | USPV | HbRH | ERoS | ASC | RCP |
|---|---|---|---|---|---|---|---|---|
| CS | 0.000000 | 0.000000 | 0.000000 | 0.006615 | 0.020590 | 0.097797 | 0.140259 | 0.064214 |
| TS | 0.000000 | 0.000000 | 0.000000 | 0.000049 | 0.029361 | 0.000871 | 0.057676 | 0.070523 |
| FS | 0.000000 | 0.000000 | 0.000000 | 0.062757 | 0.109889 | 0.227950 | 0.816382 | 0.977773 |
| USPV | 0.005912 | 0.000028 | 0.050493 | 0.000000 | 0.144487 | 0.000018 | 0.000266 | 0.000203 |
| HbRH | 0.035566 | 0.027663 | 0.120466 | 0.102983 | 0.000000 | 0.128735 | 0.846305 | 0.941104 |
| ERoS | 0.136667 | 0.001594 | 0.287837 | 0.000153 | 0.197902 | 0.000000 | 0.002056 | 0.094423 |
| ASC | 0.034789 | 0.007237 | 0.979497 | 0.000193 | 0.855884 | 0.000743 | 0.000000 | 0.000000 |
| RCP | 0.080536 | 0.073575 | 0.727705 | 0.000154 | 0.875153 | 0.038117 | 0.000000 | 0.000000 |

**Table 6.** *Cont.*

| (h) | CS | TS | FS | USPV | HbRH | ERoS | ASC | RCP |
|---|---|---|---|---|---|---|---|---|
| **CS** | 0.000000 | 0.000000 | 0.000000 | 0.006615 | 0.020590 | 0.097797 | 0.140259 | 0.064214 |
| **TS** | 0.000000 | 0.000000 | 0.000000 | 0.000049 | 0.029361 | 0.000871 | 0.057676 | 0.070523 |
| **FS** | 0.000000 | 0.000000 | 0.000000 | 0.062757 | 0.109889 | 0.227950 | 0.816382 | 0.977773 |
| **USPV** | 0.005912 | 0.000028 | 0.050493 | 0.000000 | 0.144487 | 0.000018 | 0.000266 | 0.000203 |
| **HbRH** | 0.035566 | 0.027663 | 0.120466 | 0.102983 | 0.000000 | 0.128735 | 0.846305 | 0.941104 |
| **ERoS** | 0.136667 | 0.001594 | 0.287837 | 0.000153 | 0.197902 | 0.000000 | 0.002056 | 0.094423 |
| **ASC** | 0.034789 | 0.007237 | 0.979497 | 0.000193 | 0.855884 | 0.000743 | 0.000000 | 0.000000 |
| **RCP** | 0.080536 | 0.073575 | 0.727705 | 0.000154 | 0.875153 | 0.038117 | 0.000000 | 0.000000 |

The results for the $R^2$ values of the entire model are listed in Table 7, where each sub table contains the $R^2$ values for the SLR models of all pairs of variables $(X, Y)$ when a transform is applied for $X$. Similar to Table 6, the rows are identified by $X$ and the columns are identified by $Y$ in Table 7. However, unlike the correlation values which may range from $-1$ to 1 in Table 3, the R-squared values only range from 0 to 1. Therefore, the numbers in Table 7 are dyed according to the following convention: dark green for 1, gradient green (lighter and lighter) for (less) positive values, and white for 0.

**Table 7.** SLR models' $R^2$ values for each pair of variables, by different RHS variable transform methods: (a) $X' = X^2$; (b) $X' = X^3$; (c) $X' = e^X$; (d) $X' = 2^X$; (e) $X' = \sqrt{X}$; (f) $X' = \sqrt[3]{X}$; (g) $X' = \log(X)$; (h) $X' = \lg(X)$.

| (a) | CS | TS | FS | USPV | HbRH | ERoS | ASC | RCP |
|---|---|---|---|---|---|---|---|---|
| **CS** | — | 0.737652 | 0.665950 | 0.182344 | 0.164126 | 0.094028 | 0.077766 | 0.094757 |
| **TS** | 0.775388 | — | 0.602560 | 0.366064 | 0.179988 | 0.328468 | 0.167184 | 0.106476 |
| **FS** | 0.680087 | 0.599138 | — | 0.094979 | 0.112237 | 0.041239 | 0.001035 | 0.000001 |
| **USPV** | 0.179343 | 0.367283 | 0.085371 | — | 0.048581 | 0.407865 | 0.318559 | 0.340004 |
| **HbRH** | 0.181081 | 0.178905 | 0.110312 | 0.034157 | — | 0.043164 | 0.004910 | 0.010714 |
| **ERoS** | 0.110086 | 0.369571 | 0.049006 | 0.478730 | 0.060021 | — | 0.342428 | 0.141604 |
| **ASC** | 0.037059 | 0.078054 | 0.004092 | 0.252008 | 0.008291 | 0.237607 | — | 0.559661 |
| **RCP** | 0.100227 | 0.108339 | 0.003565 | 0.307878 | 0.003547 | 0.088998 | 0.667674 | — |
| **(b)** | **CS** | **TS** | **FS** | **USPV** | **HbRH** | **ERoS** | **ASC** | **RCP** |
| **CS** | — | 0.697602 | 0.639457 | 0.171991 | 0.164775 | 0.098080 | 0.083026 | 0.090109 |
| **TS** | 0.760163 | — | 0.569927 | 0.331735 | 0.186118 | 0.320810 | 0.185304 | 0.102950 |
| **FS** | 0.676807 | 0.579680 | — | 0.093618 | 0.129642 | 0.039770 | 0.000740 | 0.000001 |
| **USPV** | 0.167293 | 0.345803 | 0.074603 | — | 0.042725 | 0.397687 | 0.312188 | 0.338609 |
| **HbRH** | 0.190586 | 0.186015 | 0.119115 | 0.018542 | — | 0.033196 | 0.008290 | 0.022518 |
| **ERoS** | 0.128374 | 0.408504 | 0.052671 | 0.526606 | 0.062689 | — | 0.381227 | 0.171407 |
| **ASC** | 0.025334 | 0.050289 | 0.004806 | 0.200628 | 0.012691 | 0.189895 | — | 0.470982 |
| **RCP** | 0.098363 | 0.108009 | 0.011236 | 0.272363 | 0.002428 | 0.068849 | 0.624378 | — |
| **(c)** | **CS** | **TS** | **FS** | **USPV** | **HbRH** | **ERoS** | **ASC** | **RCP** |
| **CS** | — | 0.103447 | 0.113169 | 0.048392 | 0.028912 | 0.017911 | 0.031003 | 0.008043 |
| **TS** | 0.351252 | — | 0.211364 | 0.078735 | 0.067590 | 0.064594 | 0.071876 | 0.022650 |
| **FS** | 0.543724 | 0.371856 | — | 0.09243 | 0.160501 | 0.020769 | 0.002057 | 0.002386 |
| **USPV** | 0.128659 | 0.277263 | 0.044524 | — | 0.027542 | 0.357041 | 0.285944 | 0.324581 |
| **HbRH** | 0.035494 | 0.055122 | 0.020455 | 0.003346 | — | 0.008304 | 0.013074 | 0.043385 |
| **ERoS** | 0.007196 | 0.085607 | 0.001299 | 0.384623 | 0.001594 | — | 0.099029 | 0.072474 |
| **ASC** | 0.021752 | 0.029362 | 0.004775 | 0.137867 | 0.020202 | 0.106588 | — | 0.348001 |
| **RCP** | 0.09981 | 0.109144 | 0.007707 | 0.291261 | 0.002210 | 0.076623 | 0.640398 | — |

**Table 7.** *Cont.*

| (d) | CS | TS | FS | USPV | HbRH | ERoS | ASC | RCP |
|---|---|---|---|---|---|---|---|---|
| CS | — | 0.104875 | 0.114209 | 0.048635 | 0.029264 | 0.018238 | 0.031402 | 0.008217 |
| TS | 0.455491 | — | 0.277951 | 0.109412 | 0.095853 | 0.109775 | 0.105039 | 0.033861 |
| FS | 0.606428 | 0.445038 | — | 0.092774 | 0.17259 | 0.026743 | 0.000293 | 0.000779 |
| USPV | 0.14779 | 0.311338 | 0.058583 | — | 0.034634 | 0.378565 | 0.300128 | 0.333176 |
| HbRH | 0.054487 | 0.078168 | 0.031428 | 0.001371 | — | 0.01239 | 0.013939 | 0.053442 |
| ERoS | 0.011655 | 0.111044 | 0.001553 | 0.409599 | 0.002444 | — | 0.131961 | 0.091879 |
| ASC | 0.02739 | 0.046491 | 0.004451 | 0.185503 | 0.015206 | 0.161951 | — | 0.442842 |
| RCP | 0.100725 | 0.109191 | 0.003984 | 0.309129 | 0.002776 | 0.087393 | 0.662126 | — |
| **(e)** | **CS** | **TS** | **FS** | **USPV** | **HbRH** | **ERoS** | **ASC** | **RCP** |
| CS | — | 0.775094 | 0.680414 | 0.194651 | 0.153738 | 0.083208 | 0.066954 | 0.097501 |
| TS | 0.749546 | — | 0.611724 | 0.391341 | 0.148188 | 0.301813 | 0.120312 | 0.099091 |
| FS | 0.677298 | 0.622835 | — | 0.097364 | 0.083323 | 0.042409 | 0.001474 | $1.05 \times 10^{-5}$ |
| USPV | 0.19681 | 0.398086 | 0.102205 | — | 0.05819 | 0.419662 | 0.325723 | 0.338846 |
| HbRH | 0.142242 | 0.149415 | 0.08187 | 0.065935 | — | 0.060992 | 0.001659 | 0.001084 |
| ERoS | 0.075509 | 0.287058 | 0.037839 | 0.382121 | 0.051764 | — | 0.270421 | 0.094167 |
| ASC | 0.100861 | 0.178268 | 0.000208 | 0.351914 | 0.000231 | 0.313327 | — | 0.70618 |
| RCP | 0.093196 | 0.09838 | 0.000857 | 0.34861 | 0.002001 | 0.117111 | 0.64189 | — |
| **(f)** | **CS** | **TS** | **FS** | **USPV** | **HbRH** | **ERoS** | **ASC** | **RCP** |
| CS | — | 0.777183 | 0.679884 | 0.195686 | 0.151897 | 0.081703 | 0.065604 | 0.097458 |
| TS | 0.742121 | — | 0.608564 | 0.39106 | 0.143045 | 0.295626 | 0.114218 | 0.097235 |
| FS | 0.676328 | 0.624921 | — | 0.097639 | 0.080025 | 0.042445 | 0.001519 | $1.39 \times 10^{-5}$ |
| USPV | 0.198676 | 0.401344 | 0.104096 | — | 0.05931 | 0.420674 | 0.326318 | 0.33846 |
| HbRH | 0.136233 | 0.144737 | 0.077815 | 0.069512 | — | 0.062918 | 0.001453 | 0.000671 |
| ERoS | 0.071601 | 0.277087 | 0.036304 | 0.370667 | 0.050622 | — | 0.262391 | 0.089311 |
| ASC | 0.111034 | 0.18866 | $7.18 \times 10^{-5}$ | 0.354477 | $1.07 \times 10^{-6}$ | 0.312048 | — | 0.70835 |
| RCP | 0.091409 | 0.096214 | 0.001577 | 0.349613 | 0.001571 | 0.118798 | 0.628015 | — |
| **(g)** | **CS** | **TS** | **FS** | **USPV** | **HbRH** | **ERoS** | **ASC** | **RCP** |
| CS | — | 0.780015 | 0.67751 | 0.197525 | 0.147843 | 0.07855 | 0.062854 | 0.097166 |
| TS | 0.724609 | — | 0.599596 | 0.388393 | 0.132067 | 0.281544 | 0.101946 | 0.093 |
| FS | 0.673959 | 0.6287 | — | 0.098187 | 0.073441 | 0.042456 | 0.001608 | $2.32 \times 10^{-5}$ |
| USPV | 0.20235 | 0.407734 | 0.107876 | — | 0.061578 | 0.422505 | 0.327379 | 0.337531 |
| HbRH | 0.123518 | 0.13472 | 0.069435 | 0.076282 | — | 0.066555 | 0.001121 | 0.000163 |
| ERoS | 0.063972 | 0.25723 | 0.033155 | 0.347959 | 0.048274 | — | 0.246804 | 0.080094 |
| ASC | 0.124503 | 0.19365 | $1.97 \times 10^{-5}$ | 0.339445 | 0.000984 | 0.287851 | — | 0.681605 |
| RCP | 0.087115 | 0.09112 | 0.003612 | 0.347684 | 0.000737 | 0.120426 | 0.592219 | — |
| **(h)** | **CS** | **TS** | **FS** | **USPV** | **HbRH** | **ERoS** | **ASC** | **RCP** |
| CS | — | 0.780015 | 0.67751 | 0.197525 | 0.147843 | 0.07855 | 0.062854 | 0.097166 |
| TS | 0.724609 | — | 0.599596 | 0.388393 | 0.132067 | 0.281544 | 0.101946 | 0.093 |
| FS | 0.673959 | 0.6287 | — | 0.098187 | 0.073441 | 0.042456 | 0.001608 | $2.32 \times 10^{-5}$ |
| USPV | 0.20235 | 0.407734 | 0.107876 | — | 0.061578 | 0.422505 | 0.327379 | 0.337531 |
| HbRH | 0.123518 | 0.13472 | 0.069435 | 0.076282 | — | 0.066555 | 0.001121 | 0.000163 |
| ERoS | 0.063972 | 0.25723 | 0.033155 | 0.347959 | 0.048274 | — | 0.246804 | 0.080094 |
| ASC | 0.124503 | 0.19365 | $1.97 \times 10^{-5}$ | 0.339445 | 0.000984 | 0.287851 | — | 0.681605 |
| RCP | 0.087115 | 0.09112 | 0.003612 | 0.347684 | 0.000737 | 0.120426 | 0.592219 | — |

To conserve space, the results of importance are shown with different shades of background colours in the table. In this manner, the SLR models with better data-model fitness can be easily identified. To evaluate the fitness, many scientific studies use the threshold: >0.4 (i.e., it has resolved more variations in the data, so using this model for prediction is therefore more accurate). The results for other observations, such as the $p$ values of the estimated parameters $\alpha^*$ and $\beta^*$, as well as the adjusted $R^2$ value for the model, are summarised in Tables A3–A5, respectively, in Appendix C. These may also be accessed on the 'web page'.

## 4. Insights and Discussions

By analysing the results, many interesting insights can be gained. For the purpose of this study, only the most critical insights are discussed in this section; other advanced information may be obtained from the detailed results in the appendices.

### 4.1. For Method Applications

#### 4.1.1. Insights for Data Pre-Processing Results

During data curation and pre-processing (see Section 3.1), the fresh properties of HPC are excluded because of data incompatibility, and only the hardened mechanical and durability property data at day 28, day 56 and day 91 are utilised for analysis due to data commonalities (i.e., they provide equal-sized data variables and have overlaps in testing time), despite the fact that the frequency of data recording for the experiment is typically predefined [39]. This experience will be useful for locating commensurable HPC testing data during any future analysis.

#### 4.1.2. Insights for Correlation Analysis Results

In the correlation analysis (see Section 3.2), the main results in Table 2 and Figure 3 reveal that:

1. There are two strongly correlated groups of variables: {CS, TS, FS} form a salient group, while {ASC, RCP} form another. All of the correlations identified are either very strong ($r > 0.8$) or near very strong ($0.6 < r < 0.8$ but $r \sim 0.8$).
2. Each group in 1. exists with respect to the same variable category, i.e., CS, TS and FS are hardened mechanical properties, and ASC (in terms of the weight loss percentage) and RCP (in terms of the coulombs measured) are both durability properties.
3. TS is a variable of interest because it forms a medium correlation group with USPV and ERoS. It has a strong ($0.6 < 0.622967 < 0.8$) positive correlation with USPV and a medium ($0.4 < 0.562644 < 0.6$) correlation with ERoS. Since USPV also has a strong positive correlation with ERoS (0.645174), {TS, USPV, ERoS} is another correlated group that overlaps both parametric categories of an HPC sample.
4. The observation of negative correlations is of interest, but there are no strong or very strong negative correlations ($< -0.6$) identified among all variables.
5. USPV has similar medium negative correlations with both ASC (−0.56893) and RCP (−0.58283); this can be explained by the fact that ASC and RCP are strongly correlated (0.817532).
6. ERoS and ASC have a medium negative correlation (−0.54312), and TS and HbRH have a medium (but near-weak) negative correlation (−0.40234). Since a negative correlation does not actually mean a 'poor relation' but rather a relation in the opposite 'direction', the observations of the above medium-negative P-Co-Cos are meaningful.

Beyond the correlations, additional information is provided. Table 3 includes eight sub tables with correlations produced when variable $X$ (of the variable pair $(X, Y)$) is converted using the eight transformation methods described in Section 2.3.4. These sub tables confirm the main results in Figure 3 and Table 2 (when no data variable is transformed), and also help identify the optimal transforms to be used for each variable (this is further discussed in Section 4.1.4). In this regard, such information is valuable.

#### 4.1.3. Insights from the Results of the Cosine-Similarity Analysis

By treating the data variables as vectors, the results obtained from the cosine-similarity analysis essentially confirm the correlations identified between pairs of variables. With few exceptions, a higher Cos-Sim index (between 0 and 1) in Table 4 indicates it has also had a relatively high P-Co-Co in Table 2 (between −1 and 1); this result can also be observed by comparing Figures 3 and 4. As such, the main results of the correlation analysis are further confirmed (and 'double checked', in addition to the confirmations provided in Section 4.1.2 by the P-Co-Cos recalculated after different variable transforms).

Another insight of this research relates to methodology and theory. This study offers new support for the claim that a cosine-similarity analysis can be used as a supplement to the traditional correlation analysis.

#### 4.1.4. Insights from the Established SLR Models

In this study, after successfully estimating the model parameters used in the datasets, 504 SLR models are established. The major insights gained from these analyses are summarised as follows.

##### The Established 'Knowledge Base' Is Novel and Benefits Future HPC Sample Testing

The information relating to the established models (see Tables A1 and A2) is valuable because it can be used to create a true 'knowledge base' for practical applications. Given this knowledge base, if one experimental variable for the HPC sample can be used to predict another variable based on the known mathematical relationship and a guarantee of prediction accuracy, it is logical that the number of testing items that are truly necessary may be reduced. This is particularly true today as stakeholders in the construction and civil engineering industries have reduced the time and resources needed to complete a project, so the time available to test the HPC samples is limited. This should be considered as the primary and original contribution of this study.

##### A Method Is Provided to Explore the Insights into the Variables That Can Practically Be Used to Predict Another Variable and to Determine How Accurate the Prediction Will Be

Table A1 lists nine SLR models (where eight other models are derived from the first 'base model') as a group. It is critical for data analysis in such a design to utilise all opportunities to identify the optimal model that offers both better predictive power and better data-model fitness simultaneously.

Examining SLR models in which variable $X$ is CS, for example (M# = {1, 2, 3, . . . , 27} in Table 5), the first nine models are established for predicting TS ($Y$) from CS ($X$). If $p$(M) < 0.5 is the threshold to confirm a model's significance, the M# = 4 ($X' = e^X$) and M# = 5 ($X' = 2^X$) models are not qualified to be effective ($p$(M) = 0.0558 and $p$(M) = 0.0540). In addition, the models are far from data-model fitness, because $R^2$ = 0.103447 for M# = 4 and $R^2$ = 0.104875 for M# = 5 (which are far below the levels of data-model fitness for seven other models), meaning that these two models may not provide good prediction accuracy. Based on these two results, it is evident that the two SLR models with the 2 'power $X$' transforms are inadequate. A negative implication of this finding could be that the true relation between CS and TS does not exist on this basis, but a positive implication could be that there have been seven models that can be recommended (M# = {1, 2, 3, 6, 7, 8, 9}) in practice or for future research.

Investigating two subsequent ($X$, $Y$) combinations, for M# = {10, 11, . . . , 18}, the models are established for predicting FS ($Y$) from CS ($X$). The resulting situation is similar to using CS to predict TS: the two 'power $X$' SLR models (M# = {13, 14}) are inadequate, and "no true relation between CS and FS exists on this basis". However, the claim can also be made that "another 7 models (M# = {10, 11, 12, 15, 16, 17, 18}) can be effective in practice or for future research".

The M# = {19, 20, . . . , 27} are also established for predicting USPV ($Y$) from CS ($X$). However, no model is qualified to build a predictive relation between these two variables. Two SLR models (M# = {22, 23}) are ineffective ($p > 0.1$), and every model's $R^2$ value is poor ($R^2 < 0.2$) meaning they provide insufficient data-model fitness. Thus, we conclude that that the value of USPV from the value of CS in the experiments performed to test HPC samples cannot be anticipated, and that further research is recommended to find a method to predict USPV using CS.

This analysis is not continued throughout the entire Table A1. The above process can be repeated for all other models in Table A1 to gain other insights regarding the variables that can be used to predict other variables and the accuracy of the prediction process. In addition to these empirical insights that are expected, the experimental design to enable such explorations is the second contribution of this work.

Another Perspective to View the Model Information Is Offered to Differentiate and Recommend the Appropriate Transforms to Be Used for a Variable

An alternative method to evaluate the information relating to the established SLR models can provide additional insights. In Section 3.5 and Appendix C, the significance of the entire model ($p$(M)), the significance of the estimated $\alpha^*$ value (i.e., $p(\alpha^*)$), the significance of the estimated $\beta^*$ value (i.e., $p(\beta^*)$), the R square value (i.e., $R^2$), and the adjusted R square value (i.e., $(R^2)^*$) of the SLR models are systematically presented as separate tables (Tables 6, 7, A1 and A2) according to the transform applied on the RHS variable in the SLR model. This provides another perspective for the model information which can be used in addition to the previous analytical viewpoint that presents a group of models derived from a base model at the same time (e.g., Tables 5, A1 and A2). Therefore, offering two complementary perspectives is another contribution of this research. The following discussion highlights the positive outcomes of this new data-viewing perspective.

Based on the results in Tables 6 and 7 (for the $p$(M) values and the $R^2$ values, respectively), the following insights are gained:

7. In most of the RHS variable transformation cases (refer to the corresponding sub tables in Table 6), the SLR model's $p$(M) value agrees with that of the 'no transform' case (shown in the 'significant' cells with red borders). This not only confirms that the 'no transform' SLR model is effective in its predictive power, but also identifies an alternative set of effective transformed SLR models.
8. In most of the RHS variable transformation cases (refer to the corresponding sub tables in Table 7), the SLR model's $R^2$ value agrees with that of the 'no transform' case (see the colour intensity of a cell). This not only confirms that the 'no transform' SLR model is able to provide accurate predictions, but also identifies an alternative set of qualified 'having transform' SLR models.
9. For most of the RHS variable transforms, the SLR model's $R^2$ value in Table 7 also concurs with the correlation identified in Figure 3 in terms of the equal P-Co-Co value for $(X, Y)$ or $(Y, X)$, and also for the counterpart model with $Y$ and $X$ exchanged as the RHS and LHS variables of the SLR model. In addition, the SLR model's $R^2$ may also concur with the Cos-Sim index calculated for $(X, Y)$ (and $(Y, X)$). In other words, if a high correlation is identified between two variables, a high data-model fitness typically exists, and vice versa. Furthermore, in such cases with high correlation and high data-model fitness, the Cos-Sim values are generally also high. These insights are critical because the relationships between these three values (P-Co-Co, Cos-Sim and $R^2$) that were previously unidentified have now been clarified for this application.
10. The 'lowest-performing RHS variable transforms' can also be identified from the results, supporting the claim that the true relationship between the two variables in an SLR model (i.e., $Y$ and the $X$ being converted using a worse transform) is far from 'a variable is transformed like that'. This claim further implies that some variable transformation methods can be ignored in the future.

Regarding the 'worse variable transforms', they can be identified by interpreting Table 6 in detail (in addition to Tables A1 and A2, as required) to determine if, for a pair of $(X, Y)$, a 'no transform' model shows good results (e.g., $p(M) < 0.05$ and/or $R^2 > 0.6$), and which model(s) with a variable transform shows poor results (e.g., $p(M) > 0.1$ and/or $R^2 < 0.2$)?

Examining the highly-correlated variable group of {CS, TS, FS} (see Section 4.1.2), the SLR models established between all pairs of these variables using every other transform (for $X$) offers a preferred data-model fitness for providing accurate predictions (i.e., $R^2 \geq 0.6$) (shown in the dark cells in the upper left of Table 7a,b,e–h); however, this is not the case using the two 'power $X$ methods' (i.e., $X' = e^X$ and $X' = 2^X$) (shown in the light cells in the upper left part of Table 7c,d).

Further evaluating Table 7 reveals that for all pairs of variables, using these two 'power $X'$' methods may depress the $R^2$ value in general and make it difficult to uncover the relevant information (e.g., to determine which model provides better data-model fitness). It is therefore recommended to neglect both 'power $X'$' methods in future studies for all pairs of test variables.

#### 4.1.5. Other Insights from the Research

In addition to the above insights, several other interesting findings are obtained from the data analysis in this study.

For example, the $p$(M) values in Table 6g,h are identical, as are the $R^2$ values in Table 7g,h; interestingly, the associated P-Co-Co values in Table 3g,h are also identical. By further inspecting Table 5 and A1, these can be explained by the fact that the SLR models will be identical if the transformation method is $X' = \log(X)$ or $X' = \lg(X)$. That means that the base (here, 10 or 2) does not change if a logarithm is used for the RHS variable transform. Therefore, it is recommended for future studies that, for each pair of variables, keeping either SLR model is sufficient.

### 4.2. *Theory-Linking*

In this section, from three theoretical perspectives, the importance of several theory-reflexive finding and insights are discussed.

Through the numerical analysis, the relationship between FS and CS identified in this study (in its mathematical form) is found to concur with their relationship as defined in ACI 318-19 [40–42] and in [43] (except for the coefficients). Therefore, this result is reflexive to the existing theories; in turn, this outcome further validates the results obtained in this study.

Next to this, the measures, data variables, and tests established for the durability of concrete are critical, as durability is an important category of parameters for HPC, particularly when it is to be used in humid areas or marine environments which are prone to acid attacks [44–46]. However, due to the time required to obtain results, true tests for durability are difficult to perform, so other indirect tools such as ASC, RCP, and ERoS should be applied. For this, a conventional theory is that, if a correlation between a hardened mechanical (i.e., strength) property and another durability property can be identified (or if their causal relationship can be established), durability can thus be anticipated by strength properties. In this research, other than very strong or strong relationships identified for the HPC parameters in the same variable category, some significant relationships (from strong to median intensities) between variables across the two categories (see Sections 3 and 4.1) are also found. In other words, these positive outcomes have confirmed the core theoretical logic.

Following these, another theoretical perspective relates to destructive and non-destructive tests (NDTs) [23,47]. Referring to Table 8, which is a simplified version of Table 2 with colour levels added (full red for 1, gradient red (lighter and lighter) for (less) positive values, gradient blue (heavier and heavier) for (more) negative values, dark blue for −1), the upper left cells containing the hardened mechanical properties show that, other than the very strong or near very strong correlations identified among the three destructive 'strength' measures, {CS, TS, FS}, USPV and HbRH are NDTs. Two interesting and important observations about these NDTs can be made.

The first observation is that USPV is an NDT that is more correlated with all destructive tests than HbRH, as evidenced by the magnitude of the *P-Co-Cos*, $|0.43176| > |-0.39808|$, $|0.62297| > |-0.40234|$, and $|0.31071| > |-0.30525|$. Thus, USPV has a medium correlation with TS, a medium correlation with CS, and a weak correlation with FS, while HbRH has a weak negative correlation with TS, a weak negative correlation with CS, and a weak negative correlation with FS. Therefore, the test results for USPV may be more associated with the results obtained from the destructive tests than that of HbRH. This supports the recommendation to keep only one NDT (i.e., USPV, because it provides links to destructive tests more effectively), which is similar to the previous finding supporting the recommendation to keep only one destructive test (i.e., either CS, or TS, or FS, because they are interchangeable; see Section 4.1).

**Table 8.** Simplified and visualised information for Table 2.

| | CS | TS | FS | USPV | HbRH | ERoS | ASC | RCP |
|---|---|---|---|---|---|---|---|---|
| **CS** | — | 0.875333 | 0.824243 | 0.43716 | −0.39808 | | | |
| **TS** | 0.875333 | — | 0.784765 | 0.622967 | −0.40234 | | | |
| **FS** | 0.824243 | 0.784765 | — | 0.310713 | −0.30525 | | | |
| **USPV** | 0.43716 | 0.622967 | 0.310713 | — | −0.23428 | | | |
| **HbRH** | −0.39808 | −0.40234 | −0.30525 | −0.23428 | — | | | |
| **ERoS** | | | | | | — | −0.54312 | −0.33091 |
| **ASC** | | | | | | −0.54312 | — | 0.817532 |
| **RCP** | | | | | | −0.33091 | 0.817532 | — |

The second observation is that there is no significant correlation between the two NDTs for hardened mechanical properties, despite the fact that the test results for USPV may have a weak negative correlation with those of HbRH (−0.23428). This means that, for two similar HPC samples (with the same admixture) tested on the same day, these two NDTs would typically produce diversified results. An evidence-based argument can be presented that the mechanisms behind these two NDTs should be intrinsically different.

Similar observations to the destructive tests and NDTs can also be made by examining the lower right cells with respect to the durability properties. Other theory-reflexive insights of interest based on these results could be gained through future studies.

## 5. Conclusions

This study proposes and applies a systematic data analytic methodology to analyse the experimental data obtained from tests of HPC samples with different admixtures. In contrast with other relevant studies aimed primarily at performing experiments (and providing rationale for the experiments based on materials, physics, or chemistry) or identifying the optimal HPC admixture(s) for grouting concrete materials to be used in the sea (e.g., for the base construction of offshore wind turbines), the purpose of this study is to perform a thorough investigation of the experimental variables related to the testing data.

To achieve this purpose, a methodological framework is proposed. In order to generate comprehensive and in-depth views of the data, numerous methods are utilised, including *P-Co-Co*, *Cos-Sim*, SLR, and heatmap or heat-based tabularised visualisation. This approach is significantly different from those of other experimental-based research or admixture-selection studies. Highlights of the research activities, results, and insights gained are presented below.

1. All variables in the dataset testing the 12 HPC samples of different admixtures were previewed. The dataset was sourced intentionally to cover as many tests for HPC as possible. Among the Cat2 and Cat3 properties, the overlapping (and compatible) data for the eight variables (i.e., the data gathered at 28, 56, and 91 days for CS, TS, FS, USPV, HbRH, ERoS, ASC, and RCP) are identified and specified to support subsequent analyses.
2. Rather than providing a descriptive analysis for the testing data, the investigation began with a correlation analysis, in which the concept of PCA was applied. Within the same variable category, variable groups were identified in which the correlation between each pair of variables was very strong {CS, TS, FS} and {ASC, RCP}. Another medium-correlation variable group was {TS, USPV, ERoS}, in which the included variables overlap the two variable categories. In addition to the positive correlations, some relatively strong negative correlations were also identified between some pairs of variables. The result for P-Co-Cos in the main correlation matrix with no variable

transformation was confirmed by another eight correlation matrices with variable $X$ in Equation (2) being converted using transforms (2)–(9) described in Section 2.3.4.

3. The correlation matrix was validated using the cosine similarity analysis. With few exceptions, a higher Cos-Sim index (between 0 and 1) meant it also had a relatively high P-Co-Co.
4. This analysis generated 504 SLR models, establishing a novel 'knowledge base' which can benefit future HPC sample-testing. By grouping these models using the $C_2^8 \times 2! = 56$ 'base models', each group consisted of nine models (associated with the nine types of RHS variable transform, including 'no transform') (see Table 5). When they are grouped using the 9 types of RHS variable transform, each stratification has 56 models. These two perspectives enabled different types of subsequent analyse.
5. With the established knowledge base, an investigation was performed to determine if a variable can be used to predict another (or not) and, if so, how accurate the prediction will be. For example, by inspecting each SLR model's significance value, $p$(M), it was shown that CS test data can safely be used to predict TS and FS for an HPC sample, but it may not be used to anticipate USPV (see Section 4.1.4). By inspecting the SLR models' $R^2$ values, CS was found to predict TS and FS accurately. These results were cross-validated with the very strong correlations observed among {CS, TS, TS} from the correlation analysis. Insights such as these implied that in future construction projects, valuable time and effort can be saved with respect to HPC testing.
6. By utilising a second perspective to view the model information (see #4 above), the appropriate RHS variable transforms in this analytical context were identified. It was also found that the two 'power $X$' transforms (i.e., $X' = e^X$ and $X' = 2^X$) are both inadequate and should be omitted, and one of the 'logarithm' transforms (i.e., $X' = \log(X)$ or $X' = \lg(X)$) is redundant. Thus, at least three out of the nine transforms can be excluded in future analyses.
7. Some of the insights gained from the results were linked to theories, including the following: (1) The results confirmed relationships found in the existing literature or standards between the test variables; (2) The results verified that the core theoretical logic of this research is effective; (3) The results of some NDTs were more related to the destructive tests (while some NDTs were less related), and different NDTs led to different results; unlike the destructive tests (particularly strength tests), these NDTs are not interchangeable.

The methods used in this study are common, and utilising them for this study was proven to be effective. Therefore, it should also be valid to utilise the methodological framework proposed in this research for similar purposes, e.g., data analysis for other concrete samples with different admixtures. From this perspective, a future story upon taking the proposed framework is perhaps clear from scratch. When a construction project is launched, one first sees if the HPC material(s) is to be used.

As can be expected, if so (using HPC) and if the planned admixtures of the HPC samples to be tested are analogous to the ones being tested in this study, efforts related to testing the materials can be reduced considerably by either policy. The first policy is to use the direct testing results for the samples in the established knowledge base (if 'no testing' is allowed legally). The second is to determine the testing item to be 'predicted' (as desired), look up the knowledge base to know what any other testing item can predict it (and also the information about whether the prediction will be effective and/or accurate), and decide whether to really but safely save the money, and time in most of the cases, for making tests by anticipating the results for the testing item (if some regulated testing items are mandatory).

However, if so (using HPC) and the planned HPC samples to be tested are different (e.g., due to a diversified purpose of use), or if not so (using the non-HPC material), theoretically, the first step is to make similar tests for these samples, followed by using the proposed framework to conduct the thorough data experiments (in order to identify the relationship and the mutual predictive power between each pair of sample parameters).

Then as can be imagined, another knowledge base for the samples included by this purpose of use can be established. Although the outcomes and the unanticipated insights gained may vary in different cases or contexts, after this one-time job, what follows is a similar policy-making problem (see the former paragraph). Anyhow, all this may support efforts to save the time required for pre-project experiments, especially when the project is urgent, which is a crucial but inevitable fact in the current A/E/C industry.

It should be noted that the applications of this framework are not limited to cases in which the source datasets for each measure are tested using conventional methods (i.e., a parameter of HPC samples may have more than one testing method). For example, the framework could be applied when a part of the data is obtained using NMR (nuclear magnetic resonance) [48,49]. In many countries, NMR equipment is legal and commercialised, and the data for some variables can be obtained [50,51]. Other than this, except for HPC, the question of 'can similar data analysis using this framework be carried out for other construction materials?' is also worth of exploration.

Future research directions are shown in Figure 5, wherein the aim of this study which is clearly differentiated from other relevant studies.

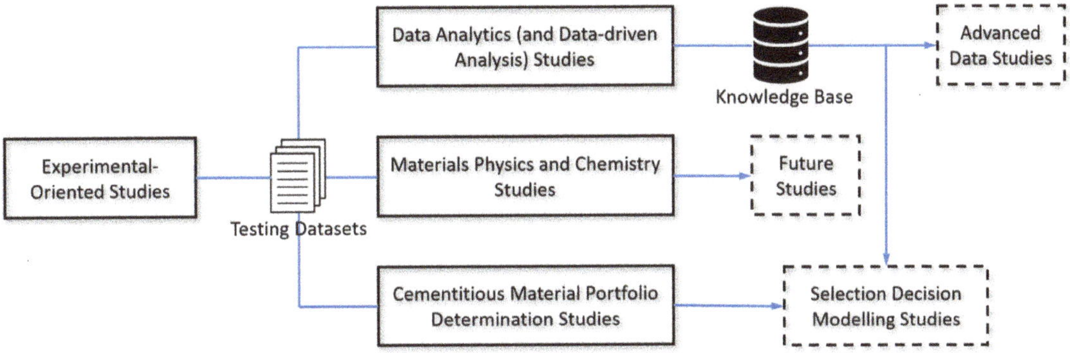


**Figure 5.** The role of this research for HPC studies.

In the figure, the experimental-oriented studies are focused on the performance of the HPC, the inclusion of parameters for developing experiments, and statistics and variability in the data. A critical element that future research directions rely upon is the testing data sets produced by this research. Following this step, there are clear boundaries among the three research directions. The materials, physics and chemistry studies research direction involves a significant amount of opportunity for future studies. However, the knowledge base constructed for the HPC testing parameters as an outcome of the data analytics studies may benefit advance data studies in the future as well as subsequent selection decision modelling studies and current cementitious material portfolio determination studies. The selection decision modelling studies aim to rank the HPC samples with different admixtures and select the optimal candidate using the scientific decision models; therefore, the knowledge base may provide a precise numerical foundation for these models (and added value from the expert knowledge of the researchers).

**Author Contributions:** Conceptualization, methodology, software, visualization, Z.-Y.Z.; data curation, investigation, supervision, validation, W.-T.K.; funding acquisition, project administration, writing—original draft preparation, writing—review and editing, Z.-Y.Z. and W.-T.K. All authors have read and agreed to the published version of the manuscript.

**Funding:** This research was funded by Ministry of Science and Technology (MOST, Taiwan, ROC), grant number MOST 110-2410-H-992-020 and 110-2637-E-992-001. The APC was funded by MDPI discount voucher and MOST 111-2410-H-992-011. Note that the institution 'Ministry of Science and Technology (MOST)' has been renamed as National Science and Technology Council (NSTC) in August 2022 in this area. However, we still keep the original funding number for these projects were approved before the transition.

**Data Availability Statement:** N/A. All related data for the results of this study are detailed in the appendix sections.

**Conflicts of Interest:** The authors declare no conflict of interest.

## Acronyms (Alphabetic Order)

| | |
|---|---|
| ACI | American Concrete Institute |
| A/E/C | Architect/Engineering/Construction |
| ASC | Anti-Sulphate Capability |
| ASTM | American Society for Testing and Materials |
| Cat1 | (Property) Category 1 (Fresh Mechanical) |
| Cat2 | (Property) Category 1 (Hardened Mechanical) |
| Cat3 | (Property) Category 3 (Durability) |
| CE | Construction Engineering |
| Cos-Sim | Cosine Similarity |
| CS | Compressive Strength |
| DDDM | Data-Driven Decision-Making |
| EE | Electrical Engineering |
| ERoS | Electricity Resistivity on Surface |
| ESG | Environmental, Social, Governance |
| FS | Flexural Strength |
| HbRH | Hardness by Rebound Hammer |
| HPC | High Performance Concrete |
| LHS | Left-Hand Side |
| NDT | Non-Destructive Test |
| NMR | Nuclear Magnetic Resonance |
| OLS | Ordinal Least Square |
| PCA | Pairwise Comparison Approach |
| P-Co-Co | Pearson Correlation Coefficient |
| RC | Reinforced Concrete |
| RCP | Rapid Chloride Permeability |
| RCPT | Rapid Chloride Permeability Test |
| RE | Renewable Energy |
| RHS | Right-Hand Side |
| SDG | Sustainable Development Goals |
| SLR | Simple Linear Regression |
| TS | Tensile Strength |
| USPV | Ultrasound Pulse Velocity |

## Appendix A

In that previous study, the data analytical process to obtain Equation (1) progressed as follows:

1. Each pair of (CS, ACP) data (i.e., ACP stands for accumulated charge passed for RCP) was treated as a data point (entry), considering the three testing time points (in terms of #days = 28, 56 and 91) and the 12 mixture portfolios of the concrete sample all together. This yields a data table of two variables, containing the 36 data points.

2. A rough trend was observed by plotting the data points (see Figure 1a; the data are sourced and replotted from the Appendix of (Kuo and Zhuang, 2021)). However, establishing a simple linear regression (SLR) model to fit them, the model did not provide a satisfactory explanation for the relationship between CS and RCP, i.e., the $p$ value of the $F$ statistic was acceptable but relatively week ($p$ = 0.06415), and the regression coefficient of the SLR model was merely $R^2$ = 0.09721, which is poor. Therefore, more solid clues about the relationship must be sought for.
3. The approach of 'dimensional alternation of the variables' was performed. The square root, square and log of the predictor variable (CS) was used as new variables, so several SLR models taking one of these predicting variables and keeping the variable being predicted (accumulated charge passed, or ACP) were reconstructed. In this process, outlier removals were also considered. However, as the case using the original data of variable CS, these trials did not give any more model that is satisfactory.
4. The second approach of 'finding a condensed set of data points and estimating a new SLR model that fits these data points' was then used. It used the K-means to cluster the data points (see Figure 1b), so the cluster centres form a condensed dataset. This process, in itself, removed the outliers to a certain extent, while keeping only the representative information of the data. Eventually, an effective SLR model: "ACP = 7966.72 + (−97.76)CS" was established.
5. The model in (4) was effective because the $R^2$ (regression for data-model fitness) of the model was as high as $R^2$ = 0.8386. The $p$ value of the model also showed a qualified but weak result, which was $p$ = 0.07071. However, since this was a model with one dependent variable and only one independent variable, a further scrutiny revealed that the correlation coefficient between these two data variables was as low as r = −0.9293, which could be justified as near totally negatively correlated.

From the above experiment, the last SLR model in step (5) can serve as an accurate (or, accurately calibrated) predictor when one knows a concrete sample's CS value and would like to postulate the value of ACP that connotes the RCP of the sample. As CS is usually treated as a 'hardened mechanical property' but RCPT is usually a 'durability property' for the high-performance concrete (HPC) materials, and as this process has revealed a relationship between two experimental variables from two parametric categories, such a predictive model should be worthy of note.

## Appendix B

The information in Table A1 is associated the 504 established SLR models. Table A2 lists these models in detail.

**Table A1.** Results from Estimating and Establishing the 504 SLR Models.

| M# | Y | X | P-Co-Co | $\alpha^*$ | $\beta^*$ | $p(\alpha^*)$ | $p(\beta^*)$ | $p$(M) | $R^2$ | $(R^2)^*$ |
|---|---|---|---|---|---|---|---|---|---|---|
| 1 | TS | CS | 0.875333 | −3.667713 | 0.278209 | 0.0466 | 0.0000 | 0.0000 | 0.766208 | 0.759331 |
| 2 | TS | CS | 0.858867 | 5.766322 | 0.001985 | 0.0000 | 0.0000 | 0.0000 | 0.737652 | 0.729936 |
| 3 | TS | CS | 0.835226 | 8.989167 | 0.000018 | 0.0000 | 0.0000 | 0.0000 | 0.697602 | 0.688708 |
| 4 | TS | CS | 0.321631 | 14.568831 | 0.000000 | 0.0000 | 0.0558 | 0.0558 | 0.103447 | 0.077077 |
| 5 | TS | CS | 0.323844 | 14.565644 | 0.000000 | 0.0000 | 0.0540 | 0.0540 | 0.104875 | 0.078548 |
| 6 | TS | CS | 0.880394 | −22.324474 | 4.575331 | 0.0000 | 0.0000 | 0.0000 | 0.775094 | 0.768479 |
| 7 | TS | CS | 0.881580 | −40.931240 | 13.814583 | 0.0000 | 0.0000 | 0.0000 | 0.777183 | 0.770630 |
| 8 | TS | CS | 0.883184 | −62.850941 | 42.785409 | 0.0000 | 0.0000 | 0.0000 | 0.780015 | 0.773545 |
| 9 | TS | CS | 0.883184 | −62.850941 | 12.879692 | 0.0000 | 0.0000 | 0.0000 | 0.780015 | 0.773545 |
| 10 | FS | CS | 0.824243 | 4.243931 | 0.104528 | 0.0000 | 0.0000 | 0.0000 | 0.679376 | 0.669946 |
| 11 | FS | CS | 0.816058 | 7.757808 | 0.000752 | 0.0000 | 0.0000 | 0.0000 | 0.665950 | 0.656125 |
| 12 | FS | CS | 0.799660 | 8.962874 | 0.000007 | 0.0000 | 0.0000 | 0.0000 | 0.639457 | 0.628852 |
| 13 | FS | CS | 0.336406 | 11.087036 | 0.000000 | 0.0000 | 0.0448 | 0.0448 | 0.113169 | 0.087086 |

**Table A1.** *Cont.*

| M# | Y | X | P-Co-Co | $\alpha^*$ | $\beta^*$ | $p(\alpha^*)$ | $p(\beta^*)$ | $p$(M) | $R^2$ | $(R^2)^*$ |
|---|---|---|---|---|---|---|---|---|---|---|
| 14 | FS | CS | 0.337948 | 11.085907 | 0.000000 | 0.0000 | 0.0438 | 0.0438 | 0.114209 | 0.088156 |
| 15 | FS | CS | 0.824872 | −2.696166 | 1.710449 | 0.1088 | 0.0000 | 0.0000 | 0.680414 | 0.671015 |
| 16 | FS | CS | 0.824551 | −9.616035 | 5.155510 | 0.0004 | 0.0000 | 0.0000 | 0.679884 | 0.670469 |
| 17 | FS | CS | 0.823110 | −17.693156 | 15.910370 | 0.0000 | 0.0000 | 0.0000 | 0.677510 | 0.668025 |
| 18 | FS | CS | 0.823110 | −17.693156 | 4.789498 | 0.0000 | 0.0000 | 0.0000 | 0.677510 | 0.668025 |
| 19 | USPV | CS | 0.437160 | 3.857498 | 0.017583 | 0.0000 | 0.0077 | 0.0077 | 0.191108 | 0.167318 |
| 20 | USPV | CS | 0.427017 | 4.456269 | 0.000125 | 0.0000 | 0.0094 | 0.0094 | 0.182344 | 0.158295 |
| 21 | USPV | CS | 0.414718 | 4.659517 | 0.000001 | 0.0000 | 0.0119 | 0.0119 | 0.171991 | 0.147638 |
| 22 | USPV | CS | 0.219982 | 5.005173 | 0.000000 | 0.0000 | 0.1973 | 0.1973 | 0.048392 | 0.020403 |
| 23 | USPV | CS | 0.220534 | 5.004976 | 0.000000 | 0.0000 | 0.1962 | 0.1962 | 0.048635 | 0.020654 |
| 24 | USPV | CS | 0.441192 | 2.670380 | 0.290147 | 0.0027 | 0.0071 | 0.0071 | 0.194651 | 0.170964 |
| 25 | USPV | CS | 0.442364 | 1.485802 | 0.877204 | 0.2362 | 0.0069 | 0.0069 | 0.195686 | 0.172030 |
| 26 | USPV | CS | 0.444438 | 0.079830 | 2.724581 | 0.9630 | 0.0066 | 0.0066 | 0.197525 | 0.173923 |
| 27 | USPV | CS | 0.444438 | 0.079830 | 0.820181 | 0.9630 | 0.0066 | 0.0066 | 0.197525 | 0.173923 |
| 28 | HBRH | CS | −0.398077 | 65.693972 | −0.390663 | 0.0000 | 0.0162 | 0.0162 | 0.158465 | 0.133714 |
| 29 | HBRH | CS | −0.405125 | 52.917551 | −0.002891 | 0.0000 | 0.0142 | 0.0142 | 0.164126 | 0.139542 |
| 30 | HBRH | CS | −0.405925 | 48.479311 | −0.000027 | 0.0000 | 0.0140 | 0.0140 | 0.164775 | 0.140210 |
| 31 | HBRH | CS | −0.170036 | 40.133608 | 0.000000 | 0.0000 | 0.3215 | 0.3215 | 0.028912 | 0.000351 |
| 32 | HBRH | CS | −0.171068 | 40.138531 | 0.000000 | 0.0000 | 0.3185 | 0.3185 | 0.029264 | 0.000713 |
| 33 | HBRH | CS | −0.392094 | 90.813888 | −6.291765 | 0.0001 | 0.0180 | 0.0180 | 0.153738 | 0.128848 |
| 34 | HBRH | CS | −0.389739 | 115.838473 | −18.857628 | 0.0007 | 0.0188 | 0.0188 | 0.151897 | 0.126953 |
| 35 | HBRH | CS | −0.384503 | 144.146418 | −57.514972 | 0.0020 | 0.0206 | 0.0206 | 0.147843 | 0.122779 |
| 36 | HBRH | CS | −0.384503 | 144.146418 | −17.313732 | 0.0020 | 0.0206 | 0.0206 | 0.147843 | 0.122779 |
| 37 | ERoS | CS | 0.295620 | 48.638331 | 0.451482 | 0.0068 | 0.0800 | 0.0800 | 0.087391 | 0.060550 |
| 38 | ERoS | CS | 0.306640 | 63.112123 | 0.003405 | 0.0000 | 0.0689 | 0.0689 | 0.094028 | 0.067382 |
| 39 | ERoS | CS | 0.313178 | 68.142433 | 0.000033 | 0.0000 | 0.0629 | 0.0629 | 0.098080 | 0.071553 |
| 40 | ERoS | CS | 0.133833 | 78.154441 | 0.000000 | 0.0000 | 0.4365 | 0.4365 | 0.017911 | −0.010974 |
| 41 | ERoS | CS | 0.135049 | 78.147143 | 0.000000 | 0.0000 | 0.4323 | 0.4323 | 0.018238 | −0.010637 |
| 42 | ERoS | CS | 0.288458 | 20.158438 | 7.203358 | 0.5500 | 0.0880 | 0.0880 | 0.083208 | 0.056243 |
| 43 | ERoS | CS | 0.285838 | −8.222433 | 21.523030 | 0.8703 | 0.0910 | 0.0910 | 0.081703 | 0.054695 |
| 44 | ERoS | CS | 0.280268 | −39.800918 | 65.241658 | 0.5712 | 0.0978 | 0.0978 | 0.078550 | 0.051449 |
| 45 | ERoS | CS | 0.280268 | −39.800918 | 19.639696 | 0.5712 | 0.0978 | 0.0978 | 0.078550 | 0.051449 |
| 46 | ASC | CS | −0.266200 | 4.951321 | −0.038650 | 0.0043 | 0.1166 | 0.1166 | 0.070862 | 0.043535 |
| 47 | ASC | CS | −0.278865 | 3.725393 | −0.000294 | 0.0001 | 0.0996 | 0.0996 | 0.077766 | 0.050641 |
| 48 | ASC | CS | −0.288142 | 3.301028 | −0.000003 | 0.0000 | 0.0883 | 0.0883 | 0.083026 | 0.056056 |
| 49 | ASC | CS | −0.176077 | 2.440935 | 0.000000 | 0.0000 | 0.3043 | 0.3043 | 0.031003 | 0.002503 |
| 50 | ASC | CS | −0.177206 | 2.441708 | 0.000000 | 0.0000 | 0.3012 | 0.3012 | 0.031402 | 0.002914 |
| 51 | ASC | CS | −0.258755 | 7.370267 | −0.614301 | 0.0276 | 0.1276 | 0.1276 | 0.066954 | 0.039512 |
| 52 | ASC | CS | −0.256134 | 9.782740 | −1.833536 | 0.0491 | 0.1316 | 0.1316 | 0.065604 | 0.038122 |
| 53 | ASC | CS | −0.250707 | 12.455418 | −5.548271 | 0.0706 | 0.1403 | 0.1403 | 0.062854 | 0.035291 |
| 54 | ASC | CS | −0.250707 | 12.455418 | −1.670196 | 0.0706 | 0.1403 | 0.1403 | 0.062854 | 0.035291 |
| 55 | RCP | CS | −0.311780 | 2.267223 | −0.011399 | 0.0000 | 0.0642 | 0.0642 | 0.097207 | 0.070654 |
| 56 | RCP | CS | −0.307826 | 1.882996 | −0.000082 | 0.0000 | 0.0678 | 0.0678 | 0.094757 | 0.068132 |
| 57 | RCP | CS | −0.300182 | 1.750786 | −0.000001 | 0.0000 | 0.0753 | 0.0753 | 0.090109 | 0.063348 |
| 58 | RCP | CS | −0.089682 | 1.518182 | 0.000000 | 0.0000 | 0.6030 | 0.6030 | 0.008043 | −0.021132 |
| 59 | RCP | CS | −0.090646 | 1.518310 | 0.000000 | 0.0000 | 0.5991 | 0.5991 | 0.008217 | −0.020953 |
| 60 | RCP | CS | −0.312252 | 3.025176 | −0.186665 | 0.0005 | 0.0637 | 0.0637 | 0.097501 | 0.070957 |
| 61 | RCP | CS | −0.312183 | 3.780745 | −0.562728 | 0.0031 | 0.0638 | 0.0638 | 0.097458 | 0.070913 |
| 62 | RCP | CS | −0.311714 | 4.663141 | −1.737056 | 0.0078 | 0.0642 | 0.0642 | 0.097166 | 0.070612 |
| 63 | RCP | CS | −0.311714 | 4.663141 | −0.522906 | 0.0078 | 0.0642 | 0.0642 | 0.097166 | 0.070612 |
| 64 | CS | TS | 0.875333 | 25.601048 | 2.754069 | 0.0000 | 0.0000 | 0.0000 | 0.766208 | 0.759331 |
| 65 | CS | TS | 0.880561 | 44.236410 | 0.094689 | 0.0000 | 0.0000 | 0.0000 | 0.775388 | 0.768782 |
| 66 | CS | TS | 0.871874 | 50.906382 | 0.003981 | 0.0000 | 0.0000 | 0.0000 | 0.760163 | 0.753109 |
| 67 | CS | TS | 0.592665 | 64.102514 | 0.000000 | 0.0000 | 0.0001 | 0.0001 | 0.351252 | 0.332171 |
| 68 | CS | TS | 0.674900 | 62.934793 | 0.000010 | 0.0000 | 0.0000 | 0.0000 | 0.455491 | 0.439476 |

**Table A1.** *Cont.*

| M# | Y | X | *P-Co-Co* | $\alpha^*$ | $\beta^*$ | $p(\alpha^*)$ | $p(\beta^*)$ | $p$(M) | $R^2$ | $(R^2)^*$ |
|---|---|---|---|---|---|---|---|---|---|---|
| 69 | CS | TS | 0.865763 | −10.632674 | 20.195417 | 0.1761 | 0.0000 | 0.0000 | 0.749546 | 0.742180 |
| 70 | CS | TS | 0.861465 | −46.679370 | 46.417521 | 0.0003 | 0.0000 | 0.0000 | 0.742121 | 0.734537 |
| 71 | CS | TS | 0.851240 | −29.233152 | 82.840374 | 0.0069 | 0.0000 | 0.0000 | 0.724609 | 0.716509 |
| 72 | CS | TS | 0.851240 | −29.233152 | 24.937437 | 0.0069 | 0.0000 | 0.0000 | 0.724609 | 0.716509 |
| 73 | FS | TS | 0.784765 | 6.546860 | 0.313125 | 0.0000 | 0.0000 | 0.0000 | 0.615856 | 0.604557 |
| 74 | FS | TS | 0.776248 | 8.707568 | 0.010586 | 0.0000 | 0.0000 | 0.0000 | 0.602560 | 0.590871 |
| 75 | FS | TS | 0.754935 | 9.483795 | 0.000437 | 0.0000 | 0.0000 | 0.0000 | 0.569927 | 0.557278 |
| 76 | FS | TS | 0.459743 | 10.957931 | 0.000000 | 0.0000 | 0.0048 | 0.0048 | 0.211364 | 0.188168 |
| 77 | FS | TS | 0.527210 | 10.840734 | 0.000001 | 0.0000 | 0.0010 | 0.0010 | 0.277951 | 0.256714 |
| 78 | FS | TS | 0.782128 | 2.360285 | 2.313706 | 0.0604 | 0.0000 | 0.0000 | 0.611724 | 0.600305 |
| 79 | FS | TS | 0.780105 | −1.800375 | 5.330575 | 0.3222 | 0.0000 | 0.0000 | 0.608564 | 0.597051 |
| 80 | FS | TS | 0.774336 | 0.153470 | 9.556441 | 0.9219 | 0.0000 | 0.0000 | 0.599596 | 0.587820 |
| 81 | FS | TS | 0.774336 | 0.153470 | 2.876775 | 0.9219 | 0.0000 | 0.0000 | 0.599596 | 0.587820 |
| 82 | USPV | TS | 0.622967 | 3.858270 | 0.078833 | 0.0000 | 0.0000 | 0.0000 | 0.388088 | 0.370090 |
| 83 | USPV | TS | 0.605032 | 4.413516 | 0.002617 | 0.0000 | 0.0001 | 0.0001 | 0.366064 | 0.347419 |
| 84 | USPV | TS | 0.575965 | 4.614242 | 0.000106 | 0.0000 | 0.0002 | 0.0002 | 0.331735 | 0.312080 |
| 85 | USPV | TS | 0.280597 | 4.981383 | 0.000000 | 0.0000 | 0.0974 | 0.0974 | 0.078735 | 0.051639 |
| 86 | USPV | TS | 0.330775 | 4.956893 | 0.000000 | 0.0000 | 0.0488 | 0.0488 | 0.109412 | 0.083218 |
| 87 | USPV | TS | 0.625573 | 2.787452 | 0.586914 | 0.0000 | 0.0000 | 0.0000 | 0.391341 | 0.373440 |
| 88 | USPV | TS | 0.625348 | 1.724675 | 1.355218 | 0.0204 | 0.0000 | 0.0000 | 0.391060 | 0.373150 |
| 89 | USPV | TS | 0.623212 | 2.210177 | 2.439321 | 0.0009 | 0.0000 | 0.0000 | 0.388393 | 0.370405 |
| 90 | USPV | TS | 0.623212 | 2.210177 | 0.734309 | 0.0009 | 0.0000 | 0.0000 | 0.388393 | 0.370405 |
| 91 | HBRH | TS | −0.402338 | 58.151339 | −1.242304 | 0.0000 | 0.0150 | 0.0150 | 0.161876 | 0.137225 |
| 92 | HBRH | TS | −0.424250 | 50.224970 | −0.044771 | 0.0000 | 0.0099 | 0.0099 | 0.179988 | 0.155870 |
| 93 | HBRH | TS | −0.431414 | 47.267905 | −0.001933 | 0.0000 | 0.0086 | 0.0086 | 0.186118 | 0.162180 |
| 94 | HBRH | TS | −0.259981 | 40.738938 | 0.000000 | 0.0000 | 0.1257 | 0.1257 | 0.067590 | 0.040166 |
| 95 | HBRH | TS | −0.309601 | 41.307878 | −0.000004 | 0.0000 | 0.0661 | 0.0661 | 0.095853 | 0.069260 |
| 96 | HBRH | TS | −0.384953 | 73.363122 | −8.812436 | 0.0000 | 0.0204 | 0.0204 | 0.148188 | 0.123135 |
| 97 | HBRH | TS | −0.378213 | 88.470940 | −19.999337 | 0.0001 | 0.0229 | 0.0229 | 0.143045 | 0.117840 |
| 98 | HBRH | TS | −0.363411 | 79.818313 | −34.707485 | 0.0001 | 0.0294 | 0.0294 | 0.132067 | 0.106540 |
| 99 | HBRH | TS | −0.363411 | 79.818313 | −10.447994 | 0.0001 | 0.0294 | 0.0294 | 0.132067 | 0.106540 |
| 100 | ERoS | TS | 0.562644 | 38.619765 | 2.703589 | 0.0007 | 0.0004 | 0.0004 | 0.316568 | 0.296467 |
| 101 | ERoS | TS | 0.573121 | 56.641239 | 0.094122 | 0.0000 | 0.0003 | 0.0003 | 0.328468 | 0.308717 |
| 102 | ERoS | TS | 0.566401 | 63.300021 | 0.003950 | 0.0000 | 0.0003 | 0.0003 | 0.320810 | 0.300834 |
| 103 | ERoS | TS | 0.254154 | 77.132815 | 0.000000 | 0.0000 | 0.1347 | 0.1347 | 0.064594 | 0.037082 |
| 104 | ERoS | TS | 0.331323 | 76.049148 | 0.000007 | 0.0000 | 0.0484 | 0.0484 | 0.109775 | 0.083592 |
| 105 | ERoS | TS | 0.549375 | 4.016119 | 19.571676 | 0.8391 | 0.0005 | 0.0005 | 0.301813 | 0.281278 |
| 106 | ERoS | TS | 0.543715 | −30.329979 | 44.742613 | 0.3022 | 0.0006 | 0.0006 | 0.295626 | 0.274909 |
| 107 | ERoS | TS | 0.530608 | −12.372872 | 78.862286 | 0.6246 | 0.0009 | 0.0009 | 0.281544 | 0.260413 |
| 108 | ERoS | TS | 0.530608 | −12.372872 | 23.739914 | 0.6246 | 0.0009 | 0.0009 | 0.281544 | 0.260413 |
| 109 | ASC | TS | −0.371395 | 4.895963 | −0.169661 | 0.0001 | 0.0257 | 0.0257 | 0.137934 | 0.112579 |
| 110 | ASC | TS | −0.408882 | 3.876248 | −0.006384 | 0.0000 | 0.0133 | 0.0133 | 0.167184 | 0.142690 |
| 111 | ASC | TS | −0.430470 | 3.492236 | −0.000285 | 0.0000 | 0.0088 | 0.0088 | 0.185304 | 0.161342 |
| 112 | ASC | TS | −0.268098 | 2.533072 | 0.000000 | 0.0000 | 0.1139 | 0.1139 | 0.071876 | 0.044579 |
| 113 | ASC | TS | −0.324097 | 2.623368 | −0.000001 | 0.0000 | 0.0538 | 0.0538 | 0.105039 | 0.078716 |
| 114 | ASC | TS | −0.346861 | 6.863957 | −1.174774 | 0.0024 | 0.0382 | 0.0382 | 0.120312 | 0.094439 |
| 115 | ASC | TS | −0.337961 | 8.824142 | −2.643973 | 0.0072 | 0.0438 | 0.0438 | 0.114218 | 0.088166 |
| 116 | ASC | TS | −0.319290 | 7.591518 | −4.511500 | 0.0074 | 0.0577 | 0.0577 | 0.101946 | 0.075533 |
| 117 | ASC | TS | −0.319290 | 7.591518 | −1.358097 | 0.0074 | 0.0577 | 0.0577 | 0.101946 | 0.075533 |
| 118 | RCP | TS | −0.321650 | 2.058247 | −0.037000 | 0.0000 | 0.0558 | 0.0558 | 0.103459 | 0.077090 |
| 119 | RCP | TS | −0.326306 | 1.810395 | −0.001283 | 0.0000 | 0.0521 | 0.0521 | 0.106476 | 0.080196 |
| 120 | RCP | TS | −0.320857 | 1.718591 | −0.000054 | 0.0000 | 0.0564 | 0.0564 | 0.102950 | 0.076566 |
| 121 | RCP | TS | −0.150499 | 1.531887 | 0.000000 | 0.0000 | 0.3810 | 0.3810 | 0.022650 | −0.006096 |
| 122 | RCP | TS | −0.184014 | 1.545030 | 0.000000 | 0.0000 | 0.2827 | 0.2827 | 0.033861 | 0.005445 |

**Table A1.** *Cont.*

| M# | Y | X | P-Co-Co | $\alpha^*$ | $\beta^*$ | $p(\alpha^*)$ | $p(\beta^*)$ | p(M) | $R^2$ | $(R^2)^*$ |
|---|---|---|---|---|---|---|---|---|---|---|
| 123 | RCP | TS | −0.314787 | 2.534156 | −0.268461 | 0.0000 | 0.0615 | 0.0615 | 0.099091 | 0.072593 |
| 124 | RCP | TS | −0.311825 | 3.006627 | −0.614281 | 0.0005 | 0.0641 | 0.0641 | 0.097235 | 0.070683 |
| 125 | RCP | TS | −0.304959 | 2.762761 | −1.085035 | 0.0002 | 0.0705 | 0.0705 | 0.093000 | 0.066324 |
| 126 | RCP | TS | −0.304959 | 2.762761 | −0.326628 | 0.0002 | 0.0705 | 0.0705 | 0.093000 | 0.066324 |
| 127 | CS | FS | 0.824243 | −6.326697 | 6.499481 | 0.4688 | 0.0000 | 0.0000 | 0.679376 | 0.669946 |
| 128 | CS | FS | 0.824674 | 29.978536 | 0.285562 | 0.0000 | 0.0000 | 0.0000 | 0.680087 | 0.670677 |
| 129 | CS | FS | 0.822683 | 42.091645 | 0.016421 | 0.0000 | 0.0000 | 0.0000 | 0.676807 | 0.667302 |
| 130 | CS | FS | 0.737376 | 60.470131 | 0.000028 | 0.0000 | 0.0000 | 0.0000 | 0.543724 | 0.530304 |
| 131 | CS | FS | 0.778735 | 57.730585 | 0.002138 | 0.0000 | 0.0000 | 0.0000 | 0.606428 | 0.594853 |
| 132 | CS | FS | 0.822981 | −78.861881 | 43.526353 | 0.0001 | 0.0000 | 0.0000 | 0.677298 | 0.667806 |
| 133 | CS | FS | 0.822392 | −151.371735 | 97.564705 | 0.0000 | 0.0000 | 0.0000 | 0.676328 | 0.666809 |
| 134 | CS | FS | 0.820950 | −107.999762 | 166.924543 | 0.0000 | 0.0000 | 0.0000 | 0.673959 | 0.664369 |
| 135 | CS | FS | 0.820950 | −107.999762 | 50.249294 | 0.0000 | 0.0000 | 0.0000 | 0.673959 | 0.664369 |
| 136 | TS | FS | 0.784765 | −7.199919 | 1.966805 | 0.0222 | 0.0000 | 0.0000 | 0.615856 | 0.604557 |
| 137 | TS | FS | 0.774040 | 3.942253 | 0.085188 | 0.0173 | 0.0000 | 0.0000 | 0.599138 | 0.587348 |
| 138 | TS | FS | 0.761367 | 7.656856 | 0.004830 | 0.0000 | 0.0000 | 0.0000 | 0.579680 | 0.567318 |
| 139 | TS | FS | 0.609800 | 13.245188 | 0.000007 | 0.0000 | 0.0001 | 0.0001 | 0.371856 | 0.353381 |
| 140 | TS | FS | 0.667112 | 12.444315 | 0.000582 | 0.0000 | 0.0000 | 0.0000 | 0.445038 | 0.428716 |
| 141 | TS | FS | 0.789199 | −29.465695 | 13.266225 | 0.0000 | 0.0000 | 0.0000 | 0.622835 | 0.611742 |
| 142 | TS | FS | 0.790519 | −51.724286 | 29.807449 | 0.0000 | 0.0000 | 0.0000 | 0.624921 | 0.613889 |
| 143 | TS | FS | 0.792906 | −38.728089 | 51.241706 | 0.0000 | 0.0000 | 0.0000 | 0.628700 | 0.617779 |
| 144 | TS | FS | 0.792906 | −38.728089 | 15.425291 | 0.0000 | 0.0000 | 0.0000 | 0.628700 | 0.617779 |
| 145 | USPV | FS | 0.310713 | 3.922078 | 0.098543 | 0.0000 | 0.0651 | 0.0651 | 0.096542 | 0.069970 |
| 146 | USPV | FS | 0.308186 | 4.477288 | 0.004292 | 0.0000 | 0.0675 | 0.0675 | 0.094979 | 0.068361 |
| 147 | USPV | FS | 0.305970 | 4.661097 | 0.000246 | 0.0000 | 0.0695 | 0.0695 | 0.093618 | 0.066959 |
| 148 | USPV | FS | 0.304023 | 4.926546 | 0.000000 | 0.0000 | 0.0714 | 0.0714 | 0.092430 | 0.065737 |
| 149 | USPV | FS | 0.304588 | 4.888412 | 0.000034 | 0.0000 | 0.0709 | 0.0709 | 0.092774 | 0.066091 |
| 150 | USPV | FS | 0.312032 | 2.809590 | 0.663750 | 0.0208 | 0.0639 | 0.0639 | 0.097364 | 0.070816 |
| 151 | USPV | FS | 0.312473 | 1.696791 | 1.490970 | 0.3353 | 0.0635 | 0.0635 | 0.097639 | 0.071099 |
| 152 | USPV | FS | 0.313347 | 2.347447 | 2.562552 | 0.1011 | 0.0628 | 0.0628 | 0.098187 | 0.071663 |
| 153 | USPV | FS | 0.313347 | 2.347447 | 0.771405 | 0.1011 | 0.0628 | 0.0628 | 0.098187 | 0.071663 |
| 154 | HBRH | FS | −0.305249 | 66.188588 | −2.362177 | 0.0000 | 0.0702 | 0.0702 | 0.093177 | 0.066505 |
| 155 | HBRH | FS | −0.335018 | 54.273515 | −0.113847 | 0.0000 | 0.0458 | 0.0458 | 0.112237 | 0.086126 |
| 156 | HBRH | FS | −0.360059 | 50.190731 | −0.007053 | 0.0000 | 0.0310 | 0.0310 | 0.129642 | 0.104044 |
| 157 | HBRH | FS | −0.400625 | 42.901148 | −0.000015 | 0.0000 | 0.0155 | 0.0155 | 0.160501 | 0.135809 |
| 158 | HBRH | FS | −0.415439 | 44.279189 | −0.001119 | 0.0000 | 0.0117 | 0.0117 | 0.172590 | 0.148254 |
| 159 | HBRH | FS | −0.288658 | 89.759913 | −14.982391 | 0.0034 | 0.0878 | 0.0878 | 0.083323 | 0.056362 |
| 160 | HBRH | FS | −0.282887 | 113.273551 | −32.935350 | 0.0122 | 0.0946 | 0.0946 | 0.080025 | 0.052967 |
| 161 | HBRH | FS | −0.271000 | 96.258836 | −54.076411 | 0.0085 | 0.1099 | 0.1099 | 0.073441 | 0.046189 |
| 162 | HBRH | FS | −0.271000 | 96.258836 | −16.278622 | 0.0085 | 0.1099 | 0.1099 | 0.073441 | 0.046189 |
| 163 | ERoS | FS | 0.205377 | 50.933952 | 2.473320 | 0.0321 | 0.2295 | 0.2295 | 0.042180 | 0.014008 |
| 164 | ERoS | FS | 0.203073 | 64.911772 | 0.107393 | 0.0000 | 0.2349 | 0.2349 | 0.041239 | 0.013040 |
| 165 | ERoS | FS | 0.199425 | 69.609113 | 0.006079 | 0.0000 | 0.2436 | 0.2436 | 0.039770 | 0.011528 |
| 166 | ERoS | FS | 0.144114 | 76.831094 | 0.000008 | 0.0000 | 0.4017 | 0.4017 | 0.020769 | −0.008032 |
| 167 | ERoS | FS | 0.163534 | 75.822904 | 0.000686 | 0.0000 | 0.3406 | 0.3406 | 0.026743 | −0.001882 |
| 168 | ERoS | FS | 0.205933 | 23.096644 | 16.633940 | 0.6135 | 0.2282 | 0.2282 | 0.042409 | 0.014244 |
| 169 | ERoS | FS | 0.206023 | −4.709195 | 37.327985 | 0.9451 | 0.2280 | 0.2280 | 0.042445 | 0.014282 |
| 170 | ERoS | FS | 0.206050 | 11.758881 | 63.985459 | 0.8305 | 0.2279 | 0.2279 | 0.042456 | 0.014293 |
| 171 | ERoS | FS | 0.206050 | 11.758881 | 19.261543 | 0.8305 | 0.2279 | 0.2279 | 0.042456 | 0.014293 |
| 172 | ASC | FS | 0.036492 | 1.922044 | 0.041780 | 0.3912 | 0.8327 | 0.8327 | 0.001332 | −0.028041 |
| 173 | ASC | FS | 0.032173 | 2.183163 | 0.001618 | 0.0631 | 0.8522 | 0.8522 | 0.001035 | −0.028346 |
| 174 | ASC | FS | 0.027196 | 2.272707 | 0.000079 | 0.0070 | 0.8749 | 0.8749 | 0.000740 | −0.028650 |
| 175 | ASC | FS | −0.045355 | 2.440934 | 0.000000 | 0.0000 | 0.7928 | 0.7928 | 0.002057 | −0.027294 |
| 176 | ASC | FS | −0.017110 | 2.416220 | −0.000007 | 0.0000 | 0.9211 | 0.9211 | 0.000293 | −0.029110 |

**Table A1.** *Cont.*

| M# | Y | X | P-Co-Co | $\alpha^*$ | $\beta^*$ | $p(\alpha^*)$ | $p(\beta^*)$ | $p$(M) | $R^2$ | $(R^2)^*$ |
|---|---|---|---|---|---|---|---|---|---|---|
| 177 | ASC | FS | 0.038388 | 1.405803 | 0.294780 | 0.7512 | 0.8241 | 0.8241 | 0.001474 | −0.027895 |
| 178 | ASC | FS | 0.038979 | 0.890954 | 0.671410 | 0.8933 | 0.8214 | 0.8214 | 0.001519 | −0.027848 |
| 179 | ASC | FS | 0.040100 | 1.152764 | 1.183835 | 0.8288 | 0.8164 | 0.8164 | 0.001608 | −0.027756 |
| 180 | ASC | FS | 0.040100 | 1.152764 | 0.356370 | 0.8288 | 0.8164 | 0.8164 | 0.001608 | −0.027756 |
| 181 | RCP | FS | −0.002053 | 1.518122 | −0.000592 | 0.0101 | 0.9905 | 0.9905 | 0.000004 | −0.029407 |
| 182 | RCP | FS | −0.000813 | 1.512816 | −0.000010 | 0.0000 | 0.9962 | 0.9962 | 0.000001 | −0.029411 |
| 183 | RCP | FS | −0.001017 | 1.512601 | −0.000001 | 0.0000 | 0.9953 | 0.9953 | 0.000001 | −0.029411 |
| 184 | RCP | FS | −0.048844 | 1.525620 | 0.000000 | 0.0000 | 0.7773 | 0.7773 | 0.002386 | −0.026956 |
| 185 | RCP | FS | −0.027904 | 1.522730 | −0.000003 | 0.0000 | 0.8717 | 0.8717 | 0.000779 | −0.028610 |
| 186 | RCP | FS | −0.003242 | 1.532411 | −0.006268 | 0.1759 | 0.9850 | 0.9850 | 0.000011 | −0.029401 |
| 187 | RCP | FS | −0.003723 | 1.547533 | −0.016148 | 0.3581 | 0.9828 | 0.9828 | 0.000014 | −0.029397 |
| 188 | RCP | FS | −0.004813 | 1.548869 | −0.035782 | 0.2535 | 0.9778 | 0.9778 | 0.000023 | −0.029388 |
| 189 | RCP | FS | −0.004813 | 1.548869 | −0.010772 | 0.2535 | 0.9778 | 0.9778 | 0.000023 | −0.029388 |
| 190 | CS | USPV | 0.437160 | 11.699855 | 10.869184 | 0.5495 | 0.0077 | 0.0077 | 0.191108 | 0.167318 |
| 191 | CS | USPV | 0.423489 | 40.273723 | 1.021887 | 0.0002 | 0.0101 | 0.0101 | 0.179343 | 0.155206 |
| 192 | CS | USPV | 0.409015 | 49.795528 | 0.126614 | 0.0000 | 0.0133 | 0.0133 | 0.167293 | 0.142802 |
| 193 | CS | USPV | 0.358691 | 58.320079 | 0.046366 | 0.0000 | 0.0317 | 0.0317 | 0.128659 | 0.103032 |
| 194 | CS | USPV | 0.384434 | 53.582094 | 0.368616 | 0.0000 | 0.0206 | 0.0206 | 0.147790 | 0.122725 |
| 195 | CS | USPV | 0.443633 | −45.426919 | 49.906242 | 0.2492 | 0.0067 | 0.0067 | 0.196810 | 0.173187 |
| 196 | CS | USPV | 0.445731 | −102.544737 | 98.687379 | 0.0870 | 0.0064 | 0.0064 | 0.198676 | 0.175108 |
| 197 | CS | USPV | 0.449834 | −25.625450 | 131.505274 | 0.4194 | 0.0059 | 0.0059 | 0.202350 | 0.178890 |
| 198 | CS | USPV | 0.449834 | −25.625450 | 39.587032 | 0.4194 | 0.0059 | 0.0059 | 0.202350 | 0.178890 |
| 199 | TS | USPV | 0.622967 | −9.951659 | 4.922893 | 0.0715 | 0.0000 | 0.0000 | 0.388088 | 0.370090 |
| 200 | TS | USPV | 0.606038 | 2.940218 | 0.464793 | 0.2864 | 0.0001 | 0.0001 | 0.367283 | 0.348673 |
| 201 | TS | USPV | 0.588051 | 7.236153 | 0.057857 | 0.0004 | 0.0002 | 0.0002 | 0.345803 | 0.326562 |
| 202 | TS | USPV | 0.526558 | 11.054730 | 0.021633 | 0.0000 | 0.0010 | 0.0010 | 0.277263 | 0.256006 |
| 203 | TS | USPV | 0.557977 | 8.911075 | 0.170046 | 0.0000 | 0.0004 | 0.0004 | 0.311338 | 0.291083 |
| 204 | TS | USPV | 0.630941 | −35.725535 | 22.558924 | 0.0020 | 0.0000 | 0.0000 | 0.398086 | 0.380383 |
| 205 | TS | USPV | 0.633517 | −61.495192 | 44.580580 | 0.0005 | 0.0000 | 0.0000 | 0.401344 | 0.383737 |
| 206 | TS | USPV | 0.638541 | −26.695574 | 59.330542 | 0.0038 | 0.0000 | 0.0000 | 0.407734 | 0.390315 |
| 207 | TS | USPV | 0.638541 | −26.695574 | 17.860273 | 0.0038 | 0.0000 | 0.0000 | 0.407734 | 0.390315 |
| 208 | FS | USPV | 0.310713 | 6.252685 | 0.979698 | 0.0215 | 0.0651 | 0.0651 | 0.096542 | 0.069970 |
| 209 | FS | USPV | 0.292183 | 8.896879 | 0.089411 | 0.0000 | 0.0838 | 0.0838 | 0.085371 | 0.058470 |
| 210 | FS | USPV | 0.273135 | 9.776372 | 0.010723 | 0.0000 | 0.1070 | 0.1070 | 0.074603 | 0.047385 |
| 211 | FS | USPV | 0.211007 | 10.578733 | 0.003459 | 0.0000 | 0.2167 | 0.2167 | 0.044524 | 0.016422 |
| 212 | FS | USPV | 0.242040 | 10.158619 | 0.029432 | 0.0000 | 0.1550 | 0.1550 | 0.058583 | 0.030895 |
| 213 | FS | USPV | 0.319695 | 0.963590 | 4.560833 | 0.8540 | 0.0573 | 0.0573 | 0.102205 | 0.075799 |
| 214 | FS | USPV | 0.322639 | −4.325082 | 9.059050 | 0.5830 | 0.0550 | 0.0550 | 0.104096 | 0.077746 |
| 215 | FS | USPV | 0.328445 | 2.662270 | 12.176713 | 0.5309 | 0.0505 | 0.0505 | 0.107876 | 0.081637 |
| 216 | FS | USPV | 0.328445 | 2.662270 | 3.665556 | 0.5309 | 0.0505 | 0.0505 | 0.107876 | 0.081637 |
| 217 | HBRH | USPV | −0.234281 | 68.508814 | −5.716477 | 0.0021 | 0.1690 | 0.1690 | 0.054887 | 0.027090 |
| 218 | HBRH | USPV | −0.220410 | 53.086186 | −0.521949 | 0.0000 | 0.1964 | 0.1964 | 0.048581 | 0.020598 |
| 219 | HBRH | USPV | −0.206701 | 47.978201 | −0.062795 | 0.0000 | 0.2265 | 0.2265 | 0.042725 | 0.014570 |
| 220 | HBRH | USPV | −0.165958 | 43.416246 | −0.021053 | 0.0000 | 0.3334 | 0.3334 | 0.027542 | −0.001060 |
| 221 | HBRH | USPV | −0.186103 | 45.834858 | −0.175122 | 0.0000 | 0.2772 | 0.2772 | 0.034634 | 0.006241 |
| 222 | HBRH | USPV | −0.241226 | 99.412808 | −26.631142 | 0.0213 | 0.1564 | 0.1564 | 0.058190 | 0.030489 |
| 223 | HBRH | USPV | −0.243537 | 130.327235 | −52.916167 | 0.0426 | 0.1523 | 0.1523 | 0.059310 | 0.031643 |
| 224 | HBRH | USPV | −0.248149 | 89.558454 | −71.193135 | 0.0112 | 0.1445 | 0.1445 | 0.061578 | 0.033977 |
| 225 | HBRH | USPV | −0.248149 | 89.558454 | −21.431269 | 0.0112 | 0.1445 | 0.1445 | 0.061578 | 0.033977 |
| 226 | ERoS | USPV | 0.645174 | −44.489856 | 24.498482 | 0.0854 | 0.0000 | 0.0000 | 0.416249 | 0.399080 |
| 227 | ERoS | USPV | 0.638643 | 18.633250 | 2.353559 | 0.1490 | 0.0000 | 0.0000 | 0.407865 | 0.390450 |
| 228 | ERoS | USPV | 0.630624 | 39.712583 | 0.298140 | 0.0001 | 0.0000 | 0.0000 | 0.397687 | 0.379972 |
| 229 | ERoS | USPV | 0.597529 | 58.274020 | 0.117962 | 0.0000 | 0.0001 | 0.0001 | 0.357041 | 0.338131 |
| 230 | ERoS | USPV | 0.615276 | 47.489651 | 0.901006 | 0.0000 | 0.0001 | 0.0001 | 0.378565 | 0.360287 |

**Table A1.** *Cont.*

| M# | Y | X | P-Co-Co | $\alpha^*$ | $\beta^*$ | $p(\alpha^*)$ | $p(\beta^*)$ | $p$(M) | $R^2$ | $(R^2)^*$ |
|---|---|---|---|---|---|---|---|---|---|---|
| 231 | ERoS | USPV | 0.647813 | −170.590922 | 111.297801 | 0.0018 | 0.0000 | 0.0000 | 0.419662 | 0.402593 |
| 232 | ERoS | USPV | 0.648594 | −296.650864 | 219.314541 | 0.0004 | 0.0000 | 0.0000 | 0.420674 | 0.403635 |
| 233 | ERoS | USPV | 0.650004 | −124.288756 | 290.210321 | 0.0044 | 0.0000 | 0.0000 | 0.422505 | 0.405520 |
| 234 | ERoS | USPV | 0.650004 | −124.288756 | 87.362012 | 0.0044 | 0.0000 | 0.0000 | 0.422505 | 0.405520 |
| 235 | ASC | USPV | −0.568934 | 12.705658 | −2.053830 | 0.0000 | 0.0003 | 0.0003 | 0.323686 | 0.303795 |
| 236 | ASC | USPV | −0.564411 | 7.424747 | −0.197743 | 0.0000 | 0.0003 | 0.0003 | 0.318559 | 0.298517 |
| 237 | ASC | USPV | −0.558738 | 5.661972 | −0.025113 | 0.0000 | 0.0004 | 0.0004 | 0.312188 | 0.291958 |
| 238 | ASC | USPV | −0.534737 | 4.115692 | −0.010036 | 0.0000 | 0.0008 | 0.0008 | 0.285944 | 0.264942 |
| 239 | ASC | USPV | −0.547840 | 5.019864 | −0.076269 | 0.0000 | 0.0005 | 0.0005 | 0.300128 | 0.279544 |
| 240 | ASC | USPV | −0.570721 | 23.257569 | −9.321819 | 0.0001 | 0.0003 | 0.0003 | 0.325723 | 0.305891 |
| 241 | ASC | USPV | −0.571243 | 33.806622 | −18.363466 | 0.0001 | 0.0003 | 0.0003 | 0.326318 | 0.306504 |
| 242 | ASC | USPV | −0.572171 | 19.365242 | −24.286354 | 0.0001 | 0.0003 | 0.0003 | 0.327379 | 0.307596 |
| 243 | ASC | USPV | −0.572171 | 19.365242 | −7.310921 | 0.0001 | 0.0003 | 0.0003 | 0.327379 | 0.307596 |
| 244 | RCP | USPV | −0.582828 | 4.172769 | −0.529796 | 0.0000 | 0.0002 | 0.0002 | 0.339688 | 0.320267 |
| 245 | RCP | USPV | −0.583099 | 2.821553 | −0.051442 | 0.0000 | 0.0002 | 0.0002 | 0.340004 | 0.320593 |
| 246 | RCP | USPV | −0.581901 | 2.369856 | −0.006586 | 0.0000 | 0.0002 | 0.0002 | 0.338609 | 0.319156 |
| 247 | RCP | USPV | −0.569720 | 1.974772 | −0.002692 | 0.0000 | 0.0003 | 0.0003 | 0.324581 | 0.304715 |
| 248 | RCP | USPV | −0.577214 | 2.209524 | −0.020235 | 0.0000 | 0.0002 | 0.0002 | 0.333176 | 0.313564 |
| 249 | RCP | USPV | −0.582105 | 6.871176 | −2.394108 | 0.0000 | 0.0002 | 0.0002 | 0.338846 | 0.319400 |
| 250 | RCP | USPV | −0.581773 | 9.568523 | −4.709275 | 0.0000 | 0.0002 | 0.0002 | 0.338460 | 0.319003 |
| 251 | RCP | USPV | −0.580974 | 5.852026 | −6.209543 | 0.0000 | 0.0002 | 0.0002 | 0.337531 | 0.318047 |
| 252 | RCP | USPV | −0.580974 | 5.852026 | −1.869259 | 0.0000 | 0.0002 | 0.0002 | 0.337531 | 0.318047 |
| 253 | CS | HBRH | −0.398077 | 82.439445 | −0.405632 | 0.0000 | 0.0162 | 0.0162 | 0.158465 | 0.133714 |
| 254 | CS | HBRH | −0.425536 | 75.405760 | −0.005286 | 0.0000 | 0.0097 | 0.0097 | 0.181081 | 0.156995 |
| 255 | CS | HBRH | −0.436562 | 72.862987 | −0.000082 | 0.0000 | 0.0078 | 0.0078 | 0.190586 | 0.166780 |
| 256 | CS | HBRH | −0.188399 | 66.902783 | 0.000000 | 0.0000 | 0.2712 | 0.2712 | 0.035494 | 0.007126 |
| 257 | CS | HBRH | −0.233424 | 67.170846 | 0.000000 | 0.0000 | 0.1706 | 0.1706 | 0.054487 | 0.026678 |
| 258 | CS | HBRH | −0.377150 | 96.047260 | −4.770542 | 0.0000 | 0.0234 | 0.0234 | 0.142242 | 0.117014 |
| 259 | CS | HBRH | −0.369097 | 109.563955 | −12.804268 | 0.0000 | 0.0267 | 0.0267 | 0.136233 | 0.110828 |
| 260 | CS | HBRH | −0.351451 | 115.311600 | −31.034230 | 0.0000 | 0.0356 | 0.0356 | 0.123518 | 0.097739 |
| 261 | CS | HBRH | −0.351451 | 115.311600 | −9.342234 | 0.0000 | 0.0356 | 0.0356 | 0.123518 | 0.097739 |
| 262 | TS | HBRH | −0.402338 | 19.962206 | −0.130303 | 0.0000 | 0.0150 | 0.0150 | 0.161876 | 0.137225 |
| 263 | TS | HBRH | −0.422972 | 17.654315 | −0.001670 | 0.0000 | 0.0102 | 0.0102 | 0.178905 | 0.154755 |
| 264 | TS | HBRH | −0.431294 | 16.838405 | −0.000026 | 0.0000 | 0.0086 | 0.0086 | 0.186015 | 0.162074 |
| 265 | TS | HBRH | −0.234780 | 15.016574 | 0.000000 | 0.0000 | 0.1681 | 0.1681 | 0.055122 | 0.027331 |
| 266 | TS | HBRH | −0.279586 | 15.109310 | 0.000000 | 0.0000 | 0.0986 | 0.0986 | 0.078168 | 0.051056 |
| 267 | TS | HBRH | −0.386542 | 24.467753 | −1.553990 | 0.0000 | 0.0199 | 0.0199 | 0.149415 | 0.124398 |
| 268 | TS | HBRH | −0.380444 | 28.951092 | −4.194722 | 0.0000 | 0.0221 | 0.0221 | 0.144737 | 0.119583 |
| 269 | TS | HBRH | −0.367043 | 31.046247 | −10.301284 | 0.0001 | 0.0277 | 0.0277 | 0.134720 | 0.109271 |
| 270 | TS | HBRH | −0.367043 | 31.046247 | −3.100995 | 0.0001 | 0.0277 | 0.0277 | 0.134720 | 0.109271 |
| 271 | FS | HBRH | −0.305249 | 12.743570 | −0.039445 | 0.0000 | 0.0702 | 0.0702 | 0.093177 | 0.066505 |
| 272 | FS | HBRH | −0.332132 | 12.075405 | −0.000523 | 0.0000 | 0.0478 | 0.0478 | 0.110312 | 0.084144 |
| 273 | FS | HBRH | −0.345131 | 11.832096 | −0.000008 | 0.0000 | 0.0393 | 0.0393 | 0.119115 | 0.093207 |
| 274 | FS | HBRH | −0.143020 | 11.232133 | 0.000000 | 0.0000 | 0.4053 | 0.4053 | 0.020455 | −0.008356 |
| 275 | FS | HBRH | −0.177279 | 11.257978 | 0.000000 | 0.0000 | 0.3010 | 0.3010 | 0.031428 | 0.002941 |
| 276 | FS | HBRH | −0.286130 | 14.036116 | −0.458979 | 0.0000 | 0.0907 | 0.0907 | 0.081870 | 0.054866 |
| 277 | FS | HBRH | −0.278953 | 15.320721 | −1.227221 | 0.0000 | 0.0994 | 0.0994 | 0.077815 | 0.050692 |
| 278 | FS | HBRH | −0.263505 | 15.834248 | −2.950814 | 0.0000 | 0.1205 | 0.1205 | 0.069435 | 0.042065 |
| 279 | FS | HBRH | −0.263505 | 15.834248 | −0.888283 | 0.0000 | 0.1205 | 0.1205 | 0.069435 | 0.042065 |
| 280 | USPV | HBRH | −0.234281 | 5.405271 | −0.009602 | 0.0000 | 0.1690 | 0.1690 | 0.054887 | 0.027090 |
| 281 | USPV | HBRH | −0.184816 | 5.182285 | −0.000092 | 0.0000 | 0.2805 | 0.2805 | 0.034157 | 0.005750 |
| 282 | USPV | HBRH | −0.136169 | 5.105547 | −0.000001 | 0.0000 | 0.4284 | 0.4284 | 0.018542 | −0.010324 |
| 283 | USPV | HBRH | 0.057846 | 5.015714 | 0.000000 | 0.0000 | 0.7375 | 0.7375 | 0.003346 | −0.025967 |

**Table A1.** *Cont.*

| M# | Y | X | *P-Co-Co* | $\alpha^*$ | $\beta^*$ | $p(\alpha^*)$ | $p(\beta^*)$ | $p$(M) | $R^2$ | $(R^2)^*$ |
|---|---|---|---|---|---|---|---|---|---|---|
| 284 | USPV | HBRH | 0.037031 | 5.017615 | 0.000000 | 0.0000 | 0.8302 | 0.8302 | 0.001371 | −0.028000 |
| 285 | USPV | HBRH | −0.256779 | 5.837827 | −0.130634 | 0.0000 | 0.1306 | 0.1306 | 0.065935 | 0.038463 |
| 286 | USPV | HBRH | −0.263651 | 6.266212 | −0.367864 | 0.0000 | 0.1203 | 0.1203 | 0.069512 | 0.042145 |
| 287 | USPV | HBRH | −0.276192 | 6.572387 | −0.980913 | 0.0000 | 0.1030 | 0.1030 | 0.076282 | 0.049114 |
| 288 | USPV | HBRH | −0.276192 | 6.572387 | −0.295284 | 0.0000 | 0.1030 | 0.1030 | 0.076282 | 0.049114 |
| 289 | ERoS | HBRH | −0.234481 | 93.091502 | −0.364904 | 0.0000 | 0.1687 | 0.1687 | 0.054981 | 0.027187 |
| 290 | ERoS | HBRH | −0.207759 | 85.361839 | −0.003941 | 0.0000 | 0.2240 | 0.2240 | 0.043164 | 0.015022 |
| 291 | ERoS | HBRH | −0.182198 | 82.755136 | −0.000052 | 0.0000 | 0.2875 | 0.2875 | 0.033196 | 0.004761 |
| 292 | ERoS | HBRH | −0.091124 | 79.017465 | 0.000000 | 0.0000 | 0.5971 | 0.5971 | 0.008304 | −0.020864 |
| 293 | ERoS | HBRH | −0.111312 | 79.206401 | 0.000000 | 0.0000 | 0.5181 | 0.5181 | 0.012390 | −0.016657 |
| 294 | ERoS | HBRH | −0.246965 | 108.321812 | −4.770826 | 0.0000 | 0.1465 | 0.1465 | 0.060992 | 0.033374 |
| 295 | ERoS | HBRH | −0.250834 | 123.476061 | −13.289413 | 0.0002 | 0.1401 | 0.1401 | 0.062918 | 0.035356 |
| 296 | ERoS | HBRH | −0.257982 | 133.518156 | −34.791338 | 0.0006 | 0.1287 | 0.1287 | 0.066555 | 0.039100 |
| 297 | ERoS | HBRH | −0.257982 | 133.518156 | −10.473236 | 0.0006 | 0.1287 | 0.1287 | 0.066555 | 0.039100 |
| 298 | ASC | HBRH | −0.049579 | 2.680781 | −0.007335 | 0.0155 | 0.7740 | 0.7740 | 0.002458 | −0.026881 |
| 299 | ASC | HBRH | −0.070073 | 2.606651 | −0.000126 | 0.0001 | 0.6847 | 0.6847 | 0.004910 | −0.024357 |
| 300 | ASC | HBRH | −0.091050 | 2.587695 | −0.000002 | 0.0000 | 0.5974 | 0.5974 | 0.008290 | −0.020878 |
| 301 | ASC | HBRH | −0.114341 | 2.442201 | 0.000000 | 0.0000 | 0.5067 | 0.5067 | 0.013074 | −0.015953 |
| 302 | ASC | HBRH | −0.118064 | 2.453005 | 0.000000 | 0.0000 | 0.4928 | 0.4928 | 0.013939 | −0.015063 |
| 303 | ASC | HBRH | −0.040729 | 2.855343 | −0.074799 | 0.1595 | 0.8136 | 0.8136 | 0.001659 | −0.027704 |
| 304 | ASC | HBRH | −0.038112 | 3.037543 | −0.191964 | 0.3075 | 0.8253 | 0.8253 | 0.001453 | −0.027917 |
| 305 | ASC | HBRH | −0.033478 | 3.066784 | −0.429225 | 0.3848 | 0.8463 | 0.8463 | 0.001121 | −0.028258 |
| 306 | ASC | HBRH | −0.033478 | 3.066784 | −0.129210 | 0.3848 | 0.8463 | 0.8463 | 0.001121 | −0.028258 |
| 307 | RCP | HBRH | −0.055419 | 1.593665 | −0.002065 | 0.0000 | 0.7482 | 0.7482 | 0.003071 | −0.026250 |
| 308 | RCP | HBRH | −0.103506 | 1.592503 | −0.000047 | 0.0000 | 0.5480 | 0.5480 | 0.010714 | −0.018383 |
| 309 | RCP | HBRH | −0.150060 | 1.594011 | −0.000001 | 0.0000 | 0.3824 | 0.3824 | 0.022518 | −0.006231 |
| 310 | RCP | HBRH | −0.208290 | 1.535961 | 0.000000 | 0.0000 | 0.2228 | 0.2228 | 0.043385 | 0.015249 |
| 311 | RCP | HBRH | −0.231175 | 1.543119 | 0.000000 | 0.0000 | 0.1749 | 0.1749 | 0.053442 | 0.025602 |
| 312 | RCP | HBRH | −0.032930 | 1.606473 | −0.015229 | 0.0029 | 0.8488 | 0.8488 | 0.001084 | −0.028295 |
| 313 | RCP | HBRH | −0.025904 | 1.622523 | −0.032854 | 0.0350 | 0.8808 | 0.8808 | 0.000671 | −0.028721 |
| 314 | RCP | HBRH | −0.012764 | 1.576586 | −0.041207 | 0.0813 | 0.9411 | 0.9411 | 0.000163 | −0.029244 |
| 315 | RCP | HBRH | −0.012764 | 1.576586 | −0.012404 | 0.0813 | 0.9411 | 0.9411 | 0.000163 | −0.029244 |
| 316 | CS | ERoS | 0.295620 | 51.089202 | 0.193566 | 0.0000 | 0.0800 | 0.0800 | 0.087391 | 0.060550 |
| 317 | CS | ERoS | 0.331792 | 56.118078 | 0.001563 | 0.0000 | 0.0481 | 0.0481 | 0.110086 | 0.083912 |
| 318 | CS | ERoS | 0.358293 | 57.830500 | 0.000015 | 0.0000 | 0.0319 | 0.0319 | 0.128374 | 0.102738 |
| 319 | CS | ERoS | 0.084827 | 65.905974 | 0.000000 | 0.0000 | 0.6228 | 0.6228 | 0.007196 | −0.022005 |
| 320 | CS | ERoS | 0.107960 | 65.737481 | 0.000000 | 0.0000 | 0.5308 | 0.5308 | 0.011655 | −0.017414 |
| 321 | CS | ERoS | 0.274790 | 40.634083 | 2.919209 | 0.0131 | 0.1048 | 0.1048 | 0.075509 | 0.048318 |
| 322 | CS | ERoS | 0.267583 | 30.025712 | 8.532478 | 0.1907 | 0.1146 | 0.1146 | 0.071601 | 0.044295 |
| 323 | CS | ERoS | 0.252926 | 19.986077 | 24.635373 | 0.5160 | 0.1367 | 0.1367 | 0.063972 | 0.036441 |
| 324 | CS | ERoS | 0.252926 | 19.986077 | 7.415986 | 0.5160 | 0.1367 | 0.1367 | 0.063972 | 0.036441 |
| 325 | TS | ERoS | 0.562644 | 5.576983 | 0.117092 | 0.0252 | 0.0004 | 0.0004 | 0.316568 | 0.296467 |
| 326 | TS | ERoS | 0.607924 | 8.848829 | 0.000910 | 0.0000 | 0.0001 | 0.0001 | 0.369571 | 0.351029 |
| 327 | TS | ERoS | 0.639143 | 9.976274 | 0.000009 | 0.0000 | 0.0000 | 0.0000 | 0.408504 | 0.391107 |
| 328 | TS | ERoS | 0.292588 | 14.347418 | 0.000000 | 0.0000 | 0.0833 | 0.0833 | 0.085607 | 0.058714 |
| 329 | TS | ERoS | 0.333232 | 14.227276 | 0.000000 | 0.0000 | 0.0470 | 0.0470 | 0.111044 | 0.084898 |
| 330 | TS | ERoS | 0.535778 | −1.126909 | 1.809041 | 0.7964 | 0.0008 | 0.0008 | 0.287058 | 0.266089 |
| 331 | TS | ERoS | 0.526391 | −7.901861 | 5.334861 | 0.2188 | 0.0010 | 0.0010 | 0.277087 | 0.255825 |
| 332 | TS | ERoS | 0.507179 | −14.739001 | 15.700921 | 0.0964 | 0.0016 | 0.0016 | 0.257230 | 0.235384 |
| 333 | TS | ERoS | 0.507179 | −14.739001 | 4.726448 | 0.0964 | 0.0016 | 0.0016 | 0.257230 | 0.235384 |
| 334 | FS | ERoS | 0.205377 | 9.833960 | 0.017054 | 0.0000 | 0.2295 | 0.2295 | 0.042180 | 0.014008 |
| 335 | FS | ERoS | 0.221374 | 10.312553 | 0.000132 | 0.0000 | 0.1944 | 0.1944 | 0.049006 | 0.021036 |
| 336 | FS | ERoS | 0.229501 | 10.486081 | 0.000001 | 0.0000 | 0.1782 | 0.1782 | 0.052671 | 0.024808 |
| 337 | FS | ERoS | 0.036046 | 11.152775 | 0.000000 | 0.0000 | 0.8347 | 0.8347 | 0.001299 | −0.028074 |

**Table A1.** *Cont.*

| M# | Y | X | P-Co-Co | α* | β* | p(α*) | p(β*) | p(M) | $R^2$ | $(R^2)^*$ |
|---|---|---|---|---|---|---|---|---|---|---|
| 338 | FS | ERoS | 0.039413 | 11.147949 | 0.000000 | 0.0000 | 0.8195 | 0.8195 | 0.001553 | −0.027813 |
| 339 | FS | ERoS | 0.194523 | 8.869969 | 0.262068 | 0.0001 | 0.2556 | 0.2556 | 0.037839 | 0.009541 |
| 340 | FS | ERoS | 0.190537 | 7.898445 | 0.770501 | 0.0103 | 0.2657 | 0.2657 | 0.036304 | 0.007960 |
| 341 | FS | ERoS | 0.182086 | 6.945734 | 2.249155 | 0.0857 | 0.2878 | 0.2878 | 0.033155 | 0.004719 |
| 342 | FS | ERoS | 0.182086 | 6.945734 | 0.677063 | 0.0857 | 0.2878 | 0.2878 | 0.033155 | 0.004719 |
| 343 | USPV | ERoS | 0.645174 | 3.688208 | 0.016991 | 0.0000 | 0.0000 | 0.0000 | 0.416249 | 0.399080 |
| 344 | USPV | ERoS | 0.691903 | 4.169383 | 0.000131 | 0.0000 | 0.0000 | 0.0000 | 0.478730 | 0.463398 |
| 345 | USPV | ERoS | 0.725676 | 4.333436 | 0.000001 | 0.0000 | 0.0000 | 0.0000 | 0.526606 | 0.512683 |
| 346 | USPV | ERoS | 0.620180 | 4.907974 | 0.000000 | 0.0000 | 0.0001 | 0.0001 | 0.384623 | 0.366523 |
| 347 | USPV | ERoS | 0.640000 | 4.889594 | 0.000000 | 0.0000 | 0.0000 | 0.0000 | 0.409599 | 0.392235 |
| 348 | USPV | ERoS | 0.618160 | 2.701183 | 0.264124 | 0.0000 | 0.0001 | 0.0001 | 0.382121 | 0.363949 |
| 349 | USPV | ERoS | 0.608824 | 1.703869 | 0.780821 | 0.0285 | 0.0001 | 0.0001 | 0.370667 | 0.352157 |
| 350 | USPV | ERoS | 0.589881 | 0.679038 | 2.310856 | 0.5109 | 0.0002 | 0.0002 | 0.347959 | 0.328781 |
| 351 | USPV | ERoS | 0.589881 | 0.679038 | 0.695637 | 0.5109 | 0.0002 | 0.0002 | 0.347959 | 0.328781 |
| 352 | HBRH | ERoS | −0.234481 | 51.632405 | −0.150674 | 0.0000 | 0.1687 | 0.1687 | 0.054981 | 0.027187 |
| 353 | HBRH | ERoS | −0.244991 | 47.170465 | −0.001133 | 0.0000 | 0.1498 | 0.1498 | 0.060021 | 0.032374 |
| 354 | HBRH | ERoS | −0.250378 | 45.600695 | −0.000010 | 0.0000 | 0.1408 | 0.1408 | 0.062689 | 0.035121 |
| 355 | HBRH | ERoS | −0.039922 | 39.974847 | 0.000000 | 0.0000 | 0.8172 | 0.8172 | 0.001594 | −0.027771 |
| 356 | HBRH | ERoS | −0.049440 | 40.045698 | 0.000000 | 0.0000 | 0.7746 | 0.7746 | 0.002444 | −0.026896 |
| 357 | HBRH | ERoS | −0.227516 | 60.646771 | −2.371986 | 0.0004 | 0.1820 | 0.1820 | 0.051764 | 0.023874 |
| 358 | HBRH | ERoS | −0.224994 | 69.724786 | −7.040810 | 0.0036 | 0.1871 | 0.1871 | 0.050622 | 0.022699 |
| 359 | HBRH | ERoS | −0.219712 | 79.274806 | −21.001758 | 0.0127 | 0.1979 | 0.1979 | 0.048274 | 0.020282 |
| 360 | HBRH | ERoS | −0.219712 | 79.274806 | −6.322159 | 0.0127 | 0.1979 | 0.1979 | 0.048274 | 0.020282 |
| 361 | ASC | ERoS | −0.543122 | 6.445815 | −0.051634 | 0.0000 | 0.0006 | 0.0006 | 0.294982 | 0.274246 |
| 362 | ASC | ERoS | −0.585174 | 4.995638 | −0.000400 | 0.0000 | 0.0002 | 0.0002 | 0.342428 | 0.323088 |
| 363 | ASC | ERoS | −0.617435 | 4.507457 | −0.000004 | 0.0000 | 0.0001 | 0.0001 | 0.381227 | 0.363027 |
| 364 | ASC | ERoS | −0.314689 | 2.599928 | 0.000000 | 0.0000 | 0.0616 | 0.0616 | 0.099029 | 0.072530 |
| 365 | ASC | ERoS | −0.363265 | 2.662620 | 0.000000 | 0.0000 | 0.0294 | 0.0294 | 0.131961 | 0.106431 |
| 366 | ASC | ERoS | −0.520020 | 9.440434 | −0.802103 | 0.0000 | 0.0012 | 0.0012 | 0.270421 | 0.248963 |
| 367 | ASC | ERoS | −0.512241 | 12.470584 | −2.371573 | 0.0001 | 0.0014 | 0.0014 | 0.262391 | 0.240697 |
| 368 | ASC | ERoS | −0.496794 | 15.596328 | −7.025658 | 0.0004 | 0.0021 | 0.0021 | 0.246804 | 0.224651 |
| 369 | ASC | ERoS | −0.496794 | 15.596328 | −2.114934 | 0.0004 | 0.0021 | 0.0021 | 0.246804 | 0.224651 |
| 370 | RCP | ERoS | −0.330914 | 2.133923 | −0.007922 | 0.0000 | 0.0487 | 0.0487 | 0.109504 | 0.083313 |
| 371 | RCP | ERoS | −0.376303 | 1.933609 | −0.000065 | 0.0000 | 0.0237 | 0.0237 | 0.141604 | 0.116357 |
| 372 | RCP | ERoS | −0.414013 | 1.869216 | −0.000001 | 0.0000 | 0.0121 | 0.0121 | 0.171407 | 0.147036 |
| 373 | RCP | ERoS | −0.269210 | 1.556967 | 0.000000 | 0.0000 | 0.1123 | 0.1123 | 0.072474 | 0.045194 |
| 374 | RCP | ERoS | −0.303115 | 1.569021 | 0.000000 | 0.0000 | 0.0723 | 0.0723 | 0.091879 | 0.065169 |
| 375 | RCP | ERoS | −0.306866 | 2.559308 | −0.119186 | 0.0001 | 0.0687 | 0.0687 | 0.094167 | 0.067525 |
| 376 | RCP | ERoS | −0.298850 | 2.992587 | −0.348403 | 0.0008 | 0.0766 | 0.0766 | 0.089311 | 0.062527 |
| 377 | RCP | ERoS | −0.283008 | 3.406065 | −1.007805 | 0.0040 | 0.0944 | 0.0944 | 0.080094 | 0.053038 |
| 378 | RCP | ERoS | −0.283008 | 3.406065 | −0.303379 | 0.0040 | 0.0944 | 0.0944 | 0.080094 | 0.053038 |
| 379 | CS | ASC | −0.266200 | 70.677622 | −1.833423 | 0.0000 | 0.1166 | 0.1166 | 0.070862 | 0.043535 |
| 380 | CS | ASC | −0.192506 | 68.271436 | −0.225307 | 0.0000 | 0.2607 | 0.2607 | 0.037059 | 0.008737 |
| 381 | CS | ASC | −0.159165 | 67.549397 | −0.032803 | 0.0000 | 0.3538 | 0.3538 | 0.025334 | −0.003333 |
| 382 | CS | ASC | −0.147486 | 67.191905 | −0.017965 | 0.0000 | 0.3907 | 0.3907 | 0.021752 | −0.007020 |
| 383 | CS | ASC | −0.165498 | 67.710319 | −0.124547 | 0.0000 | 0.3347 | 0.3347 | 0.027390 | −0.001217 |
| 384 | CS | ASC | −0.317586 | 75.050290 | −6.179200 | 0.0000 | 0.0591 | 0.0591 | 0.100861 | 0.074416 |
| 385 | CS | ASC | −0.333218 | 79.074048 | −10.387030 | 0.0000 | 0.0470 | 0.0470 | 0.111034 | 0.084888 |
| 386 | CS | ASC | −0.352850 | 67.936551 | −8.663600 | 0.0000 | 0.0348 | 0.0348 | 0.124503 | 0.098753 |
| 387 | CS | ASC | −0.352850 | 67.936551 | −2.608003 | 0.0000 | 0.0348 | 0.0348 | 0.124503 | 0.098753 |
| 388 | TS | ASC | −0.371395 | 16.719103 | −0.812997 | 0.0000 | 0.0257 | 0.0257 | 0.137934 | 0.112579 |
| 389 | TS | ASC | −0.279381 | 15.687327 | −0.103926 | 0.0000 | 0.0989 | 0.0989 | 0.078054 | 0.050938 |
| 390 | TS | ASC | −0.224252 | 15.337423 | −0.014689 | 0.0000 | 0.1886 | 0.1886 | 0.050289 | 0.022356 |
| 391 | TS | ASC | −0.171353 | 15.107116 | −0.006634 | 0.0000 | 0.3177 | 0.3177 | 0.029362 | 0.000814 |
| 392 | TS | ASC | −0.215619 | 15.361860 | −0.051573 | 0.0000 | 0.2066 | 0.2066 | 0.046491 | 0.018447 |

**Table A1.** *Cont.*

| M# | Y | X | *P-Co-Co* | $\alpha^*$ | $\beta^*$ | $p(\alpha^*)$ | $p(\beta^*)$ | $p$(M) | $R^2$ | $(R^2)^*$ |
|---|---|---|---|---|---|---|---|---|---|---|
| 393 | TS | ASC | −0.422217 | 18.475277 | −2.610992 | 0.0000 | 0.0103 | 0.0103 | 0.178268 | 0.154099 |
| 394 | TS | ASC | −0.434350 | 20.070081 | −4.303288 | 0.0000 | 0.0081 | 0.0081 | 0.188660 | 0.164797 |
| 395 | TS | ASC | −0.440057 | 15.426529 | −3.434117 | 0.0000 | 0.0072 | 0.0072 | 0.193650 | 0.169934 |
| 396 | TS | ASC | −0.440057 | 15.426529 | −1.033772 | 0.0000 | 0.0072 | 0.0072 | 0.193650 | 0.169934 |
| 397 | FS | ASC | 0.036492 | 11.097746 | 0.031874 | 0.0000 | 0.8327 | 0.8327 | 0.001332 | −0.028041 |
| 398 | FS | ASC | 0.063966 | 11.090721 | 0.009494 | 0.0000 | 0.7109 | 0.7109 | 0.004092 | −0.025200 |
| 399 | FS | ASC | 0.069324 | 11.104756 | 0.001812 | 0.0000 | 0.6879 | 0.6879 | 0.004806 | −0.024465 |
| 400 | FS | ASC | 0.069100 | 11.120763 | 0.001067 | 0.0000 | 0.6888 | 0.6888 | 0.004775 | −0.024497 |
| 401 | FS | ASC | 0.066715 | 11.101677 | 0.006367 | 0.0000 | 0.6991 | 0.6991 | 0.004451 | −0.024830 |
| 402 | FS | ASC | 0.014438 | 11.123430 | 0.035624 | 0.0000 | 0.9334 | 0.9334 | 0.000208 | −0.029197 |
| 403 | FS | ASC | 0.008473 | 11.132687 | 0.033496 | 0.0000 | 0.9609 | 0.9609 | 0.000072 | −0.029338 |
| 404 | FS | ASC | 0.004440 | 11.171274 | 0.013825 | 0.0000 | 0.9795 | 0.9795 | 0.000020 | −0.029391 |
| 405 | FS | ASC | 0.004440 | 11.171274 | 0.004162 | 0.0000 | 0.9795 | 0.9795 | 0.000020 | −0.029391 |
| 406 | USPV | ASC | −0.568934 | 5.399677 | −0.157601 | 0.0000 | 0.0003 | 0.0003 | 0.323686 | 0.303795 |
| 407 | USPV | ASC | −0.502004 | 5.230189 | −0.023631 | 0.0000 | 0.0018 | 0.0018 | 0.252008 | 0.230008 |
| 408 | USPV | ASC | −0.447915 | 5.164850 | −0.003713 | 0.0000 | 0.0062 | 0.0062 | 0.200628 | 0.177117 |
| 409 | USPV | ASC | −0.371304 | 5.113722 | −0.001819 | 0.0000 | 0.0258 | 0.0258 | 0.137867 | 0.112510 |
| 410 | USPV | ASC | −0.430700 | 5.171037 | −0.013036 | 0.0000 | 0.0087 | 0.0087 | 0.185503 | 0.161547 |
| 411 | USPV | ASC | −0.593223 | 5.680741 | −0.464229 | 0.0000 | 0.0001 | 0.0001 | 0.351914 | 0.332853 |
| 412 | USPV | ASC | −0.595380 | 5.941333 | −0.746448 | 0.0000 | 0.0001 | 0.0001 | 0.354477 | 0.335491 |
| 413 | USPV | ASC | −0.582619 | 5.132017 | −0.575354 | 0.0000 | 0.0002 | 0.0002 | 0.339445 | 0.320016 |
| 414 | USPV | ASC | −0.582619 | 5.132017 | −0.173199 | 0.0000 | 0.0002 | 0.0002 | 0.339445 | 0.320016 |
| 415 | HBRH | ASC | −0.049579 | 40.594427 | −0.335109 | 0.0000 | 0.7740 | 0.7740 | 0.002458 | −0.026881 |
| 416 | HBRH | ASC | −0.091052 | 40.710013 | −0.104582 | 0.0000 | 0.5974 | 0.5974 | 0.008291 | −0.020877 |
| 417 | HBRH | ASC | −0.112654 | 40.663258 | −0.022785 | 0.0000 | 0.5130 | 0.5130 | 0.012691 | −0.016348 |
| 418 | HBRH | ASC | −0.142134 | 40.639521 | −0.016991 | 0.0000 | 0.4083 | 0.4083 | 0.020202 | −0.008616 |
| 419 | HBRH | ASC | −0.123312 | 40.826770 | −0.091071 | 0.0000 | 0.4737 | 0.4737 | 0.015206 | −0.013759 |
| 420 | HBRH | ASC | −0.015192 | 40.204772 | −0.290080 | 0.0000 | 0.9299 | 0.9299 | 0.000231 | −0.029174 |
| 421 | HBRH | ASC | −0.001036 | 39.832882 | −0.031701 | 0.0000 | 0.9952 | 0.9952 | 0.000001 | −0.029411 |
| 422 | HBRH | ASC | 0.031369 | 39.650913 | 0.755863 | 0.0000 | 0.8559 | 0.8559 | 0.000984 | −0.028399 |
| 423 | HBRH | ASC | 0.031369 | 39.650913 | 0.227537 | 0.0000 | 0.8559 | 0.8559 | 0.000984 | −0.028399 |
| 424 | ERoS | ASC | −0.543122 | 92.218082 | −5.712918 | 0.0000 | 0.0006 | 0.0006 | 0.294982 | 0.274246 |
| 425 | ERoS | ASC | −0.487450 | 86.202996 | −0.871296 | 0.0000 | 0.0026 | 0.0026 | 0.237607 | 0.215184 |
| 426 | ERoS | ASC | −0.435769 | 83.803982 | −0.137159 | 0.0000 | 0.0079 | 0.0079 | 0.189895 | 0.166068 |
| 427 | ERoS | ASC | −0.326479 | 81.593360 | −0.060736 | 0.0000 | 0.0520 | 0.0520 | 0.106588 | 0.080311 |
| 428 | ERoS | ASC | −0.402432 | 83.816290 | −0.462527 | 0.0000 | 0.0150 | 0.0150 | 0.161951 | 0.137303 |
| 429 | ERoS | ASC | −0.559757 | 102.130604 | −16.633197 | 0.0000 | 0.0004 | 0.0004 | 0.313327 | 0.293131 |
| 430 | ERoS | ASC | −0.558613 | 111.281384 | −26.593706 | 0.0000 | 0.0004 | 0.0004 | 0.312048 | 0.291814 |
| 431 | ERoS | ASC | −0.536518 | 82.376110 | −20.118595 | 0.0000 | 0.0007 | 0.0007 | 0.287851 | 0.266906 |
| 432 | ERoS | ASC | −0.536518 | 82.376110 | −6.056301 | 0.0000 | 0.0007 | 0.0007 | 0.287851 | 0.266906 |
| 433 | RCP | ASC | 0.817532 | 1.019731 | 0.205859 | 0.0000 | 0.0000 | 0.0000 | 0.668359 | 0.658605 |
| 434 | RCP | ASC | 0.748105 | 1.231091 | 0.032011 | 0.0000 | 0.0000 | 0.0000 | 0.559661 | 0.546710 |
| 435 | RCP | ASC | 0.686281 | 1.314202 | 0.005171 | 0.0000 | 0.0000 | 0.0000 | 0.470982 | 0.455422 |
| 436 | RCP | ASC | 0.589916 | 1.380754 | 0.002627 | 0.0000 | 0.0002 | 0.0002 | 0.348001 | 0.328825 |
| 437 | RCP | ASC | 0.665464 | 1.303851 | 0.018309 | 0.0000 | 0.0000 | 0.0000 | 0.442842 | 0.426455 |
| 438 | RCP | ASC | 0.840345 | 0.664786 | 0.597778 | 0.0000 | 0.0000 | 0.0000 | 0.706180 | 0.697538 |
| 439 | RCP | ASC | 0.841635 | 0.331701 | 0.959175 | 0.0202 | 0.0000 | 0.0000 | 0.708350 | 0.699772 |
| 440 | RCP | ASC | 0.825594 | 1.371320 | 0.741115 | 0.0000 | 0.0000 | 0.0000 | 0.681605 | 0.672241 |
| 441 | RCP | ASC | 0.825594 | 1.371320 | 0.223098 | 0.0000 | 0.0000 | 0.0000 | 0.681605 | 0.672241 |
| 442 | CS | RCP | −0.311780 | 79.187609 | −8.527803 | 0.0000 | 0.0642 | 0.0642 | 0.097207 | 0.070654 |
| 443 | CS | RCP | −0.316587 | 73.215514 | −2.791395 | 0.0000 | 0.0599 | 0.0599 | 0.100227 | 0.073764 |
| 444 | CS | RCP | −0.313630 | 70.953828 | −1.074032 | 0.0000 | 0.0625 | 0.0625 | 0.098363 | 0.071845 |
| 445 | CS | RCP | −0.315927 | 74.539445 | −1.649554 | 0.0000 | 0.0605 | 0.0605 | 0.099810 | 0.073334 |
| 446 | CS | RCP | −0.317372 | 78.505386 | −4.086523 | 0.0000 | 0.0593 | 0.0593 | 0.100725 | 0.074276 |
| 447 | CS | RCP | −0.305280 | 90.302318 | −19.752229 | 0.0000 | 0.0702 | 0.0702 | 0.093196 | 0.066525 |
| 448 | CS | RCP | −0.302340 | 101.206498 | −30.737982 | 0.0000 | 0.0731 | 0.0731 | 0.091409 | 0.064686 |

**Table A1.** *Cont.*

| M# | Y | X | P-Co-Co | $\alpha^*$ | $\beta^*$ | $p(\alpha^*)$ | $p(\beta^*)$ | $p$(M) | $R^2$ | $(R^2)^*$ |
|---|---|---|---|---|---|---|---|---|---|---|
| 449 | CS | RCP | −0.295152 | 70.232348 | −24.866765 | 0.0000 | 0.0805 | 0.0805 | 0.087115 | 0.060265 |
| 450 | CS | RCP | −0.295152 | 70.232348 | −7.485642 | 0.0000 | 0.0805 | 0.0805 | 0.087115 | 0.060265 |
| 451 | TS | RCP | −0.321650 | 19.003440 | −2.796214 | 0.0000 | 0.0558 | 0.0558 | 0.103459 | 0.077090 |
| 452 | TS | RCP | −0.329149 | 17.062865 | −0.922398 | 0.0000 | 0.0500 | 0.0500 | 0.108339 | 0.082114 |
| 453 | TS | RCP | −0.328647 | 16.327649 | −0.357708 | 0.0000 | 0.0503 | 0.0503 | 0.108009 | 0.081774 |
| 454 | TS | RCP | −0.330369 | 17.516156 | −0.548248 | 0.0000 | 0.0491 | 0.0491 | 0.109144 | 0.082942 |
| 455 | TS | RCP | −0.330440 | 18.816681 | −1.352311 | 0.0000 | 0.0490 | 0.0490 | 0.109191 | 0.082990 |
| 456 | TS | RCP | −0.313657 | 22.615719 | −6.450167 | 0.0000 | 0.0625 | 0.0625 | 0.098380 | 0.071862 |
| 457 | TS | RCP | −0.310183 | 26.159902 | −10.022973 | 0.0001 | 0.0656 | 0.0656 | 0.096214 | 0.069632 |
| 458 | TS | RCP | −0.301861 | 16.055905 | −8.083126 | 0.0000 | 0.0736 | 0.0736 | 0.091120 | 0.064389 |
| 459 | TS | RCP | −0.301861 | 16.055905 | −2.433263 | 0.0000 | 0.0736 | 0.0736 | 0.091120 | 0.064389 |
| 460 | FS | RCP | −0.002053 | 11.184655 | −0.007123 | 0.0000 | 0.9905 | 0.9905 | 0.000004 | −0.029407 |
| 461 | FS | RCP | −0.059707 | 11.339342 | −0.066763 | 0.0000 | 0.7294 | 0.7294 | 0.003565 | −0.025742 |
| 462 | FS | RCP | −0.106002 | 11.373457 | −0.046035 | 0.0000 | 0.5384 | 0.5384 | 0.011236 | −0.017845 |
| 463 | FS | RCP | −0.087791 | 11.464327 | −0.058131 | 0.0000 | 0.6107 | 0.6107 | 0.007707 | −0.021478 |
| 464 | FS | RCP | −0.063122 | 11.481795 | −0.103072 | 0.0000 | 0.7146 | 0.7146 | 0.003984 | −0.025310 |
| 465 | FS | RCP | 0.029273 | 10.881983 | 0.240196 | 0.0000 | 0.8654 | 0.8654 | 0.000857 | −0.028530 |
| 466 | FS | RCP | 0.039716 | 10.592340 | 0.512068 | 0.0002 | 0.8181 | 0.8181 | 0.001577 | −0.027788 |
| 467 | FS | RCP | 0.060098 | 11.072289 | 0.642116 | 0.0000 | 0.7277 | 0.7277 | 0.003612 | −0.025694 |
| 468 | FS | RCP | 0.060098 | 11.072289 | 0.193296 | 0.0000 | 0.7277 | 0.7277 | 0.003612 | −0.025694 |
| 469 | USPV | RCP | −0.582828 | 5.992315 | −0.641168 | 0.0000 | 0.0002 | 0.0002 | 0.339688 | 0.320267 |
| 470 | USPV | RCP | −0.554868 | 5.510830 | −0.196771 | 0.0000 | 0.0004 | 0.0004 | 0.307878 | 0.287522 |
| 471 | USPV | RCP | −0.521885 | 5.334800 | −0.071882 | 0.0000 | 0.0011 | 0.0011 | 0.272363 | 0.250962 |
| 472 | USPV | RCP | −0.539686 | 5.589440 | −0.113335 | 0.0000 | 0.0007 | 0.0007 | 0.291261 | 0.270415 |
| 473 | USPV | RCP | −0.555993 | 5.883336 | −0.287937 | 0.0000 | 0.0004 | 0.0004 | 0.309129 | 0.288809 |
| 474 | USPV | RCP | −0.590432 | 6.890460 | −1.536494 | 0.0000 | 0.0002 | 0.0002 | 0.348610 | 0.329451 |
| 475 | USPV | RCP | −0.591281 | 7.769025 | −2.417779 | 0.0000 | 0.0001 | 0.0001 | 0.349613 | 0.330484 |
| 476 | USPV | RCP | −0.589648 | 5.339331 | −1.998062 | 0.0000 | 0.0002 | 0.0002 | 0.347684 | 0.328499 |
| 477 | USPV | RCP | −0.589648 | 5.339331 | −0.601477 | 0.0000 | 0.0002 | 0.0002 | 0.347684 | 0.328499 |
| 478 | HBRH | RCP | −0.055419 | 42.042378 | −1.487582 | 0.0000 | 0.7482 | 0.7482 | 0.003071 | −0.026250 |
| 479 | HBRH | RCP | −0.059554 | 41.070969 | −0.515318 | 0.0000 | 0.7301 | 0.7301 | 0.003547 | −0.025761 |
| 480 | HBRH | RCP | −0.049270 | 40.511711 | −0.165583 | 0.0000 | 0.7754 | 0.7754 | 0.002428 | −0.026913 |
| 481 | HBRH | RCP | −0.047011 | 40.997438 | −0.240888 | 0.0000 | 0.7854 | 0.7854 | 0.002210 | −0.027137 |
| 482 | HBRH | RCP | −0.052685 | 41.782675 | −0.665751 | 0.0000 | 0.7602 | 0.7602 | 0.002776 | −0.026554 |
| 483 | HBRH | RCP | −0.044729 | 43.245445 | −2.840126 | 0.0027 | 0.7956 | 0.7956 | 0.002001 | −0.027352 |
| 484 | HBRH | RCP | −0.039640 | 44.285542 | −3.955010 | 0.0298 | 0.8185 | 0.8185 | 0.001571 | −0.027794 |
| 485 | HBRH | RCP | −0.027139 | 40.148932 | −2.243900 | 0.0000 | 0.8752 | 0.8752 | 0.000737 | −0.028654 |
| 486 | HBRH | RCP | −0.027139 | 40.148932 | −0.675481 | 0.0000 | 0.8752 | 0.8752 | 0.000737 | −0.028654 |
| 487 | ERoS | RCP | −0.330914 | 99.464458 | −13.823229 | 0.0000 | 0.0487 | 0.0487 | 0.109504 | 0.083313 |
| 488 | ERoS | RCP | −0.298326 | 88.526158 | −4.017212 | 0.0000 | 0.0772 | 0.0772 | 0.088998 | 0.062204 |
| 489 | ERoS | RCP | −0.262390 | 84.519693 | −1.372315 | 0.0000 | 0.1221 | 0.1221 | 0.068849 | 0.041462 |
| 490 | ERoS | RCP | −0.276809 | 89.598988 | −2.207320 | 0.0000 | 0.1022 | 0.1022 | 0.076623 | 0.049465 |
| 491 | ERoS | RCP | −0.295622 | 95.936765 | −5.813376 | 0.0000 | 0.0800 | 0.0800 | 0.087393 | 0.060551 |
| 492 | ERoS | RCP | −0.342215 | 119.666621 | −33.816057 | 0.0000 | 0.0411 | 0.0411 | 0.117111 | 0.091144 |
| 493 | ERoS | RCP | −0.344671 | 139.348934 | −53.516849 | 0.0000 | 0.0395 | 0.0395 | 0.118798 | 0.092881 |
| 494 | ERoS | RCP | −0.347025 | 85.635639 | −44.651839 | 0.0000 | 0.0381 | 0.0381 | 0.120426 | 0.094556 |
| 495 | ERoS | RCP | −0.347025 | 85.635639 | −13.441543 | 0.0000 | 0.0381 | 0.0381 | 0.120426 | 0.094556 |
| 496 | ASC | RCP | 0.817532 | −2.518481 | 3.246674 | 0.0003 | 0.0000 | 0.0000 | 0.668359 | 0.658605 |
| 497 | ASC | RCP | 0.817113 | −0.203489 | 1.046058 | 0.5736 | 0.0000 | 0.0000 | 0.667674 | 0.657900 |
| 498 | ASC | RCP | 0.790176 | 0.685673 | 0.392889 | 0.0247 | 0.0000 | 0.0000 | 0.624378 | 0.613330 |
| 499 | ASC | RCP | 0.800249 | −0.642195 | 0.606666 | 0.1436 | 0.0000 | 0.0000 | 0.640398 | 0.629821 |
| 500 | ASC | RCP | 0.813711 | −2.155522 | 1.521251 | 0.0008 | 0.0000 | 0.0000 | 0.662126 | 0.652188 |
| 501 | ASC | RCP | 0.801181 | −6.757952 | 7.526514 | 0.0000 | 0.0000 | 0.0000 | 0.641890 | 0.631358 |
| 502 | ASC | RCP | 0.792474 | −10.896336 | 11.697965 | 0.0000 | 0.0000 | 0.0000 | 0.628015 | 0.617075 |
| 503 | ASC | RCP | 0.769558 | 0.899398 | 9.413692 | 0.0034 | 0.0000 | 0.0000 | 0.592219 | 0.580226 |
| 504 | ASC | RCP | 0.769558 | 0.899398 | 2.833804 | 0.0034 | 0.0000 | 0.0000 | 0.592219 | 0.580226 |

**Table A2.** All SLR Models in Detail.

| M# | SLR Model | M# | SLR Model |
|---|---|---|---|
| 1 | TS = −3.66771333146045 + 0.278209291384233 × CS | 253 | CS = 82.4394445799055 + −0.405631800581188 × HbRH |
| 2 | TS = 5.76632220397327 + 0.0019846216045102 × CS ^2 | 254 | CS = 75.4057601873381 + −0.00528600953754371 × HbRH ^2 |
| 3 | TS = 8.98916688640433 + 1.80409811683411e-05 × CS ^3 | 255 | CS = 72.8629873306788 + −8.21782427238269e-05 × HbRH ^3 |
| 4 | TS = 14.5688311121099 + 2.68866314762229e-41 × e^ CS | 256 | CS = 66.9027834245253 + −5.14345524118181e-26 × e^ HbRH |
| 5 | TS = 14.5656439363982 + 1.41265796681029e-28 × 2^ CS | 257 | CS = 67.1708457607848 + −7.48294418841369e-18 × 2^ HbRH |
| 6 | TS = −22.3244737238133 + 4.57533122714702 × CS ^(1/2) | 258 | CS = 96.0472599467819 + −4.77054201210767 × HbRH ^(1/2) |
| 7 | TS = −40.9312402935274 + 13.814582720438 × CS ^(1/3) | 259 | CS = 109.56395500385 + −12.8042675489701 × HbRH ^(1/3) |
| 8 | TS = −62.8509407538376 + 42.7854090996044 × log(CS) | 260 | CS = 115.311599942547 + −31.0342296205957 × log(HbRH) |
| 9 | TS = −62.8509407538376 + 12.8796915157356 × lg(CS) | 261 | CS = 115.311599942547 + −9.34223400812292 × lg(HbRH) |
| 10 | FS = 4.24393079882703 + 0.104527758280078 × CS | 262 | TS = 19.9622058974958 + −0.1303029585153 × HbRH |
| 11 | FS = 7.75780828627991 + 0.000752403905718667 × CS ^2 | 263 | TS = 17.6543146320032 + −0.00166994242748144 × HbRH ^2 |
| 12 | FS = 8.96287429220537 + 6.89191529938516e-06 × CS ^3 | 264 | TS = 16.83840458393 + −2.58037721938791e-05 × HbRH ^3 |
| 13 | FS = 11.0870361916049 + 1.12206961389585e-41 × e^ CS | 265 | TS = 15.0165743978521 + −2.03721394407596e-26 × e^ HbRH |
| 14 | FS = 11.0859072057969 + 5.88205048357479e-29 × 2^ CS | 266 | TS = 15.1093101981532 + −2.84866062385744e-18 × 2^ HbRH |
| 15 | FS = −2.69616622448087 + 1.71044934171127 × CS ^(1/2) | 267 | TS = 24.4677526180197 + −1.55399032698069 × HbRH ^(1/2) |
| 16 | FS = −9.61603490429178 + 5.15551033198057 × CS ^(1/3) | 268 | TS = 28.9510915589449 + −4.19472168720189 × HbRH ^(1/3) |
| 17 | FS = −17.6931562993013 + 15.9103695111459 × log(CS) | 269 | TS = 31.0462473657117 + −10.3012836038027 × log(HbRH) |
| 18 | FS = −17.6931562993013 + 4.7894984649526 × lg(CS) | 270 | TS = 31.0462473657117 + −3.10099535858616 × lg(HbRH) |
| 19 | USPV = 3.8574983286046 + 0.0175825945943531 × CS | 271 | FS = 12.7435704060995 + −0.039445290747868 × HbRH |
| 20 | USPV = 4.45626900695745 + 0.000124865325500488 × CS ^2 | 272 | FS = 12.0754048065075 + −0.000523213761785206 × HbRH ^2 |
| 21 | USPV = 4.6595172220698 + 1.13358339686721e-06 × CS ^3 | 273 | FS = 11.832096250984 + −8.23893341545536e-06 × HbRH ^3 |
| 22 | USPV = 5.00517282814257 + 2.32706905364138e-42 × e^ CS | 274 | FS = 11.2321332373504 + −4.95164302968338e-27 × e^ HbRH |
| 23 | USPV = 5.00497635203039 + 1.21736498704813e-29 × 2^ CS | 275 | FS = 11.2579775085298 + −7.20711857382233e-19 × 2^ HbRH |
| 24 | USPV = 2.67038043865855 + 0.290146899587109 × CS ^(1/2) | 276 | FS = 14.0361157995126 + −0.458978534407627 × HbRH ^(1/2) |
| 25 | USPV = 1.4858023636118 + 0.877204473839206 × CS ^(1/3) | 277 | FS = 15.3207207454225 + −1.22722061378818 × HbRH ^(1/3) |
| 26 | USPV = 0.0798296243915567 + 2.72458142347606 × log(CS) | 278 | FS = 15.8342481716803 + −2.9508137441398 × log(HbRH) |
| 27 | USPV = 0.0798296243915644 + 0.820180734095162 × lg(CS) | 279 | FS = 15.8342481716803 + −0.888283448603619 × lg(HbRH) |
| 28 | HbRH = 65.6939723416543 + −0.39066291994853 × CS | 280 | USPV = 5.40527092455439 + −0.00960161616381508 × HbRH |
| 29 | HbRH = 52.9175508566686 + −0.00289053323693466 × CS ^2 | 281 | USPV = 5.18228451266245 + −9.23365939017983e-05 × HbRH ^2 |
| 30 | HbRH = 48.4793106810452 + −2.70731778165387e-05 × CS ^3 | 282 | USPV = 5.105547006142 + −1.03094075674674e-06 × HbRH ^3 |
| 31 | HbRH = 40.1336082855694 + −4.38891161917149e-41 × e^ CS | 283 | USPV = 5.01571398268165 + 6.35172119916297e-28 × e^ HbRH |
| 32 | HbRH = 40.1385314994 + −2.30412190648458e-28 × 2^ CS | 284 | USPV = 5.01761451645092 + 4.7746220119381e-20 × 2^ HbRH |
| 33 | HbRH = 90.8138878413038 + −6.29176473409599 × CS ^(1/2) | 285 | USPV = 5.83782737563726 + −0.130633680633248 × HbRH ^(1/2) |
| 34 | HbRH = 115.838473020169 + −18.8576275257236 × CS ^(1/3) | 286 | USPV = 6.26621225603992 + −0.367863560422559 × HbRH ^(1/3) |
| 35 | HbRH = 144.146417510672 + −57.5149717946972 × log(CS) | 287 | USPV = 6.57238703587405 + −0.980912750035412 × log(HbRH) |
| 36 | HbRH = 144.146417510672 + −17.3137317099717 × lg(CS) | 288 | USPV = 6.57238703587405 + −0.295284160889903 × lg(HbRH) |
| 37 | ERoS = 48.638330809566 + 0.451481569206116 × CS | 289 | ERoS = 93.0915021332562 + −0.364903933321156 × HbRH |
| 38 | ERoS = 63.1121233323246 + 0.00340477469183933 × CS ^2 | 290 | ERoS = 85.3618387083167 + −0.00394146430058687 × HbRH ^2 |
| 39 | ERoS = 68.1424331649685 + 3.25053196642623e-05 × CS ^3 | 291 | ERoS = 82.755136193978 + −5.23793613335234e-05 × HbRH ^3 |
| 40 | ERoS = 78.1544409313211 + 5.37587881956385e-41 × e^ CS | 292 | ERoS = 79.0174650656451 + −3.79940100453875e-26 × e^ HbRH |
| 41 | ERoS = 78.1471425518886 + 2.83074450890616e-28 × 2^ CS | 293 | ERoS = 79.2064011656095 + −5.44974424110028e-18 × 2^ HbRH |
| 42 | ERoS = 20.1584377318609 + 7.2033577129187 × CS ^(1/2) | 294 | ERoS = 108.32181180241 + −4.77082649816592 × HbRH ^(1/2) |
| 43 | ERoS = −8.22243286374305 + 21.5230297614624 × CS ^(1/3) | 295 | ERoS = 123.47606121203 + −13.2894132486659 × HbRH ^(1/3) |
| 44 | ERoS = −39.8009179828865 + 65.241657789968 × log(CS) | 296 | ERoS = 133.518155940357 + −34.7913378742457 × log(HbRH) |
| 45 | ERoS = −39.800917982886 + 19.6396959616249 × lg(CS) | 297 | ERoS = 133.518155940357 + −10.4732362894283 × lg(HbRH) |
| 46 | ASC = 4.95132054829306 + −0.0386503401063175 × CS | 298 | ASC = 2.68078131101273 + −0.00733510672804191 × HbRH |
| 47 | ASC = 3.72539269504536 + −0.000294369718030655 × CS ^2 | 299 | ASC = 2.60665094129842 + −0.000126382796341767 × HbRH ^2 |
| 48 | ASC = 3.30102796537324 + −2.84321291493011e-06 × CS ^3 | 300 | ASC = 2.58769478660714 + −2.48849929099358e-06 × HbRH ^3 |
| 49 | ASC = 2.44093528214272 + −6.72399110323773e-42 × e^ CS | 301 | ASC = 2.44220142607004 + −4.53236511525499e-27 × e^ HbRH |
| 50 | ASC = 2.44170772986293 + −3.53122466148145e-29 × 2^ CS | 302 | ASC = 2.45300492295719 + −5.49529605773755e-19 × 2^ HbRH |
| 51 | ASC = 7.37026715497592 + −0.614301465019437 × CS ^(1/2) | 303 | ASC = 2.85534341592287 + −0.0747993160120245 × HbRH ^(1/2) |
| 52 | ASC = 9.78274044602861 + −1.83353621086711 × CS ^(1/3) | 304 | ASC = 3.03754257943617 + −0.191963698502796 × HbRH ^(1/3) |
| 53 | ASC = 12.4554180798691 + −5.54827132735972 × log(CS) | 305 | ASC = 3.06678420213484 + −0.429225020225469 × log(HbRH) |
| 54 | ASC = 12.4554180798691 + −1.67019609361768 × lg(CS) | 306 | ASC = 3.06678420213485 + −0.129209605977347 × lg(HbRH) |
| 55 | RCP = 2.2672230225556 + −0.0113988195146484 × CS | 307 | RCP = 1.59366458546644 + −0.00206458751119777 × HbRH |
| 56 | RCP = 1.88299647072557 + −8.18219613510574e-05 × CS ^2 | 308 | RCP = 1.59250269958169 + −4.7007782059367e-05 × HbRH ^2 |
| 57 | RCP = 1.75078592624905 + −7.4585338236417e-07 × CS ^3 | 309 | RCP = 1.5940113480029 + −1.03273072317156e-06 × HbRH ^3 |
| 58 | RCP = 1.51818175249647 + −8.62375516317953e-43 × e^ CS | 310 | RCP = 1.53596131120061 + −2.07901551267993e-27 × e^ HbRH |
| 59 | RCP = 1.51830999498204 + −4.54842382692243e-30 × 2^ CS | 311 | RCP = 1.54311874344824 + −2.70943115773082e-19 × 2^ HbRH |
| 60 | RCP = 3.0251758982295 + −0.186665056505132 × CS ^(1/2) | 312 | RCP = 1.60647332079248 + −0.0152285890543703 × HbRH ^(1/2) |
| 61 | RCP = 3.78074488916298 + −0.562728438143532 × CS ^(1/3) | 313 | RCP = 1.62252334316423 + −0.0328544818159167 × HbRH ^(1/3) |
| 62 | RCP = 4.663140971559 + −1.73705575957867 × log(CS) | 314 | RCP = 1.5765860616204 + −0.0412065467018546 × log(HbRH) |
| 63 | RCP = 4.66314097155899 + −0.522905887774061 × lg(CS) | 315 | RCP = 1.5765860616204 + −0.0124044065749873 × lg(HbRH) |
| 64 | CS = 25.6010477687945 + 2.75406936542173 × TS | 316 | CS = 51.089202219025 + 0.193565839661132 × ERoS |
| 65 | CS = 44.236410022525 + 0.0946888189494539 × TS ^2 | 317 | CS = 56.1180779980333 + 0.00156343014265882 × ERoS ^2 |
| 66 | CS = 50.9063816671094 + 0.00398118874295442 × TS ^3 | 318 | CS = 57.8305000886192 + 1.5170484861907e-05 × ERoS ^3 |
| 67 | CS = 64.102513981059 + 1.08352527227825e-08 × e^ TS | 319 | CS = 65.905973876004 + 5.56007324663602e-43 × e^ ERoS |
| 68 | CS = 62.93479281879 + 9.97078792156074e-06 × 2^ TS | 320 | CS = 65.7374812451052 + 9.51622799059243e-30 × 2^ ERoS |

**Table A2.** *Cont.*

| M# | SLR Model |
|---|---|
| 69 | CS = −10.6326736740597 + 20.195416541604 × TS ^(1/2) |
| 70 | CS = −46.6793703465627 + 46.4175211250117 × TS ^(1/3) |
| 71 | CS = −29.2331517456234 + 82.8403738530719 × log(TS) |
| 72 | CS = −29.2331517456234 + 24.9374373817928 × lg(TS) |
| 73 | FS = 6.54685987581198 + 0.313124883867076 × TS |
| 74 | FS = 8.70756838739812 + 0.0105856072945152 × TS ^2 |
| 75 | FS = 9.48379504325131 + 0.000437165189201065 × TS ^3 |
| 76 | FS = 10.9579310971029 + 1.06591164804576e-09 × e^ TS |
| 77 | FS = 10.8407343563029 + 9.87757373295459e-07 × 2^ TS |
| 78 | FS = 2.36028455472909 + 2.31370552755273 × TS ^(1/2) |
| 79 | FS = −1.80037496774033 + 5.33057504676748 × TS ^(1/3) |
| 80 | FS = 0.153469672168541 + 9.55644054219172 × log(TS) |
| 81 | FS = 0.153469672168538 + 2.87677525497907 × lg(TS) |
| 82 | USPV = 3.8582700632675 + 0.0788332945887362 × TS |
| 83 | USPV = 4.41351597695613 + 0.00261674011715778 × TS ^2 |
| 84 | USPV = 4.61424167992712 + 0.000105778685478559 × TS ^3 |
| 85 | USPV = 4.98138269039791 + 2.06326729117493e-10 × e^ TS |
| 86 | USPV = 4.95689326088985 + 1.965467135128e-07 × 2^ TS |
| 87 | USPV = 2.78745215424647 + 0.586914035384849 × TS ^(1/2) |
| 88 | USPV = 1.72467454463204 + 1.35521823819063 × TS ^(1/3) |
| 89 | USPV = 2.21017728190999 + 2.43932132966721 × log(TS) |
| 90 | USPV = 2.21017728191 + 0.734308889292778 × lg(TS) |
| 91 | HbRH = 58.1513390501432 + −1.24230352426858 × TS |
| 92 | HbRH = 50.2249700104166 + −0.044770875619361 × TS ^2 |
| 93 | HbRH = 47.2679048675656 + −0.00193325336213942 × TS ^3 |
| 94 | HbRH = 40.7389382402033 + −4.66451848397386e-09 × e^ TS |
| 95 | HbRH = 41.3078778310346 + −4.48876900783616e-06 × 2^ TS |
| 96 | HbRH = 73.3631224275738 + −8.81243566756599 × TS ^(1/2) |
| 97 | HbRH = 88.4709404052577 + −19.9993370745879 × TS ^(1/3) |
| 98 | HbRH = 79.8183132517281 + −34.7074847278588 × log(TS) |
| 99 | HbRH = 79.8183132517281 + −10.447993977135 × lg(TS) |
| 100 | ERoS = 38.619764851316 + 2.70358942299871 × TS |
| 101 | ERoS = 56.6412391630876 + 0.0941220459496348 × TS ^2 |
| 102 | ERoS = 63.3000205646993 + 0.00394992641131165 × TS ^3 |
| 103 | ERoS = 77.1328153398415 + 7.09631280316041e-09 × e^ TS |
| 104 | ERoS = 76.0491484044899 + 7.47562545588949e-06 × 2^ TS |
| 105 | ERoS = 4.01611935228142 + 19.5716763093765 × TS ^(1/2) |
| 106 | ERoS = −30.329978647475 + 44.7426132701713 × TS ^(1/3) |
| 107 | ERoS = −12.3728718803923 + 78.8622864433468 × log(TS) |
| 108 | ERoS = −12.3728718803923 + 23.7399137460923 × lg(TS) |
| 109 | ASC = 4.89596300636296 + −0.16966119937039 × TS |
| 110 | ASC = 3.87624826801362 + −0.00638384276646585 × TS ^2 |
| 111 | ASC = 3.49223640961992 + −0.000285395470139084 × TS ^3 |
| 112 | ASC = 2.53307238173374 + −7.11652324318657e-10 × e^ TS |
| 113 | ASC = 2.62336847692134 + −6.95199732594167e-07 × 2^ TS |
| 114 | ASC = 6.86395722224124 + −1.17477367334536 × TS ^(1/2) |
| 115 | ASC = 8.82414184541521 + −2.64397265299706 × TS ^(1/3) |
| 116 | ASC = 7.59151813198557 + −4.51149961240018 × log(TS) |
| 117 | ASC = 7.59151813198557 + −1.35809670875888 × lg(TS) |
| 118 | RCP = 2.0582468256364 + −0.0369995440023129 × TS |
| 119 | RCP = 1.8103952102763 + −0.00128284918583967 × TS ^2 |
| 120 | RCP = 1.71859132373958 + −5.35652053686268e-05 × TS ^3 |
| 121 | RCP = 1.53188741984192 + −1.00594344665696e-10 × e^ TS |
| 122 | RCP = 1.54503002030309 + −9.93922733664546e-08 × 2^ TS |
| 123 | RCP = 2.53415581702009 + −0.268461007718213 × TS ^(1/2) |
| 124 | RCP = 3.0066266949075 + −0.614281461688656 × TS ^(1/3) |
| 125 | RCP = 2.76276107851835 + −1.08503484360649 × log(TS) |
| 126 | RCP = 2.76276107851835 + −0.326628034266129 × lg(TS) |
| 127 | CS = −6.32669651116645 + 6.49948062049916 × FS |
| 128 | CS = 29.9785358256873 + 0.28556239381448 × FS ^2 |
| 129 | CS = 42.0916447762494 + 0.0164212042293632 × FS ^3 |
| 130 | CS = 60.4701313902147 + 2.77820273664116e-05 × e^ FS |
| 131 | CS = 57.7305850492271 + 0.00213782451902893 × 2^ FS |
| 132 | CS = −78.8618812949019 + 43.5263527058322 × FS ^(1/2) |
| 133 | CS = −151.371735363849 + 97.5647049792941 × FS ^(1/3) |
| 134 | CS = −107.999762435771 + 166.924543040243 × log(FS) |
| 135 | CS = −107.999762435771 + 50.2492944676166 × lg(FS) |
| 136 | TS = −7.19991865837233 + 1.96680522970567 × FS |
| 321 | CS = 40.6340826309488 + 2.9192088069464 × ERoS ^(1/2) |
| 322 | CS = 30.0257116537128 + 8.53247838099428 × ERoS ^(1/3) |
| 323 | CS = 19.9860767860859 + 24.6353731468881 × log(ERoS) |
| 324 | CS = 19.9860767860861 + 7.41598627158826 × lg(ERoS) |
| 325 | TS = 5.57698297874872 + 0.117091719673417 × ERoS |
| 326 | TS = 8.84882889224645 + 0.000910458534534851 × ERoS ^2 |
| 327 | TS = 9.97627368064185 + 8.60117098114261e-06 × ERoS ^3 |
| 328 | TS = 14.3474181010307 + 6.0953908828585e-43 × e^ ERoS |
| 329 | TS = 14.2272764535642 + 9.33570995950449e-30 × 2^ ERoS |
| 330 | TS = −1.1269089945942 + 1.80904069963451 × ERoS ^(1/2) |
| 331 | TS = −7.90186096886077 + 5.33486061240443 × ERoS ^(1/3) |
| 332 | TS = −14.7390012786689 + 15.7009205255199 × log(ERoS) |
| 333 | TS = −14.7390012786689 + 4.72644803771776 × lg(ERoS) |
| 334 | FS = 9.83395980893246 + 0.0170538323228349 × ERoS |
| 335 | FS = 10.3125531496547 + 0.000132286637799236 × ERoS ^2 |
| 336 | FS = 10.486081411843 + 1.23231731631889e-06 × ERoS ^3 |
| 337 | FS = 11.1527750704836 + 2.99625338895325e-44 × e^ ERoS |
| 338 | FS = 11.1479488279815 + 4.40573016768686e-31 × 2^ ERoS |
| 339 | FS = 8.86996914221672 + 0.262067593014333 × ERoS ^(1/2) |
| 340 | FS = 7.89844511181229 + 0.770500724179212 × ERoS ^(1/3) |
| 341 | FS = 6.94573366509112 + 2.24915473694021 × log(ERoS) |
| 342 | FS = 6.94573366509116 + 0.677063040708727 × lg(ERoS) |
| 343 | USPV = 3.68820846972565 + 0.0169908027073602 × ERoS |
| 344 | USPV = 4.16938309261067 + 0.000131129610994454 × ERoS ^2 |
| 345 | USPV = 4.33343560349138 + 1.23579707391254e-06 × ERoS ^3 |
| 346 | USPV = 4.90797369481416 + 1.63496289148364e-43 × e^ ERoS |
| 347 | USPV = 4.88959412661251 + 2.26894827628225e-30 × 2^ ERoS |
| 348 | USPV = 2.70118294373709 + 0.2641244615984 × ERoS ^(1/2) |
| 349 | USPV = 1.70386942486284 + 0.780820994806742 × ERoS ^(1/3) |
| 350 | USPV = 0.67903797185686 + 2.31085638391061 × log(ERoS) |
| 351 | USPV = 0.679037971856871 + 0.695637087228694 × lg(ERoS) |
| 352 | HbRH = 51.6324046062446 + −0.150673692373029 × ERoS |
| 353 | HbRH = 47.1704653947156 + −0.00113291769977207 × ERoS ^2 |
| 354 | HbRH = 45.600695422352 + −1.04038244458094e-05 × ERoS ^3 |
| 355 | HbRH = 39.9748474387209 + −2.56797542626781e-43 × e^ ERoS |
| 356 | HbRH = 40.0456981734223 + −4.27679705661863e-30 × 2^ ERoS |
| 357 | HbRH = 60.6467709227773 + −2.37198566049296 × ERoS ^(1/2) |
| 358 | HbRH = 69.7247860734917 + −7.04081020027556 × ERoS ^(1/3) |
| 359 | HbRH = 79.2748060130706 + −21.0017577568586 × log(ERoS) |
| 360 | HbRH = 79.2748060130704 + −6.32215904648311 × lg(ERoS) |
| 361 | ASC = 6.4458151840564 + −0.0516341811061645 × ERoS |
| 362 | ASC = 4.99563817395329 + −0.000400352711259067 × ERoS ^2 |
| 363 | ASC = 4.50745712993463 + −3.79575450452663e-06 × ERoS ^3 |
| 364 | ASC = 2.59992789799722 + −2.99484600134824e-43 × e^ ERoS |
| 365 | ASC = 2.66261962836728 + −4.64911698183225e-30 × 2^ ERoS |
| 366 | ASC = 9.44043438324327 + −0.802103266576792 × ERoS ^(1/2) |
| 367 | ASC = 12.47058422184 + −2.37157285658742 × ERoS ^(1/3) |
| 368 | ASC = 15.5963281401096 + −7.02565846862435 × log(ERoS) |
| 369 | ASC = 15.5963281401096 + −2.1149339383466 × lg(ERoS) |
| 370 | RCP = 2.13392259412586 + −0.00792174587898038 × ERoS |
| 371 | RCP = 1.93360893463405 + −6.48277942505363e-05 × ERoS ^2 |
| 372 | RCP = 1.86921630500704 + −6.40894224684878e-07 × ERoS ^3 |
| 373 | RCP = 1.55696749364964 + −6.45133418255133e-44 × e^ ERoS |
| 374 | RCP = 1.56902059329398 + −9.76832901653834e-31 × 2^ ERoS |
| 375 | RCP = 2.55930807240737 + −0.11918592439322 × ERoS ^(1/2) |
| 376 | RCP = 2.99258729570875 + −0.348402784873324 × ERoS ^(1/3) |
| 377 | RCP = 3.40606522122726 + −1.00780487661076 × log(ERoS) |
| 378 | RCP = 3.40606522122725 + −0.303379497636274 × lg(ERoS) |
| 379 | CS = 70.6776219375967 + −1.83342313666841 × ASC |
| 380 | CS = 68.2714363423197 + −0.225306775541211 × ASC ^2 |
| 381 | CS = 67.549396985626 + −0.0328026961997557 × ASC ^3 |
| 382 | CS = 67.1919053071617 + −0.0179652802360796 × e^ ASC |
| 383 | CS = 67.7103192435025 + −0.12454655051729 × 2^ ASC |
| 384 | CS = 75.0502895075756 + −6.17920002596685 × ASC ^(1/2) |
| 385 | CS = 79.0740475713104 + −10.3870301811374 × ASC ^(1/3) |
| 386 | CS = 67.936550509068 + −8.66359978801819 × log(ASC) |
| 387 | CS = 67.936550509068 + −2.60800340662158 × lg(ASC) |
| 388 | TS = 16.7191027509542 + −0.812996500399415 × ASC |

**Table A2.** *Cont.*

| M# | SLR Model |
|---|---|
| 137 | TS = 3.94225264358756 + 0.085188466515247 × FS ^2 |
| 138 | TS = 7.656856290228 + 0.0048301982685163 × FS ^3 |
| 139 | TS = 13.2451876094529 + 7.3023151163122e-06 × e^ FS |
| 140 | TS = 12.4443150069096 + 0.000582075431634922 × 2^ FS |
| 141 | TS = −29.4656951691208 + 13.2662252636793 × FS ^(1/2) |
| 142 | TS = −51.7242857085425 + 29.8074489486029 × FS ^(1/3) |
| 143 | TS = −38.7280892344743 + 51.2417059142874 × log(FS) |
| 144 | TS = −38.7280892344744 + 15.425290509193 × lg(FS) |
| 145 | USPV = 3.9220780594136 + 0.0985428823674005 × FS |
| 146 | USPV = 4.47728767296606 + 0.00429215529920093 × FS ^2 |
| 147 | USPV = 4.66109666398569 + 0.000245637379412544 × FS ^3 |
| 148 | USPV = 4.92654609777185 + 4.60706117880123e-07 × e^ FS |
| 149 | USPV = 4.88841164928704 + 3.36308959813418e-05 × 2^ FS |
| 150 | USPV = 2.8095900623364 + 0.663750016435372 × FS ^(1/2) |
| 151 | USPV = 1.69679136870262 + 1.49096966175817 × FS ^(1/3) |
| 152 | USPV = 2.34744679463731 + 2.56255150271304 × log(FS) |
| 153 | USPV = 2.3474467946373 + 0.771404867750439 × lg(FS) |
| 154 | HbRH = 66.1885884598235 + −2.36217666323683 × FS |
| 155 | HbRH = 54.2735150191536 + −0.113846998920061 × FS ^2 |
| 156 | HbRH = 50.1907307977067 + −0.00705311601461987 × FS ^3 |
| 157 | HbRH = 42.9011476603809 + −1.48131754198992e-05 × e^ FS |
| 158 | HbRH = 44.2791893797267 + −0.00111924473609312 × 2^ FS |
| 159 | HbRH = 89.7599134019201 + −14.9823912521973 × FS ^(1/2) |
| 160 | HbRH = 113.273550501112 + −32.9353495751625 × FS ^(1/3) |
| 161 | HbRH = 96.2588355860703 + −54.0764110249104 × log(FS) |
| 162 | HbRH = 96.2588355860704 + −16.2786217763525 × lg(FS) |
| 163 | ERoS = 50.9339523136868 + 2.47332003357847 × FS |
| 164 | ERoS = 64.9117719074914 + 0.107393071702334 × FS ^2 |
| 165 | ERoS = 69.6091128616802 + 0.00607935520541711 × FS ^3 |
| 166 | ERoS = 76.8310941676581 + 8.292501066042e-06 × e^ FS |
| 167 | ERoS = 75.8229039136237 + 0.000685638485789691 × 2^ FS |
| 168 | ERoS = 23.0966441893402 + 16.6339398116732 × FS ^(1/2) |
| 169 | ERoS = −4.70919525255609 + 37.3279849854577 × FS ^(1/3) |
| 170 | ERoS = 11.7588814417093 + 63.9854593331807 × log(FS) |
| 171 | ERoS = 11.7588814417082 + 19.2615425456256 × lg(FS) |
| 172 | ASC = 1.92204377986897 + 0.0417800027959956 × FS |
| 173 | ASC = 2.18316320247149 + 0.00161753154319612 × FS ^2 |
| 174 | ASC = 2.27270746633942 + 7.88163424212467e-05 × FS ^3 |
| 175 | ASC = 2.44093390317411 + −2.48113202997887e-07 × e^ FS |
| 176 | ASC = 2.41621956362761 + −6.81999208250955e-06 × 2^ FS |
| 177 | ASC = 1.40580253441745 + 0.294779993824526 × FS ^(1/2) |
| 178 | ASC = 0.89095417108623 + 0.671410326192849 × FS ^(1/3) |
| 179 | ASC = 1.15276439502583 + 1.18383492977662 × log(FS) |
| 180 | ASC = 1.15276439502588 + 0.356369823777509 × lg(FS) |
| 181 | RCP = 1.51812154765212 + −0.000591998745777089 × FS |
| 182 | RCP = 1.51281584514026 + −1.02938718884825e-05 × FS ^2 |
| 183 | RCP = 1.51260094503923 + −7.42379796240544e-07 × FS ^3 |
| 184 | RCP = 1.5256197518961 + −6.72812668995411e-08 × e^ FS |
| 185 | RCP = 1.52273012331417 + −2.80067024547618e-06 × 2^ FS |
| 186 | RCP = 1.53241072190053 + −0.0062681280895472 × FS ^(1/2) |
| 187 | RCP = 1.54753285970626 + −0.0161478263626681 × FS ^(1/3) |
| 188 | RCP = 1.5488693295213 + −0.0357822221626458 × log(FS) |
| 189 | RCP = 1.5488693295213 + −0.0107715221824692 × lg(FS) |
| 190 | CS = 11.6998550741678 + 10.8691835328367 × USPV |
| 191 | CS = 40.2737226314715 + 1.02188686322694 × USPV ^2 |
| 192 | CS = 49.7955275674617 + 0.126614263542547 × USPV ^3 |
| 193 | CS = 58.3200794662332 + 0.0463658521197105 × e^ USPV |
| 194 | CS = 53.5820935995622 + 0.368615720449898 × 2^ USPV |
| 195 | CS = −45.4269190022254 + 49.9062423711336 × USPV ^(1/2) |
| 196 | CS = −102.544736756829 + 98.6873794949362 × USPV ^(1/3) |
| 197 | CS = −25.625449801709 + 131.505274123587 × log(USPV) |
| 198 | CS = −25.6254498017091 + 39.587032099214 × lg(USPV) |
| 199 | TS = −9.95165895269854 + 4.92289295131928 × USPV |
| 200 | TS = 2.94021780561976 + 0.464792876737356 × USPV ^2 |
| 201 | TS = 7.23615323870508 + 0.0578570626959677 × USPV ^3 |
| 202 | TS = 11.054729627453 + 0.0216332650123754 × e^ USPV |
| 203 | TS = 8.91107549629212 + 0.170046021754159 × 2^ USPV |
| 204 | TS = −35.7255348185222 + 22.5589242403911 × USPV ^(1/2) |
| 389 | TS = 15.6873265109676 + −0.10392640936173 × ASC ^2 |
| 390 | TS = 15.3374234564199 + −0.014689150374878 × ASC ^3 |
| 391 | TS = 15.1071164105457 + −0.00663398866735903 × e^ ASC |
| 392 | TS = 15.3618601830077 + −0.0515731674779195 × 2^ ASC |
| 393 | TS = 18.4752766762153 + −2.61099160196387 × ASC ^(1/2) |
| 394 | TS = 20.0700807482682 + −4.30328784764094 × ASC ^(1/3) |
| 395 | TS = 15.4265286526908 + −3.43411719112025 × log(ASC) |
| 396 | TS = 15.4265286526908 + −1.03377228315253 × lg(ASC) |
| 397 | FS = 11.0977464494996 + 0.031873579279226 × ASC |
| 398 | FS = 11.0907213145279 + 0.00949415382452752 × ASC ^2 |
| 399 | FS = 11.1047564029815 + 0.00181183855164129 × ASC ^3 |
| 400 | FS = 11.1207633178236 + 0.00106742689440128 × e^ ASC |
| 401 | FS = 11.1016774194187 + 0.00636702684383551 × 2^ ASC |
| 402 | FS = 11.1234296660006 + 0.0356237890342107 × ASC ^(1/2) |
| 403 | FS = 11.1326874840045 + 0.0334964933392047 × ASC ^(1/3) |
| 404 | FS = 11.1712737674145 + 0.0138252031036824 × log(ASC) |
| 405 | FS = 11.1712737674145 + 0.00416180083035513 × lg(ASC) |
| 406 | USPV = 5.39967741586927 + −0.157601360131323 × ASC |
| 407 | USPV = 5.23018907761182 + −0.0236309154497613 × ASC ^2 |
| 408 | USPV = 5.16485024635992 + −0.00371278492484695 × ASC ^3 |
| 409 | USPV = 5.11372151021348 + −0.00181910156412128 × e^ ASC |
| 410 | USPV = 5.17103704429333 + −0.0130363801376633 × 2^ ASC |
| 411 | USPV = 5.6807412003617 + −0.464228977845544 × ASC ^(1/2) |
| 412 | USPV = 5.94133254828393 + −0.746448185862817 × ASC ^(1/3) |
| 413 | USPV = 5.13201708600292 + −0.575354478791353 × log(ASC) |
| 414 | USPV = 5.13201708600292 + −0.173198956255813 × lg(ASC) |
| 415 | HbRH = 40.5944269259985 + −0.335108945766807 × ASC |
| 416 | HbRH = 40.7100131929842 + −0.104581925495593 × ASC ^2 |
| 417 | HbRH = 40.6632578629035 + −0.0227846026661095 × ASC ^3 |
| 418 | HbRH = 40.6395211240528 + −0.0169908872975303 × e^ ASC |
| 419 | HbRH = 40.8267699220799 + −0.0910711458036205 × 2^ ASC |
| 420 | HbRH = 40.2047715378215 + −0.290079711211349 × ASC ^(1/2) |
| 421 | HbRH = 39.8328823193541 + −0.0317014234712671 × ASC ^(1/3) |
| 422 | HbRH = 39.6509127800971 + 0.755863068849545 × log(ASC) |
| 423 | HbRH = 39.6509127800971 + 0.227537456338342 × lg(ASC) |
| 424 | ERoS = 92.2180819190403 + −5.71291801262153 × ASC |
| 425 | ERoS = 86.2029960263802 + −0.871295867930876 × ASC ^2 |
| 426 | ERoS = 83.8039817684898 + −0.137158721334941 × ASC ^3 |
| 427 | ERoS = 81.593359750612 + −0.0607357705453962 × e^ ASC |
| 428 | ERoS = 83.8162898301957 + −0.462526675987894 × 2^ ASC |
| 429 | ERoS = 102.130604184649 + −16.6331971432884 × ASC ^(1/2) |
| 430 | ERoS = 111.281383625304 + −26.593705666928 × ASC ^(1/3) |
| 431 | ERoS = 82.3761104513835 + −20.1185948235959 × log(ASC) |
| 432 | ERoS = 82.3761104513835 + −6.05630051251246 × lg(ASC) |
| 433 | RCP = 1.01973122036864 + 0.205859469380567 × ASC |
| 434 | RCP = 1.23109107218656 + 0.0320113741552666 × ASC ^2 |
| 435 | RCP = 1.31420187113397 + 0.00517100383009007 × ASC ^3 |
| 436 | RCP = 1.38075364835372 + 0.00262715741339878 × e^ ASC |
| 437 | RCP = 1.30385075127291 + 0.0183094250349436 × 2^ ASC |
| 438 | RCP = 0.664785578422722 + 0.597777968220892 × ASC ^(1/2) |
| 439 | RCP = 0.331701338198878 + 0.95917456833655 × ASC ^(1/3) |
| 440 | RCP = 1.37132011432189 + 0.741115441374556 × log(ASC) |
| 441 | RCP = 1.37132011432189 + 0.223097978103492 × lg(ASC) |
| 442 | CS = 79.1876086152576 + −8.52780310166127 × RCP |
| 443 | CS = 73.2155143001305 + −2.79139452268105 × RCP ^2 |
| 444 | CS = 70.9538278880991 + −1.07403228395841 × RCP ^3 |
| 445 | CS = 74.5394446838858 + −1.64955391286155 × e^ RCP |
| 446 | CS = 78.5053861753618 + −4.08652317822277 × 2^ RCP |
| 447 | CS = 90.3023182739201 + −19.7522294845498 × RCP ^(1/2) |
| 448 | CS = 101.206498442361 + −30.737981999028 × RCP ^(1/3) |
| 449 | CS = 70.2323475564202 + −24.8667652877232 × log(RCP) |
| 450 | CS = 70.2323475564202 + −7.48564224674054 × lg(RCP) |
| 451 | TS = 19.0034404993042 + −2.79621405588897 × RCP |
| 452 | TS = 17.0628654215869 + −0.922398139081904 × RCP ^2 |
| 453 | TS = 16.3276494077446 + −0.357708176247271 × RCP ^3 |
| 454 | TS = 17.5161562873897 + −0.548248026177606 × e^ RCP |
| 455 | TS = 18.8166813396968 + −1.35231061799441 × 2^ RCP |
| 456 | TS = 22.6157194894507 + −6.45016652543744 × RCP ^(1/2) |

Table A2. *Cont.*

| M# | SLR Model | M# | SLR Model |
|---|---|---|---|
| 205 | TS = −61.4951920610503 + 44.5805803174326 × USPV ^(1/3) | 457 | TS = 26.1599023833134 + −10.0229727575101 × RCP ^(1/3) |
| 206 | TS = −26.695573877673 + 59.3305417374464 × log(USPV) | 458 | TS = 16.0559054395538 + −8.08312589858581 × log(RCP) |
| 207 | TS = −26.695573877673 + 17.8602727219652 × lg(USPV) | 459 | TS = 16.0559054395538 + −2.4332633542027 × lg(RCP) |
| 208 | FS = 6.25268451510252 + 0.979697960884986 × USPV | 460 | FS = 11.1846550558515 + −0.00712280503711107 × RCP |
| 209 | FS = 8.8968791736595 + 0.0894113658440846 × USPV ^2 | 461 | FS = 11.3393422008688 + −0.0667625121783373 × RCP ^2 |
| 210 | FS = 9.77637198554798 + 0.0107225118544231 × USPV ^3 | 462 | FS = 11.3734572783345 + −0.0460353492860872 × RCP ^3 |
| 211 | FS = 10.5787331781917 + 0.00345900541644401 × e^ USPV | 463 | FS = 11.4643267491817 + −0.0581305838184308 × e^ RCP |
| 212 | FS = 10.1586186111251 + 0.0294317301093728 × 2^ USPV | 464 | FS = 11.4817948094672 + −0.103072169435291 × 2^ RCP |
| 213 | FS = 0.963589896916147 + 4.56083275104788 × USPV ^(1/2) | 465 | FS = 10.8819830772816 + 0.240195832103802 × RCP ^(1/2) |
| 214 | FS = −4.32508243714884 + 9.05905049595972 × USPV ^(1/3) | 466 | FS = 10.5923404701954 + 0.512067600446326 × RCP ^(1/3) |
| 215 | FS = 2.6622701356421 + 12.1767129685834 × log(USPV) | 467 | FS = 11.0722894827807 + 0.642115587523053 × log(RCP) |
| 216 | FS = 2.66227013564208 + 3.66555585213421 × lg(USPV) | 468 | FS = 11.0722894827807 + 0.19329605252784 × lg(RCP) |
| 217 | HbRH = 68.5088143827834 + −5.71647747514457 × USPV | 469 | USPV = 5.99231505990734 + −0.641168070098011 × RCP |
| 218 | HbRH = 53.0861857332109 + −0.52194876820546 × USPV ^2 | 470 | USPV = 5.51082951115173 + −0.196770639884694 × RCP ^2 |
| 219 | HbRH = 47.9782012695454 + −0.0627945081108502 × USPV ^3 | 471 | USPV = 5.33480011717637 + −0.0718816141889639 × RCP ^3 |
| 220 | HbRH = 43.4162462402041 + −0.0210529000617608 × e^ USPV | 472 | USPV = 5.58943973451332 + −0.113334749562755 × e^ RCP |
| 221 | HbRH = 45.8348577834842 + −0.175122004454279 × 2^ USPV | 473 | USPV = 5.88333633268943 + −0.287937416420146 × 2^ RCP |
| 222 | HbRH = 99.4128084239354 + −26.6311418511156 × USPV ^(1/2) | 474 | USPV = 6.89045968374231 + −1.53649400550554 × RCP ^(1/2) |
| 223 | HbRH = 130.327235060512 + −52.9161669690371 × USPV ^(1/3) | 475 | USPV = 7.76902470819704 + −2.417778749196 × RCP ^(1/3) |
| 224 | HbRH = 89.5584537535433 + −71.1931349289109 × log(USPV) | 476 | USPV = 5.33933071656506 + −1.99806201577142 × log(RCP) |
| 225 | HbRH = 89.5584537535434 + −21.4312690989553 × lg(USPV) | 477 | USPV = 5.33933071656506 + −0.601476599944037 × lg(RCP) |
| 226 | ERoS = −44.4898561732773 + 24.4984815259042 × USPV | 478 | HbRH = 42.0423782293484 + −1.48758153721214 × RCP |
| 227 | ERoS = 18.6332502605124 + 2.35355883446629 × USPV ^2 | 479 | HbRH = 41.0709689631091 + −0.515318025294051 × RCP ^2 |
| 228 | ERoS = 39.7125834810522 + 0.298139553955767 × USPV ^3 | 480 | HbRH = 40.5117106435288 + −0.165583213312471 × RCP ^3 |
| 229 | ERoS = 58.2740201323325 + 0.117962114287251 × e^ USPV | 481 | HbRH = 40.9974382176738 + −0.24088810276325 × e^ RCP |
| 230 | ERoS = 47.4896511762263 + 0.901006174693199 × 2^ USPV | 482 | HbRH = 41.7826752623931 + −0.665750517805537 × 2^ RCP |
| 231 | ERoS = −170.590922018834 + 111.297801182333 × USPV ^(1/2) | 483 | HbRH = 43.2454450048409 + −2.8401263707606 × RCP ^(1/2) |
| 232 | ERoS = −296.650864119786 + 219.314541365346 × USPV ^(1/3) | 484 | HbRH = 44.2855416269705 + −3.95501004850947 × RCP ^(1/3) |
| 233 | ERoS = −124.288756274984 + 290.210321300233 × log(USPV) | 485 | HbRH = 40.1489323608852 + −2.24390039617392 × log(RCP) |
| 234 | ERoS = −124.288756274984 + 87.3620117626516 × lg(USPV) | 486 | HbRH = 40.1489323608852 + −0.675481326530641 × lg(RCP) |
| 235 | ASC = 12.7056584538935 + −2.05383018831603 × USPV | 487 | ERoS = 99.4644577413937 + −13.8232290332395 × RCP |
| 236 | ASC = 7.42474725865513 + −0.197743106683585 × USPV ^2 | 488 | ERoS = 88.526157863181 + −4.01721193944016 × RCP ^2 |
| 237 | ASC = 5.66197246714162 + −0.0251128822731464 × USPV ^3 | 489 | ERoS = 84.5196933817461 + −1.37231471647666 × RCP ^3 |
| 238 | ASC = 4.11569164408351 + −0.01003606279162 × e^ USPV | 490 | ERoS = 89.5989881995607 + −2.20731976041961 × e^ RCP |
| 239 | ASC = 5.01986392805905 + −0.0762694909654092 × 2^ USPV | 491 | ERoS = 95.9367653724852 + −5.81337610312006 × 2^ RCP |
| 240 | ASC = 23.2575691266695 + −9.32181911366594 × USPV ^(1/2) | 492 | ERoS = 119.666620550924 + −33.8160568760288 × RCP ^(1/2) |
| 241 | ASC = 33.8066223340896 + −18.3634660495521 × USPV ^(1/3) | 493 | ERoS = 139.34893446531 + −53.5168485493523 × RCP ^(1/3) |
| 242 | ASC = 19.3652424955155 + −24.2863538785987 × log(USPV) | 494 | ERoS = 85.6356391257711 + −44.6518386412545 × log(RCP) |
| 243 | ASC = 19.3652424955155 + −7.31092100276846 × lg(USPV) | 495 | ERoS = 85.6356391257711 + −13.4415427925656 × lg(RCP) |
| 244 | RCP = 4.17276937555526 + −0.529795858393688 × USPV | 496 | ASC = −2.51848109170778 + 3.24667448853144 × RCP |
| 245 | RCP = 2.8215531818394 + −0.0514416129538915 × USPV ^2 | 497 | ASC = −0.203489407868408 + 1.04605755869761 × RCP ^2 |
| 246 | RCP = 2.36985609246673 + −0.00658572528011425 × USPV ^3 | 498 | ASC = 0.685672739591328 + 0.392888626101718 × RCP ^3 |
| 247 | RCP = 1.97477200398597 + −0.00269246761070794 × e^ USPV | 499 | ASC = −0.642194562790037 + 0.606665572012053 × e^ RCP |
| 248 | RCP = 2.2095241687276 + −0.0202348720060608 × 2^ USPV | 500 | ASC = −2.15552198342141 + 1.52125131772239 × 2^ RCP |
| 249 | RCP = 6.87117560609697 + −2.39410754458009 × USPV ^(1/2) | 501 | ASC = −6.75795197625392 + 7.52651357171335 × RCP ^(1/2) |
| 250 | RCP = 9.56852307231565 + −4.70927504528767 × USPV ^(1/3) | 502 | ASC = −10.8963360640752 + 11.6979653703423 × RCP ^(1/3) |
| 251 | RCP = 5.85202639937843 + −6.20954309949019 × log(USPV) | 503 | ASC = 0.899397697886484 + 9.41369194818746 × log(RCP) |
| 252 | RCP = 5.85202639937842 + −1.86925873231483 × lg(USPV) | 504 | ASC = 0.899397697886484 + 2.83380364634493 × lg(RCP) |

## Appendix C

This appendix supplements Section 3.5 by viewing the $p(\alpha^*)$ (i.e., significance of the estimated $\alpha^*$ value), $p(\beta^*)$ (i.e., significance of the estimated $\beta^*$ value) and $(R^2)^*$ (i.e., the adjusted R square value) of the SLR models in Table 5 from another perspective. Tables 4, A3 and A5 re-sort the results by the transform method applied on the RHS independent variable in each SLR model. For each specific transform, the results for all pairs of variables are aggregated and summarised for clearer inspections.

The results are provided directly without any further marks for inspections. In addition, the order of the sub tables, from (a) to (h), follows the same order in Tables 3, 6 and 7 for the different transformation method.

**Table A3.** SLR Models' $p(\alpha^*)$ Values for Each Pair of Variables, by Transform Method.

| (a) | CS | TS | FS | USPV | HbRH | ERoS | ASC | RCP |
|---|---|---|---|---|---|---|---|---|
| **CS** | 0.000000 | 0.000001 | 0.000000 | 0.000000 | 0.000000 | 0.000000 | 0.000092 | 0.000000 |
| **TS** | 0.000000 | 0.000000 | 0.000000 | 0.000000 | 0.000000 | 0.000000 | 0.000001 | 0.000000 |
| **FS** | 0.000000 | 0.017322 | 0.000000 | 0.000000 | 0.000000 | 0.000003 | 0.063099 | 0.000007 |
| **USPV** | 0.000217 | 0.286449 | 0.000000 | 0.000000 | 0.000011 | 0.149037 | 0.000002 | 0.000000 |
| **HbRH** | 0.000000 | 0.000000 | 0.000000 | 0.000000 | 0.000000 | 0.000000 | 0.000147 | 0.000000 |
| **ERoS** | 0.000000 | 0.000000 | 0.000000 | 0.000000 | 0.000000 | 0.000000 | 0.000000 | 0.000000 |
| **ASC** | 0.000000 | 0.000000 | 0.000000 | 0.000000 | 0.000000 | 0.000000 | 0.000000 | 0.000000 |
| **RCP** | 0.000000 | 0.000000 | 0.000000 | 0.000000 | 0.000000 | 0.000000 | 0.573584 | 0.000000 |
| **(b)** | **CS** | **TS** | **FS** | **USPV** | **HbRH** | **ERoS** | **ASC** | **RCP** |
| **CS** | 0.000000 | 0.000000 | 0.000000 | 0.000000 | 0.000000 | 0.000000 | 0.000003 | 0.000000 |
| **TS** | 0.000000 | 0.000000 | 0.000000 | 0.000000 | 0.000000 | 0.000000 | 0.000000 | 0.000000 |
| **FS** | 0.000000 | 0.000000 | 0.000000 | 0.000000 | 0.000000 | 0.000000 | 0.006972 | 0.000000 |
| **USPV** | 0.000000 | 0.000433 | 0.000000 | 0.000000 | 0.000000 | 0.000050 | 0.000000 | 0.000000 |
| **HbRH** | 0.000000 | 0.000000 | 0.000000 | 0.000000 | 0.000000 | 0.000000 | 0.000005 | 0.000000 |
| **ERoS** | 0.000000 | 0.000000 | 0.000000 | 0.000000 | 0.000000 | 0.000000 | 0.000000 | 0.000000 |
| **ASC** | 0.000000 | 0.000000 | 0.000000 | 0.000000 | 0.000000 | 0.000000 | 0.000000 | 0.000000 |
| **RCP** | 0.000000 | 0.000000 | 0.000000 | 0.000000 | 0.000000 | 0.000000 | 0.024659 | 0.000000 |
| **(c)** | **CS** | **TS** | **FS** | **USPV** | **HbRH** | **ERoS** | **ASC** | **RCP** |
| **CS** | 0.000000 | 0.000000 | 0.000000 | 0.000000 | 0.000000 | 0.000000 | 0.000000 | 0.000000 |
| **TS** | 0.000000 | 0.000000 | 0.000000 | 0.000000 | 0.000000 | 0.000000 | 0.000000 | 0.000000 |
| **FS** | 0.000000 | 0.000000 | 0.000000 | 0.000000 | 0.000000 | 0.000000 | 0.000000 | 0.000000 |
| **USPV** | 0.000000 | 0.000000 | 0.000000 | 0.000000 | 0.000000 | 0.000000 | 0.000000 | 0.000000 |
| **HbRH** | 0.000000 | 0.000000 | 0.000000 | 0.000000 | 0.000000 | 0.000000 | 0.000000 | 0.000000 |
| **ERoS** | 0.000000 | 0.000000 | 0.000000 | 0.000000 | 0.000000 | 0.000000 | 0.000000 | 0.000000 |
| **ASC** | 0.000000 | 0.000000 | 0.000000 | 0.000000 | 0.000000 | 0.000000 | 0.000000 | 0.000000 |
| **RCP** | 0.000000 | 0.000000 | 0.000000 | 0.000000 | 0.000000 | 0.000000 | 0.143616 | 0.000000 |
| **(d)** | **CS** | **TS** | **FS** | **USPV** | **HbRH** | **ERoS** | **ASC** | **RCP** |
| **CS** | 0.000000 | 0.000000 | 0.000000 | 0.000000 | 0.000000 | 0.000000 | 0.000000 | 0.000000 |
| **TS** | 0.000000 | 0.000000 | 0.000000 | 0.000000 | 0.000000 | 0.000000 | 0.000000 | 0.000000 |
| **FS** | 0.000000 | 0.000000 | 0.000000 | 0.000000 | 0.000000 | 0.000000 | 0.000001 | 0.000000 |
| **USPV** | 0.000000 | 0.000003 | 0.000000 | 0.000000 | 0.000000 | 0.000000 | 0.000000 | 0.000000 |
| **HbRH** | 0.000000 | 0.000000 | 0.000000 | 0.000000 | 0.000000 | 0.000000 | 0.000000 | 0.000000 |
| **ERoS** | 0.000000 | 0.000000 | 0.000000 | 0.000000 | 0.000000 | 0.000000 | 0.000000 | 0.000000 |
| **ASC** | 0.000000 | 0.000000 | 0.000000 | 0.000000 | 0.000000 | 0.000000 | 0.000000 | 0.000000 |
| **RCP** | 0.000000 | 0.000000 | 0.000000 | 0.000000 | 0.000001 | 0.000000 | 0.000769 | 0.000000 |
| **(e)** | **CS** | **TS** | **FS** | **USPV** | **HbRH** | **ERoS** | **ASC** | **RCP** |
| **CS** | 0.000000 | 0.000000 | 0.108751 | 0.002671 | 0.000100 | 0.550016 | 0.027612 | 0.000549 |
| **TS** | 0.176117 | 0.000000 | 0.060424 | 0.000002 | 0.000008 | 0.839067 | 0.002417 | 0.000036 |
| **FS** | 0.000060 | 0.000018 | 0.000000 | 0.020769 | 0.003388 | 0.613544 | 0.751245 | 0.175856 |
| **USPV** | 0.249243 | 0.001985 | 0.853972 | 0.000000 | 0.021310 | 0.001780 | 0.000073 | 0.000006 |
| **HbRH** | 0.000000 | 0.000001 | 0.000000 | 0.000000 | 0.000000 | 0.000006 | 0.159489 | 0.002875 |
| **ERoS** | 0.013146 | 0.796416 | 0.000097 | 0.000007 | 0.000396 | 0.000000 | 0.000040 | 0.000064 |
| **ASC** | 0.000000 | 0.000000 | 0.000000 | 0.000000 | 0.000000 | 0.000000 | 0.000000 | 0.000000 |
| **RCP** | 0.000000 | 0.000004 | 0.000000 | 0.000000 | 0.002721 | 0.000001 | 0.000002 | 0.000000 |

**Table A3.** *Cont.*

| (f) | CS | TS | FS | USPV | HbRH | ERoS | ASC | RCP |
|---|---|---|---|---|---|---|---|---|
| CS | 0.000000 | 0.000000 | 0.000403 | 0.236202 | 0.000655 | 0.870334 | 0.049127 | 0.003082 |
| TS | 0.000265 | 0.000000 | 0.322155 | 0.020386 | 0.000131 | 0.302175 | 0.007208 | 0.000522 |
| FS | 0.000001 | 0.000001 | 0.000000 | 0.335280 | 0.012176 | 0.945119 | 0.893291 | 0.358140 |
| USPV | 0.086979 | 0.000500 | 0.582955 | 0.000000 | 0.042606 | 0.000400 | 0.000113 | 0.000020 |
| HbRH | 0.000001 | 0.000025 | 0.000000 | 0.000000 | 0.000000 | 0.000221 | 0.307528 | 0.034954 |
| ERoS | 0.190682 | 0.218798 | 0.010250 | 0.028473 | 0.003633 | 0.000000 | 0.000142 | 0.000813 |
| ASC | 0.000000 | 0.000000 | 0.000000 | 0.000000 | 0.000001 | 0.000000 | 0.000000 | 0.020166 |
| RCP | 0.000006 | 0.000118 | 0.000183 | 0.000000 | 0.029770 | 0.000024 | 0.000001 | 0.000000 |
| (g) | CS | TS | FS | USPV | HbRH | ERoS | ASC | RCP |
| CS | 0.000000 | 0.000000 | 0.000010 | 0.963047 | 0.001980 | 0.571159 | 0.070590 | 0.007804 |
| TS | 0.006873 | 0.000000 | 0.921876 | 0.000922 | 0.000073 | 0.624640 | 0.007369 | 0.000243 |
| FS | 0.000010 | 0.000004 | 0.000000 | 0.101100 | 0.008491 | 0.830472 | 0.828827 | 0.253466 |
| USPV | 0.419449 | 0.003784 | 0.530899 | 0.000000 | 0.011174 | 0.004408 | 0.000051 | 0.000003 |
| HbRH | 0.000012 | 0.000109 | 0.000005 | 0.000000 | 0.000000 | 0.000624 | 0.384850 | 0.081312 |
| ERoS | 0.515965 | 0.096413 | 0.085698 | 0.510925 | 0.012697 | 0.000000 | 0.000392 | 0.004012 |
| ASC | 0.000000 | 0.000000 | 0.000000 | 0.000000 | 0.000000 | 0.000000 | 0.000000 | 0.000000 |
| RCP | 0.000000 | 0.000000 | 0.000000 | 0.000000 | 0.000000 | 0.000000 | 0.003399 | 0.000000 |
| (h) | CS | TS | FS | USPV | HbRH | ERoS | ASC | RCP |
| CS | 0.000000 | 0.000000 | 0.000010 | 0.963047 | 0.001980 | 0.571159 | 0.070590 | 0.007804 |
| TS | 0.006873 | 0.000000 | 0.921876 | 0.000922 | 0.000073 | 0.624640 | 0.007369 | 0.000243 |
| FS | 0.000010 | 0.000004 | 0.000000 | 0.101100 | 0.008491 | 0.830472 | 0.828827 | 0.253466 |
| USPV | 0.419449 | 0.003784 | 0.530899 | 0.000000 | 0.011174 | 0.004408 | 0.000051 | 0.000003 |
| HbRH | 0.000012 | 0.000109 | 0.000005 | 0.000000 | 0.000000 | 0.000624 | 0.384850 | 0.081312 |
| ERoS | 0.515965 | 0.096413 | 0.085698 | 0.510925 | 0.012697 | 0.000000 | 0.000392 | 0.004012 |
| ASC | 0.000000 | 0.000000 | 0.000000 | 0.000000 | 0.000000 | 0.000000 | 0.000000 | 0.000000 |
| RCP | 0.000000 | 0.000000 | 0.000000 | 0.000000 | 0.000000 | 0.000000 | 0.003399 | 0.000000 |

**Table A4.** SLR Models' $p(\beta^*)$ Values for Each Pair of Variables, by Transform Method.

| (a) | CS | TS | FS | USPV | HbRH | ERoS | ASC | RCP |
|---|---|---|---|---|---|---|---|---|
| CS | 0.000000 | 0.000000 | 0.000000 | 0.009394 | 0.014243 | 0.068910 | 0.099559 | 0.067789 |
| TS | 0.000000 | 0.000000 | 0.000000 | 0.000093 | 0.009916 | 0.000259 | 0.013286 | 0.052104 |
| FS | 0.000000 | 0.000000 | 0.000000 | 0.067451 | 0.045792 | 0.234888 | 0.852231 | 0.996245 |
| USPV | 0.010063 | 0.000090 | 0.083775 | 0.000000 | 0.196446 | 0.000028 | 0.000336 | 0.000190 |
| HbRH | 0.009670 | 0.010165 | 0.047811 | 0.280548 | 0.000000 | 0.224030 | 0.684669 | 0.548015 |
| ERoS | 0.048053 | 0.000084 | 0.194446 | 0.000003 | 0.149833 | 0.000000 | 0.000178 | 0.023698 |
| ASC | 0.260660 | 0.098907 | 0.710912 | 0.001811 | 0.597407 | 0.002567 | 0.000000 | 0.000000 |
| RCP | 0.059945 | 0.049972 | 0.729410 | 0.000445 | 0.730079 | 0.077182 | 0.000000 | 0.000000 |
| (b) | CS | TS | FS | USPV | HbRH | ERoS | ASC | RCP |
| CS | 0.000000 | 0.000000 | 0.000000 | 0.011907 | 0.014034 | 0.062907 | 0.088345 | 0.075272 |
| TS | 0.000000 | 0.000000 | 0.000000 | 0.000237 | 0.008612 | 0.000317 | 0.008775 | 0.056392 |
| FS | 0.000000 | 0.000000 | 0.000000 | 0.069549 | 0.031002 | 0.243582 | 0.874895 | 0.995302 |
| USPV | 0.013253 | 0.000162 | 0.107008 | 0.000000 | 0.226450 | 0.000037 | 0.000397 | 0.000197 |
| HbRH | 0.007769 | 0.008633 | 0.039253 | 0.428430 | 0.000000 | 0.287537 | 0.597416 | 0.382366 |
| ERoS | 0.031897 | 0.000027 | 0.178153 | 0.000001 | 0.140798 | 0.000000 | 0.000060 | 0.012067 |
| ASC | 0.353814 | 0.188561 | 0.687871 | 0.006154 | 0.513013 | 0.007894 | 0.000000 | 0.000004 |
| RCP | 0.062508 | 0.050343 | 0.538354 | 0.001097 | 0.775369 | 0.122102 | 0.000000 | 0.000000 |

**Table A4.** *Cont.*

| (c) | CS | TS | FS | USPV | HbRH | ERoS | ASC | RCP |
|---|---|---|---|---|---|---|---|---|
| CS | 0.000000 | 0.055766 | 0.044846 | 0.197340 | 0.321462 | 0.436468 | 0.304312 | 0.602958 |
| TS | 0.000140 | 0.000000 | 0.004788 | 0.097387 | 0.125694 | 0.134707 | 0.113898 | 0.380960 |
| FS | 0.000000 | 0.000079 | 0.000000 | 0.071435 | 0.015465 | 0.401719 | 0.792810 | 0.777263 |
| USPV | 0.031694 | 0.000970 | 0.216709 | 0.000000 | 0.333375 | 0.000119 | 0.000780 | 0.000287 |
| HbRH | 0.271163 | 0.168105 | 0.405339 | 0.737545 | 0.000000 | 0.597118 | 0.506679 | 0.222821 |
| ERoS | 0.622799 | 0.083328 | 0.834675 | 0.000055 | 0.817184 | 0.000000 | 0.061580 | 0.112349 |
| ASC | 0.390676 | 0.317673 | 0.688828 | 0.025775 | 0.408285 | 0.051972 | 0.000000 | 0.000153 |
| RCP | 0.060509 | 0.049079 | 0.610654 | 0.000681 | 0.785420 | 0.102184 | 0.000000 | 0.000000 |

| (d) | CS | TS | FS | USPV | HbRH | ERoS | ASC | RCP |
|---|---|---|---|---|---|---|---|---|
| CS | 0.000000 | 0.054008 | 0.043813 | 0.196189 | 0.318493 | 0.432274 | 0.301173 | 0.599052 |
| TS | 0.000006 | 0.000000 | 0.000954 | 0.048784 | 0.066138 | 0.048389 | 0.053810 | 0.282676 |
| FS | 0.000000 | 0.000009 | 0.000000 | 0.070884 | 0.011745 | 0.340585 | 0.921102 | 0.871662 |
| USPV | 0.020615 | 0.000406 | 0.154958 | 0.000000 | 0.277152 | 0.000065 | 0.000543 | 0.000228 |
| HbRH | 0.170646 | 0.098650 | 0.300968 | 0.830220 | 0.000000 | 0.518074 | 0.492845 | 0.174923 |
| ERoS | 0.530833 | 0.047033 | 0.819475 | 0.000026 | 0.774611 | 0.000000 | 0.029431 | 0.072328 |
| ASC | 0.334737 | 0.206599 | 0.699058 | 0.008735 | 0.473680 | 0.014964 | 0.000000 | 0.000009 |
| RCP | 0.059279 | 0.049027 | 0.714567 | 0.000430 | 0.760238 | 0.080032 | 0.000000 | 0.000000 |

| (e) | CS | TS | FS | USPV | HbRH | ERoS | ASC | RCP |
|---|---|---|---|---|---|---|---|---|
| CS | 0.000000 | 0.000000 | 0.000000 | 0.007071 | 0.018024 | 0.087981 | 0.127552 | 0.063732 |
| TS | 0.000000 | 0.000000 | 0.000000 | 0.000045 | 0.020430 | 0.000520 | 0.038214 | 0.061495 |
| FS | 0.000000 | 0.000000 | 0.000000 | 0.063928 | 0.087751 | 0.228218 | 0.824097 | 0.985030 |
| USPV | 0.006726 | 0.000037 | 0.057342 | 0.000000 | 0.156394 | 0.000019 | 0.000278 | 0.000196 |
| HbRH | 0.023359 | 0.019873 | 0.090691 | 0.130591 | 0.000000 | 0.146475 | 0.813553 | 0.848792 |
| ERoS | 0.104815 | 0.000758 | 0.255601 | 0.000059 | 0.182040 | 0.000000 | 0.001151 | 0.068695 |
| ASC | 0.059099 | 0.010315 | 0.933397 | 0.000137 | 0.929925 | 0.000386 | 0.000000 | 0.000000 |
| RCP | 0.070213 | 0.062484 | 0.865421 | 0.000150 | 0.795613 | 0.041055 | 0.000000 | 0.000000 |

| (f) | CS | TS | FS | USPV | HbRH | ERoS | ASC | RCP |
|---|---|---|---|---|---|---|---|---|
| CS | 0.000000 | 0.000000 | 0.000000 | 0.006903 | 0.018790 | 0.091034 | 0.131594 | 0.063793 |
| TS | 0.000000 | 0.000000 | 0.000000 | 0.000045 | 0.022941 | 0.000610 | 0.043804 | 0.064114 |
| FS | 0.000000 | 0.000000 | 0.000000 | 0.063534 | 0.094571 | 0.228012 | 0.821431 | 0.982807 |
| USPV | 0.006440 | 0.000034 | 0.054960 | 0.000000 | 0.152343 | 0.000019 | 0.000274 | 0.000198 |
| HbRH | 0.026739 | 0.022083 | 0.099447 | 0.120253 | 0.000000 | 0.140053 | 0.825341 | 0.880793 |
| ERoS | 0.114620 | 0.000975 | 0.265660 | 0.000081 | 0.187065 | 0.000000 | 0.001404 | 0.076639 |
| ASC | 0.047043 | 0.008122 | 0.960882 | 0.000128 | 0.995214 | 0.000399 | 0.000000 | 0.000000 |
| RCP | 0.073097 | 0.065603 | 0.818108 | 0.000146 | 0.818453 | 0.039532 | 0.000000 | 0.000000 |

| (g) | CS | TS | FS | USPV | HbRH | ERoS | ASC | RCP |
|---|---|---|---|---|---|---|---|---|
| CS | 0.000000 | 0.000000 | 0.000000 | 0.006615 | 0.020590 | 0.097797 | 0.140259 | 0.064214 |
| TS | 0.000000 | 0.000000 | 0.000000 | 0.000049 | 0.029361 | 0.000871 | 0.057676 | 0.070523 |
| FS | 0.000000 | 0.000000 | 0.000000 | 0.062757 | 0.109889 | 0.227950 | 0.816382 | 0.977773 |
| USPV | 0.005912 | 0.000028 | 0.050493 | 0.000000 | 0.144487 | 0.000018 | 0.000266 | 0.000203 |
| HbRH | 0.035566 | 0.027663 | 0.120466 | 0.102983 | 0.000000 | 0.128735 | 0.846305 | 0.941104 |
| ERoS | 0.136667 | 0.001594 | 0.287837 | 0.000153 | 0.197902 | 0.000000 | 0.002056 | 0.094423 |
| ASC | 0.034789 | 0.007237 | 0.979497 | 0.000193 | 0.855884 | 0.000743 | 0.000000 | 0.000000 |
| RCP | 0.080536 | 0.073575 | 0.727705 | 0.000154 | 0.875153 | 0.038117 | 0.000000 | 0.000000 |

| (h) | CS | TS | FS | USPV | HbRH | ERoS | ASC | RCP |
|---|---|---|---|---|---|---|---|---|
| CS | 0.000000 | 0.000000 | 0.000000 | 0.006615 | 0.020590 | 0.097797 | 0.140259 | 0.064214 |
| TS | 0.000000 | 0.000000 | 0.000000 | 0.000049 | 0.029361 | 0.000871 | 0.057676 | 0.070523 |
| FS | 0.000000 | 0.000000 | 0.000000 | 0.062757 | 0.109889 | 0.227950 | 0.816382 | 0.977773 |
| USPV | 0.005912 | 0.000028 | 0.050493 | 0.000000 | 0.144487 | 0.000018 | 0.000266 | 0.000203 |
| HbRH | 0.035566 | 0.027663 | 0.120466 | 0.102983 | 0.000000 | 0.128735 | 0.846305 | 0.941104 |
| ERoS | 0.136667 | 0.001594 | 0.287837 | 0.000153 | 0.197902 | 0.000000 | 0.002056 | 0.094423 |
| ASC | 0.034789 | 0.007237 | 0.979497 | 0.000193 | 0.855884 | 0.000743 | 0.000000 | 0.000000 |
| RCP | 0.080536 | 0.073575 | 0.727705 | 0.000154 | 0.875153 | 0.038117 | 0.000000 | 0.000000 |

**Table A5.** SLR Models' $(R^2)^*$ Values for Each Pair of Variables, by Transform Method.

| (a) | CS | TS | FS | USPV | HbRH | ERoS | ASC | RCP |
|---|---|---|---|---|---|---|---|---|
| CS | 0.000000 | 0.729936 | 0.656125 | 0.158295 | 0.139542 | 0.067382 | 0.050641 | 0.068132 |
| TS | 0.768782 | 0.000000 | 0.590871 | 0.347419 | 0.155870 | 0.308717 | 0.142690 | 0.080196 |
| FS | 0.670677 | 0.587348 | 0.000000 | 0.068361 | 0.086126 | 0.013040 | −0.028346 | −0.029411 |
| USPV | 0.155206 | 0.348673 | 0.058470 | 0.000000 | 0.020598 | 0.390450 | 0.298517 | 0.320593 |
| HbRH | 0.156995 | 0.154755 | 0.084144 | 0.005750 | 0.000000 | 0.015022 | −0.024357 | −0.018383 |
| ERoS | 0.083912 | 0.351029 | 0.021036 | 0.463398 | 0.032374 | 0.000000 | 0.323088 | 0.116357 |
| ASC | 0.008737 | 0.050938 | −0.025200 | 0.230008 | −0.020877 | 0.215184 | 0.000000 | 0.546710 |
| RCP | 0.073764 | 0.082114 | −0.025742 | 0.287522 | −0.025761 | 0.062204 | 0.657900 | 0.000000 |

| (b) | CS | TS | FS | USPV | HbRH | ERoS | ASC | RCP |
|---|---|---|---|---|---|---|---|---|
| CS | 0.000000 | 0.688708 | 0.628852 | 0.147638 | 0.140210 | 0.071553 | 0.056056 | 0.063348 |
| TS | 0.753109 | 0.000000 | 0.557278 | 0.312080 | 0.162180 | 0.300834 | 0.161342 | 0.076566 |
| FS | 0.667302 | 0.567318 | 0.000000 | 0.066959 | 0.104044 | 0.011528 | −0.028650 | −0.029411 |
| USPV | 0.142802 | 0.326562 | 0.047385 | 0.000000 | 0.014570 | 0.379972 | 0.291958 | 0.319156 |
| HbRH | 0.166780 | 0.162074 | 0.093207 | −0.010324 | 0.000000 | 0.004761 | −0.020878 | −0.006231 |
| ERoS | 0.102738 | 0.391107 | 0.024808 | 0.512683 | 0.035121 | 0.000000 | 0.363027 | 0.147036 |
| ASC | −0.003333 | 0.022356 | −0.024465 | 0.177117 | −0.016348 | 0.166068 | 0.000000 | 0.455422 |
| RCP | 0.071845 | 0.081774 | −0.017845 | 0.250962 | −0.026913 | 0.041462 | 0.613330 | 0.000000 |

| (c) | CS | TS | FS | USPV | HbRH | ERoS | ASC | RCP |
|---|---|---|---|---|---|---|---|---|
| CS | 0.000000 | 0.077077 | 0.087086 | 0.020403 | 0.000351 | −0.010974 | 0.002503 | −0.021132 |
| TS | 0.332171 | 0.000000 | 0.188168 | 0.051639 | 0.040166 | 0.037082 | 0.044579 | −0.006096 |
| FS | 0.530304 | 0.353381 | 0.000000 | 0.065737 | 0.135809 | −0.008032 | −0.027294 | −0.026956 |
| USPV | 0.103032 | 0.256006 | 0.016422 | 0.000000 | −0.001060 | 0.338131 | 0.264942 | 0.304715 |
| HbRH | 0.007126 | 0.027331 | −0.008356 | −0.025967 | 0.000000 | −0.020864 | −0.015953 | 0.015249 |
| ERoS | −0.022005 | 0.058714 | −0.028074 | 0.366523 | −0.027771 | 0.000000 | 0.072530 | 0.045194 |
| ASC | −0.007020 | 0.000814 | −0.024497 | 0.112510 | −0.008616 | 0.080311 | 0.000000 | 0.328825 |
| RCP | 0.073334 | 0.082942 | −0.021478 | 0.270415 | −0.027137 | 0.049465 | 0.629821 | 0.000000 |

| (d) | CS | TS | FS | USPV | HbRH | ERoS | ASC | RCP |
|---|---|---|---|---|---|---|---|---|
| CS | 0.000000 | 0.078548 | 0.088156 | 0.020654 | 0.000713 | −0.010637 | 0.002914 | −0.020953 |
| TS | 0.439476 | 0.000000 | 0.256714 | 0.083218 | 0.069260 | 0.083592 | 0.078716 | 0.005445 |
| FS | 0.594853 | 0.428716 | 0.000000 | 0.066091 | 0.148254 | −0.001882 | −0.029110 | −0.028610 |
| USPV | 0.122725 | 0.291083 | 0.030895 | 0.000000 | 0.006241 | 0.360287 | 0.279544 | 0.313564 |
| HbRH | 0.026678 | 0.051056 | 0.002941 | −0.028000 | 0.000000 | −0.016657 | −0.015063 | 0.025602 |
| ERoS | −0.017414 | 0.084898 | −0.027813 | 0.392235 | −0.026896 | 0.000000 | 0.106431 | 0.065169 |
| ASC | −0.001217 | 0.018447 | −0.024830 | 0.161547 | −0.013759 | 0.137303 | 0.000000 | 0.426455 |
| RCP | 0.074276 | 0.082990 | −0.025310 | 0.288809 | −0.026554 | 0.060551 | 0.652188 | 0.000000 |

| (e) | CS | TS | FS | USPV | HbRH | ERoS | ASC | RCP |
|---|---|---|---|---|---|---|---|---|
| CS | 0.000000 | 0.768479 | 0.671015 | 0.170964 | 0.128848 | 0.056243 | 0.039512 | 0.070957 |
| TS | 0.742180 | 0.000000 | 0.600305 | 0.373440 | 0.123135 | 0.281278 | 0.094439 | 0.072593 |
| FS | 0.667806 | 0.611742 | 0.000000 | 0.070816 | 0.056362 | 0.014244 | −0.027895 | −0.029401 |
| USPV | 0.173187 | 0.380383 | 0.075799 | 0.000000 | 0.030489 | 0.402593 | 0.305891 | 0.319400 |
| HbRH | 0.117014 | 0.124398 | 0.054866 | 0.038463 | 0.000000 | 0.033374 | −0.027704 | −0.028295 |
| ERoS | 0.048318 | 0.266089 | 0.009541 | 0.363949 | 0.023874 | 0.000000 | 0.248963 | 0.067525 |
| ASC | 0.074416 | 0.154099 | −0.029197 | 0.332853 | −0.029174 | 0.293131 | 0.000000 | 0.697538 |
| RCP | 0.066525 | 0.071862 | −0.028530 | 0.329451 | −0.027352 | 0.091144 | 0.631358 | 0.000000 |

| (f) | CS | TS | FS | USPV | HbRH | ERoS | ASC | RCP |
|---|---|---|---|---|---|---|---|---|
| CS | 0.000000 | 0.770630 | 0.670469 | 0.172030 | 0.126953 | 0.054695 | 0.038122 | 0.070913 |
| TS | 0.734537 | 0.000000 | 0.597051 | 0.373150 | 0.117840 | 0.274909 | 0.088166 | 0.070683 |
| FS | 0.666809 | 0.613889 | 0.000000 | 0.071099 | 0.052967 | 0.014282 | −0.027848 | −0.029397 |
| USPV | 0.175108 | 0.383737 | 0.077746 | 0.000000 | 0.031643 | 0.403635 | 0.306504 | 0.319003 |
| HbRH | 0.110828 | 0.119583 | 0.050692 | 0.042145 | 0.000000 | 0.035356 | −0.027917 | −0.028721 |
| ERoS | 0.044295 | 0.255825 | 0.007960 | 0.352157 | 0.022699 | 0.000000 | 0.240697 | 0.062527 |
| ASC | 0.084888 | 0.164797 | −0.029338 | 0.335491 | −0.029411 | 0.291814 | 0.000000 | 0.699772 |
| RCP | 0.064686 | 0.069632 | −0.027788 | 0.330484 | −0.027794 | 0.092881 | 0.617075 | 0.000000 |

**Table A5.** *Cont.*

| (g) | CS | TS | FS | USPV | HbRH | ERoS | ASC | RCP |
|---|---|---|---|---|---|---|---|---|
| **CS** | 0.000000 | 0.773545 | 0.668025 | 0.173923 | 0.122779 | 0.051449 | 0.035291 | 0.070612 |
| **TS** | 0.716509 | 0.000000 | 0.587820 | 0.370405 | 0.106540 | 0.260413 | 0.075533 | 0.066324 |
| **FS** | 0.664369 | 0.617779 | 0.000000 | 0.071663 | 0.046189 | 0.014293 | −0.027756 | −0.029388 |
| **USPV** | 0.178890 | 0.390315 | 0.081637 | 0.000000 | 0.033977 | 0.405520 | 0.307596 | 0.318047 |
| **HbRH** | 0.097739 | 0.109271 | 0.042065 | 0.049114 | 0.000000 | 0.039100 | −0.028258 | −0.029244 |
| **ERoS** | 0.036441 | 0.235384 | 0.004719 | 0.328781 | 0.020282 | 0.000000 | 0.224651 | 0.053038 |
| **ASC** | 0.098753 | 0.169934 | −0.029391 | 0.320016 | −0.028399 | 0.266906 | 0.000000 | 0.672241 |
| **RCP** | 0.060265 | 0.064389 | −0.025694 | 0.328499 | −0.028654 | 0.094556 | 0.580226 | 0.000000 |
| **(h)** | **CS** | **TS** | **FS** | **USPV** | **HbRH** | **ERoS** | **ASC** | **RCP** |
| **CS** | 0.000000 | 0.773545 | 0.668025 | 0.173923 | 0.122779 | 0.051449 | 0.035291 | 0.070612 |
| **TS** | 0.716509 | 0.000000 | 0.587820 | 0.370405 | 0.106540 | 0.260413 | 0.075533 | 0.066324 |
| **FS** | 0.664369 | 0.617779 | 0.000000 | 0.071663 | 0.046189 | 0.014293 | −0.027756 | −0.029388 |
| **USPV** | 0.178890 | 0.390315 | 0.081637 | 0.000000 | 0.033977 | 0.405520 | 0.307596 | 0.318047 |
| **HbRH** | 0.097739 | 0.109271 | 0.042065 | 0.049114 | 0.000000 | 0.039100 | −0.028258 | −0.029244 |
| **ERoS** | 0.036441 | 0.235384 | 0.004719 | 0.328781 | 0.020282 | 0.000000 | 0.224651 | 0.053038 |
| **ASC** | 0.098753 | 0.169934 | −0.029391 | 0.320016 | −0.028399 | 0.266906 | 0.000000 | 0.672241 |
| **RCP** | 0.060265 | 0.064389 | −0.025694 | 0.328499 | −0.028654 | 0.094556 | 0.580226 | 0.000000 |

## References


1. Kuo, W.T.; Zhuang, Z.Y. A Comprehensive Study of the Mechanical and Durability Properties of High-Performance Concrete Materials for Grouting Underwater Foundations of Offshore Wind Turbines. *Materials* **2021**, *14*, 5968. [CrossRef]
2. Lee, C.C. Energy consumption and GDP in developing countries: A cointegrated panel analysis. *Energy Econ.* **2005**, *27*, 415–427. [CrossRef]
3. Stern, D.I. Economic growth and energy. In *Encyclopedia of Energy*; Cleveland, C.J., Ed.; Elsevier: Philadelphia, PA, USA, 2004; Volume 2, pp. 35–51.
4. Brundtland, G.H.; Khalid, M.; Agnelli, S.; Al-Athel, S.; Chidzero, B. Report of the World Commission on Environment and Development: Our common future. *U. N. Doc. Coop. Circ.* **1987**, *10*, 1–223.
5. Inter-Agency Task Force on Financing for Development, UN. *Financing for Sustainable Development Report 2021*; United Nations: New York, NY, USA, 2021. Available online: https://developmentfinance.un.org/fsdr2021 (accessed on 15 July 2022).
6. Herzog, A.V.; Lipman, T.E.; Kammen, D.M. Renewable Energy Sources. Encyclopedia of Life Support Systems (EOLSS) Forerunner Volume-Perspectives and Overview of Life Support Systems and Sustainable Development. Google Scholar. 2001. Available online: https://rael.berkeley.edu/old_drupal/sites/default/files/old-site-files/2001/Herzog-Lipman-Kammen-RenewableEnergy-2001.pdf (accessed on 20 July 2022).
7. Panwar, N.L.; Kaushik, S.C.; Kothari, S. Role of renewable energy sources in environmental protection: A review. *Renew. Sustain. Energy Rev.* **2011**, *15*, 1513–1524. [CrossRef]
8. Sinsel, S.R.; Riemke, R.L.; Hoffmann, V.H. Challenges and solution technologies for the integration of variable renewable energy sources—A review. *Renew. Energy* **2020**, *145*, 2271–2285. [CrossRef]
9. Ahmadi, S.A.; Mirlohi, S.M.; Ahmadi, M.H.; Ameri, M. Portfolio optimization of power plants by using renewable energy in Iran. *Int. J. Low Carbon. Technol.* **2021**, *16*, 463–475.
10. Zhuang, Z.Y.; Hocine, A.; Kouaissah, N.; Kiker, G.A. Optimising sustainable renewable energy portfolios using a multi-tolerance fuzzy goal programming approach. *Int. J. Green Energy* **2022**, 1–16. [CrossRef]
11. Wu, Y.; Xu, C.; Ke, Y.; Tao, Y.; Li, X. Portfolio optimization of renewable energy projects under type-2 fuzzy environment with sustainability perspective. *Comput. Ind. Eng.* **2019**, *133*, 69–82. [CrossRef]
12. Hocine, A.; Zhuang, Z.Y.; Kouaissah, N.; Li, D.C. Weighted-additive fuzzy multi-choice goal programming (WA-FMCGP) for supporting renewable energy site selection decisions. *Eur. J. Oper. Res.* **2020**, *285*, 642–654. [CrossRef]
13. Gerwick, B.C., Jr. *Construction of Marine and Offshore Structures*, 3rd ed.; CRC Press: Boca Raton, FL, USA, 2007; pp. 1–840.
14. Huang, T.T. Largest Offshore Wind Farm in Taiwan Starts Generating Electricity: Orsted-Operated Farm Off West Coast of Taiwan to Power 1 Million Households. Taiwan News. 2022. Available online: https://www.taiwannews.com.tw/en/news/4514439 (accessed on 12 April 2022).
15. Lee, C.L. Policy of Offshore Wind in Taiwan. France: The Official Website of Ministère de l'Économie, des Finances et de la Relance. 2019. Available online: https://www.tresor.economie.gouv.fr/Articles/f27ed9b3-e7db-40bf-880c-5f1351c0a826/files/f4318f81-7d7b-44cc-a462-193ca3840d20 (accessed on 19 July 2021).
16. Almaktar, M.; Shaaban, M. Prospects of renewable energy as a non-rivalry energy alternative in Libya. *Renew. Sustain. Energy Rev.* **2021**, *143*, 110852.

17. Ahmed, A.S. Technical and economic feasibility of the first wind farm on the coast of Mediterranean Sea. *Ain. Shams. Eng. J.* **2021**, *12*, 2145–2151. [CrossRef]
18. Zhuang, Z.Y.; Hocine, A. Meta goal programming approach for solving multi-criteria de Novo programming problem. *Eur. J. Oper. Res.* **2018**, *265*, 228–238. [CrossRef]
19. Chen, S.K.; Wu, Y.M.; Hsu, Y.J.; Chan, Y.C. Current crustal deformation of the Taiwan orogen reassessed by cGPS strain-rate estimation and focal mechanism stress inversion. *Geophys. J. Int.* **2017**, *210*, 228–239. [CrossRef]
20. Yang, Y.R.; Johnson, K.M. Crustal stress state in Taiwan: Moderately strong crust supporting gravitational and flexural loading. *J. Geophysl. Res. Solid Earth* **2020**, *125*, e2020JB019530. [CrossRef]
21. Tsai, K.C.; Hsiao, C.P.; Bruneau, M. Overview of building damages in 921 Chi-Chi earthquake. *Earthq. Eng. Eng. Seismol.* **2000**, *2*, 93–108.
22. Fan, M.F. Disaster governance and community resilience: Reflections on Typhoon Morakot in Taiwan. *J. Environ. Plan. Manag.* **2015**, *58*, 24–38. [CrossRef]
23. Yoshida, N.; Atarashi, D.; Kinose, T.; Okumira, Y. (吉田夏樹, 新大軌, 木野瀬透, 奥村勇馬). A Study on Chemical Change of Fire-damaged Concrete (火災による熱を受けたコンクリートの化学的変化に関する検討). *GBRC Gen. Build. Res. Corp.* **2019**, *44*, 22–28.
24. Lee, T.Y.; Wu, Y.T.; Kueh, M.T.; Lin, C.Y.; Lin, Y.Y.; Sheng, Y.F. Impacts of offshore wind farms on the atmospheric environment over Taiwan Strait during an extreme weather typhoon event. *Sci. Rep.* **2022**, *12*, 823. [CrossRef]
25. Wu, C.R.; Chao, S.Y.; Hsu, C. Transient, seasonal and interannual variability of the Taiwan Strait current. *J. Oceanogr.* **2007**, *63*, 821–833. [CrossRef]
26. Hung, C.C.; Wang, C.C.; Wang, H.Y. Establishment of the controlled low-strength desulfurization slag prediction model for compressive strength and surface resistivity. *Appl. Sci.* **2020**, *10*, 5674.
27. Wang, H.Y.; Wang, W.C.; Wang, J.C.; Chen, Y.W. Evaluation of the engineering properties and durability of mortar produced using ground granulated blast-furnace slag and stainless-steel reduced slag. *Constr. Build Mater.* **2021**, *280*, 122498. [CrossRef]
28. Wang, H.Y.; Tsai, S.L.; Hung, C.C.; Jian, T.Y. Research on Engineering Properties of Cement Mortar adding Stainless Steel Reduction Slag and Pozzolanic materials. *Case Stud. Constr. Mater.* **2022**, *16*, e01144. [CrossRef]
29. *ASTM*; C39 Committee Test Method for Compressive Strength of Cylindrical Concrete Specimens. ASTM International: West Conshohocken, PA, USA, 2021.
30. *ASTM*; C496 Standard Test Method for Splitting Tensile Strength of Cylindrical Concrete Specimens. ASTM International: West Conshohocken, PA, USA, 2017.
31. *ASTM*; C78 Standard Test Method for Flexural Strength of Concrete (Using Simple Beam with Third-Point Loading). ASTM International: West Conshohocken, PA, USA, 2022.
32. *ASTM*; C805 Standard Test Method for Rebound Number of Hardened Concrete. ASTM International: West Conshohocken, PA, USA, 2019.
33. *ASTM*; C597 Standard Test Method for Pulse Velocity Through Concrete. ASTM International: West Conshohocken, PA, USA, 2016.
34. *ASTM*; C1012 Standard Test Method for Length Change of Hydraulic-Cement Mortars Exposed to a Sulfate Solution. ASTM International: West Conshohocken, PA, USA, 2019.
35. *ASTM*; C1202 Standard Test Method for Electrical Indication of Concrete's Ability to Resist Chloride Ion Penetration. ASTM International: West Conshohocken, PA, USA, 2019.
36. *ASTM*; C876 Standard Test Method for Corrosion Potentials of Uncoated Reinforcing Steel in Concrete. ASTM International: West Conshohocken, PA, USA, 2016.
37. Zhang, H.Y.; Sneyd, A.; Stevenson, M. Robustness and reliability of gender bias assessment in word embeddings: The role of base pairs. In Proceedings of the 1st Conference of the Asia-Pacific Chapter of the Association for Computational Linguistics and the 10th International Joint Conference on Natural Language Processing, Suzhou, China, 4–7 December 2020; pp. 759–769.
38. Hsu, C.C.; Zhuang, Z.Y. Competition among the world's main technological powers to develop IPs: Cross-national longitudinal patentography over a 9-year time Span. *Appl. Sci.* **2019**, *9*, 2432. [CrossRef]
39. Sheen, Y.N.; Wang, H.Y.; Lin, R.Y.; Kuo, W.T. *Concrete Technologies*; Chuan Hwa Book Co., Ltd.: New Taipei City, Taiwan, 2011; pp. 1–496.
40. Thien, M.V.; Thanh, T.V. Design interaction diagrams for reinforced concrete circular columns following ACI 319-19. *J. Mat. Constr.* **2021**, *8*, 26–33. [CrossRef]
41. Moehle, J.P. Key changes in the 2019 edition of the ACI Building Code (ACI 318-19). *Concr. Int.* **2019**, *41*, 21–27.
42. American Concrete Institute (ACI) Committee 318. *Commentary on Building Code Requirements for Structural Concrete (ACI 318R-19)*; ACI: Detroit, MI, USA, 2019.
43. Sata, V.; Wongsa, A.; Chindaprasirt, P. Properties of pervious geopolymer concrete using recycled aggregates. *Constr. Build. Mater.* **2013**, *42*, 33–39. [CrossRef]
44. Hoshino, T.; Uomoto, T.; Kobayashi, K. (星野富夫, 魚本健人, 小林一輔). Corrosion Prevention Guidelines for Marine Concrete Structures (海洋コンクリト造物の防食指針(案), JCI-R1, 1983). *J. Jpn. Soc. Civ. Eng. (土木学会論文集)* **1998**, *592*, 107–120.
45. Patel, V.; Shah, N. The durability of normal strength concrete: An experimental study. *Mater Charact VII* **2015**, *90*, 195–203.
46. Atsbha, T.G.; Zhutovsky, S. The effect of external curing methods on the development of mechanical and durability-related properties of normal-strength concrete. *Constr. Build. Mater.* **2022**, *324*, 126706. [CrossRef]

47. Public Works Research Institute (PWRI); The Japanese Society for Non-Destructive Inspection (JSNDI); LIC Co., Ltd.; Zenitaka Corporate; Maeda Infroneer Holdings Inc.; JDC Corporation; Chiba Institute of Technology; Toda Corporation. (独立行政法人土木研究所技術推進本部構造物ﾏﾈｼﾞﾒﾝﾄ技術ﾁｰﾑ, (社)日本非破壊検査協会, リック(株), (株)錢高組, 前田建設工業(株), 日本国土開発(株), 千葉工業大学, 戸田建設(株)); *Joint Research Report on Quality Inspection of Concrete Structures by Non-Destructive/Local Destructive Tests (7)–XVIII Field Experiment in 2005 (非破壊·局部破壊試験によるコンクリート構造物の品質検査に関する共同研究報告書(7)–XVIII2005年度現場実験)*; Public Works Research Institute Joint Research Report No. 355 (土木研究所共同研究報告書第355号); Public Works Research Institute (国立研究開発法人土木研究所): Tsukubashi, Japan, 2006; pp. 1–153.
48. Justnes, H.; Meland, I.; Bjoergum, J.O.; Krane, J.; Skjetne, T. Nuclear magnetic resonance (NMR)—A powerful tool in cement and concrete research. *Adv. Cem. Res.* **1990**, *3*, 105–110. [CrossRef]
49. Sheen, Y.N.; Hwang, C.L. The study of hydration model and strength characteristics of cement paste by NMR. *J. Technol.* **1995**, *10*, 151–156.
50. Takahashi, T.; Ohkubo, T.; Kanehashi, K. (高橋貴文, 大窪貴洋, 金橋康二). Analysis of slag and cement materials by solid-state nuclear magnetic resonance (NMR) (固体核磁気共鳴 (NMR) によるスラグ·セメント材料の解析). *Concr. Eng. (コンクリート工学)* **2018**, *56*, 454–459.
51. Yonemura, M.; Kitagaki, R.; Ohkubo, T.; Kim, J. (米村美紀, 北垣亮馬, 大窪貴洋, 金志訓). Pore structure analysis of hardened cement using 1H NMR (1H NMR を用いたセメント硬化体の細孔構造分析). *Proc. Jpn. Concr. Inst. (コンクリート工学年次論文集)* **2015**, *37*, 511–516.

buildings

MDPI

Article

# An Intelligent Detection Logic for Fan-Blade Damage to Wind Turbines Based on Mounted-Accelerometer Data

Ming-Hung Hsu [1] and Zheng-Yun Zhuang [2,*]

[1] Department of Electrical Engineering, National Penghu University of Science and Technology, Penghu 880, Taiwan
[2] Department of Civil Engineering, National Kaohsiung University of Science and Technology, Kaohsiung 807, Taiwan
* Correspondence: wayne@nkust.edu.tw; Tel.: +886-7-3814526 (ext. 15240)

**Abstract:** Many wind turbines operate in harsh marine or shore environments. This study assists industry by establishing a real-time condition-monitoring and fault-detection system, with rules for recognizing a wind turbine's abnormal operation mainly caused by different types of fan-blade damage. This system can ensure ideal wind turbine operation by monitoring the health status of the blades, detecting sudden anomalies, and performing maintenance almost in real time. This is especially significant for wind farms in areas subject to frequent natural disasters (e.g., earthquakes and typhoons). Turbines might fail to endure these because the manufacturers have built them according to the standards developed for areas less prone to natural disasters. The system's rules are established by utilising concepts and methods from data analytics, digital signal processing (DSP) and statistics to analyse data from the accelerometer, which measures the vibration signals in three dimensions on the platform of the wind turbine's base. The patterns for those cases involving fan-blade damage are found to establish the rules. With the anomalies detected and reported effectively, repairs and maintenance can be carried out on the faulty wind turbines. This enables 'maintenance by prediction' actions for unplanned maintenance as a supplement to the 'predictive maintenance' tasks for regular planned maintenance.

**Keywords:** energy generation; green energy; wind turbines; fan-blade damage; mechanical condition monitoring system; fault diagnosis; maintenance by prediction; accelerometer; vibration data; big-data analytics

**Citation:** Hsu, M.-H.; Zhuang, Z.-Y. An Intelligent Detection Logic for Fan-Blade Damage to Wind Turbines Based on Mounted-Accelerometer Data. *Buildings* **2022**, *12*, 1588. https://doi.org/10.3390/buildings12101588

Academic Editor: Zhenjun Ma

Received: 17 August 2022
Accepted: 28 September 2022
Published: 1 October 2022



## 1. Introduction

### 1.1. Research Background

Improving the operating efficiency of wind turbines and reducing maintenance costs are goals common to all wind farm operators. However, similar to the erosional effects caused by seawater and weather conditions on the base constructs of offshore or onshore wind turbines, machines and equipment are also prone to failure, leading to a loss of efficiency in power generation operations. However, an integrated perspective to examine these issues is seldom discussed [1–3].

The industry developing wind turbines has adopted a conservative attitude, i.e., it lacks integration with related upstream and downstream industries. In some countries, for example, blade manufacturers concentrate only on the R&D (research and development) of the blade shape and materials, while the generator manufacturers care only about the design and manufacture of the generator bodies. The power-converter manufacturers alone bear the final responsibility for power generation efficiency. However, converter manufacturers do not usually have the expertise to integrate blades and generators. Moreover, the international market currently lacks protocols or standard operating procedures (SOPs) for these integrative matters. This is critical for turbine maintenance.

Although avoiding the replacement of faulty major parts is the most efficient way to reduce operating costs, an alternative logic for turbine maintenance entails detecting faults efficiently and replacing the necessary part promptly, in addition to scheduled preventive maintenance (e.g., predictive maintenance). Within the entire power generation system utilising wind as a resource, the blade is an important component. To ensure the normal operation of a wind turbine, the key factors are the durability, safety and reliability of the wind turbine blades, and these have become a concern for wind farm operators and manufacturers. However, to the authors' knowledge, most fan blade health diagnoses today still rely on subjective human efforts to identify whether the blades are damaged or not, i.e., acoustic or vision-based inspections performed by professionals, and currently there is no rapid and objective method for blade diagnosis [1–3].

Benbouzid et al. [4] reviewed the current progress in condition monitoring of wind turbines, from traditional condition monitoring and signal-processing tools to machine learning-based condition monitoring and predictive maintenance using big-data mining. Their systematic review of signal-based and data-driven modelling approaches using intelligence and machine learning approaches examined recent developments in the field and their use in diagnosis, prognosis, health assessment, and predictive maintenance of wind turbines and farms [4]. Natilli et al. [5] developed a multi-scale method for wind turbine gearbox fault detection and tested it in real-world test cases. The novelty of this work is that the detection method was developed using industrial datasets provided by standard SCADA (supervisory control and data acquisition) and TCM (turbine condition monitoring) systems [5].

Papi [6] proposed using an uncertainty-quantification method to model the effect of blade damage on the performance of multi-megawatt wind turbines. The proposed method aims to overcome some of the problems associated with evaluating individual test cases. In fact, treating blade damage as a random phenomenon avoids biases due to specific test cases of combinations of blade damage factors and allows for more general conclusions [6]. This point is critical for this study. Santolamazza et al. [7] presented an approach based on machine learning techniques using data from SCADA systems. Since these systems are usually already implemented on most wind turbines, they provide a lot of data without the need for additional sensors. In particular, they used artificial neural networks (ANNs) to develop models to describe the behaviour of some of the main components of wind turbines, such as gearboxes and generators, and predict operational anomalies. The proposed method was tested on real wind turbines in Italy to verify its effectiveness and applicability and proved to be able to provide important assistance in the maintenance of wind farms [7].

Knowing the status of the blades at any time should be critical to the normal operation of turbines at the lowest cost, because wind has become one of the most popular sources of green power generation (solar being another popular energy source), and more and more wind farms are operating or being constructed worldwide. The offshore equipment is usually expensive and produces electricity almost constantly; however, it may fail, incurring significant operational losses if it cannot be repaired quickly during the downtime [1–3].

The repair and subsequent maintenance tasks involve considerable effort and costs, thus exacerbating the problem. Moreover, in some places, such as Penghu (or the Pescadores Islands) and Taiwan, most wind turbines are maintained by non-native manufacturers, who regard internal equipment maintenance as a business secret (and do not disclose it easily). However, a discussion about this critical problem is beyond the scope of this paper. In any case, there is a need to detect and diagnose problems related to offshore wind turbines quickly, in order to launch the subsequent unplanned yet necessary repairs [1–3].

### 1.2. This Study: An Overview

In this study, a complete functioning mechanical condition monitoring system (CMS) is established with an initial stable performance, which measures parameters such as the vibrations and noises of a turbine during operation. The recorded data are stored and reported remotely (from the remote-side CMS). They are organised as datasets (on the server-side CMS), and the characteristics and patterns of/in the signal are then analysed using the established rules and classified as having different levels of fan-blade damage severity. This process automatically detects the health status of a turbine by inspecting its vibration states, rather than through human efforts.

Thus, if effective rules can be established to program the CMS (which embeds these rules) to detect the different severity levels of turbine fan-blade damage based on a real-time database that is regularly imported and updated by the (remote) CMS, the performance of the server-side CMS relies heavily on the effectiveness of the rules. Once an anomaly is detected, the CMS serves to guide the subsequent fan maintenance and repair actions. In this study, we call this 'maintenance by prediction', in contrast to 'predictive maintenance', which is a type of 'preventive maintenance' [1–3].

The CMS can thus warn of blade damage almost in real time, allowing wind farm operators to launch and arrange repair operations for maintenance by prediction at the earliest moment. This reduces the turbine's downtime. As such, this is different from predictive maintenance, which aims to avoid wasting time and money by preventing serious damage by performing regular maintenance in advance. Maintenance by prediction operates by giving sudden fan-blade status warnings (in the event of damage) for launching the actions as soon as possible.

An example may help to illustrate the difference. During predictive maintenance, the engineers are recommended to replace some parts of the turbine, but they may be replaced improperly, ultimately resulting in a machine shutdown. By contrast, the proposed maintenance by prediction mechanism works to detect, warn of, and support addressing an incident in almost real time.

Theoretically, the mechanism proposed by this study to detect wind turbine blade damage in wind turbines aims to support unplanned maintenance and repairs, which are carried out for the faulty wind turbine(s) using rule-based predictions (i.e., the maintenance by prediction actions). If the CMS is programmed with this detection mechanism, most of the equipment manager's efforts can be focused on the machine's normal operations, e.g., acquiring or preparing the parts required for equipment repairs, i.e., preventive maintenance. In operational practice, such a CMS can improve the safety (effective operation) rate (or operating efficiency) of the power system, reduce maintenance costs and increase the reliability of the wind turbine [1–3,8–12].

Section 2 focuses on the design of the CMS for wind turbines, describing how our research led us to the fault-diagnosis functions for fan blades from the traditional functions used for power generators, while reviewing the relevant literature. Section 3 presents the results from the data analysis upon taking real sampling data and introduces the methods used 'by example' (on the fly). Although the methods being applied are individually common in data processing, statistics, and digital signal processing (DSP), hybridising them for the purpose of the application described in this study is novel. The rules are established in Section 4 based on the analytical results, followed by a discussion of key findings. Section 5 concludes this study.

## 2. Condition Monitoring System (CMS) and Methods

This section describes the development of the CMS, which involves remote-side and server-side subsystems. Because the remote-side CMS is mounted on the structure of a wind turbine on the base platform, we start from an overview of the wind turbines.

### 2.1. Wind Turbines

A wind turbine is mainly composed of control systems, transmission mechanisms, generators, converters, blades, towers, power cables, converters and transformers. The internal transmission mechanisms include gearboxes, hubs and steering systems. The performance of wind turbines varies from one model to another.

As discussed previously, the installation site's climate conditions and the natural disasters which may occur affect wind turbines' operation, resulting in various causes of sudden failure, which is to be resolved by maintenance by prediction. In addition to these external factors, some wind turbines may have inherent (internal) drawbacks, such as inconsistent design specifications, low manufacturing quality and inappropriate design specifications for the sites on which they operate. All of these factors could cause them to operate in an unstable manner and accelerate the possible damage to their components, including the blades.

The reasons for fan-blade damage are manifold, but each can be categorised as either internal or external. Internal damage may be caused by kinetic or mechanical reasons. For example, machine wear due to operation over a long period will result in the deformation of parts due to stress. Also, during fan operation on normal days, the tower column will produce different deformations and displacements due to different operating conditions, tower conditions, and conditions of the base. External damage, by contrast, can result from the sudden occurrence of natural disasters, e.g., large typhoons and earthquakes. These are more harmful to the blades, and the consequences are usually unexpected. While the events causing damage are all occasional, they meet the condition that the tool provided by this study can address. See these further in Section 4.3.

### 2.2. CMS Hardware

The central component of the remote-side CMS hardware uses Adlink's USB-2405, with which each channel can be connected to an accelerometer or a microphone. The USB-2405 is a 24-bit USB interface for dynamic signal data acquisition. The USB-2405 has a USB-interface power supply and BNC connector; there are four analogue input channels, and each channel provides up to 128 kS/s of simultaneous sampling. The USB-2405 also features selectable AC or DC coupled inputs and a built-in high-precision 2 mA excitation current measurement integrated electronic piezoelectric (IEPE) sensor. Each channel can be connected to an accelerometer or microphone for measuring vibrations or noises. In the measurement system, each channel was connected to a PCB 601A01 accelerometers with a sensitivity of 100 mV/g and a frequency band range of 0.27 to 10,000 Hz.

Relatedly, the CMS can be divided into vibration and noise detection subsystems, and each subsystem has its effective fault detection and diagnostic methods. The integration of technologies inspecting the wind turbine near the accelerometer or microphone enables the operation status monitoring functions and the preliminary remote-side fault-diagnosis functions. However, it is difficult to derive effective rules that can be used for such fault diagnosis on the remote side. This is discussed in the next subsection.

### 2.3. Fault-Diagnosis Functions for Wind Turbines

This subsection illustrates the fault-diagnosis process and the analysis functions of the CMS for a wind turbine. When the machine experiences abnormal vibration, the diagnostic analysis can be performed according to the following guidelines [13–19]:

(A) Understand the machine's construction. For an abnormally functioning machine, it is necessary to understand its primary structure, component transmission methods and even an overview of the system to determine the possible failure modes and the appropriate measurement method and position.

(B) Observe abnormalities. When an exception occurs in a machine, it is necessary to understand the machine's operational conditions at that moment and to observe the abnormal phenomenon that is causing the device to operate abnormally.

(C) Abnormal phenomenon measurement. It is necessary to select appropriate sensors and instrumentation equipment, measure the vibration signal of the unusual machine phenomenon, and process the measured signal through spectrum analysis.

(D) Abnormal signal analysis. The abnormal signals measured should be analysed to determine the most severe locations and conditions and the components that are causing the problems. The reasons for these problems should be investigated. If necessary, further measurements should be made to confirm the causes of the issues.

(E) Verification of the results. After confirming the cause (or causes) of the abnormality, and after repairing or replacing the machine's faulty components, it is necessary to observe or measure again to determine whether the abnormality still exists and whether there are any signs of improvement. If the situation has not improved, the above steps must be repeated until the problem is resolved.

The following list summarises the judging criteria for abnormal situations that have traditionally been applied for fault diagnosis on the remote side (or 'device side', 'client side', or 'site side') CMS for the power generator component(s). Usually, a subset of these criteria is sufficient to program the diagnostic function. Appendix A [1–3,17,18] presents detailed descriptions.

- Vibrations generated by the generator caused by an uneven magnetic force acting on the rotor or the stator.
- Vibrations generated by the generator or vibrations caused by an abnormal rotor or rotor coils.
- Vibrations generated by the generator or vibrations caused by an abnormal stator or stator coils [1–3,20].
- Unbalanced weight of the shaft of the driveshaft [1–3].
- Bent shaft of the driveshaft.
- Improper installation of the shaft of the driveshaft [1–3].
- Misalignment of the driveshaft with the coupling connecting the machine during installation, causing the machine's severe vibration [1–3,13–15].
- Shaft diameter too small during bearing installation, resulting in the bearing's movement relative to the inner ring of the bearing [1–3].
- Poor shoulder angle during bearing installation, resulting in the centre of the inner and outer rings no longer being on the same straight line [1–3].
- Ball bearing causing temperature rise and vibration during mounting of the bearing due to insufficient lubrication or improper lubrication viscosity [1–3,13–15].
- Gearbox [1–3,13–15,20,21].
- Bearing damage [22,23].

While the faults listed above are mainly mechanical problems that can be judged using vibration signals (with the means of judging these detailed in Appendix A), the acoustic interface is another main interface used to understand whether a wind turbine is faulty, specifically through noise measurements. Appendix B provides further details about this process [1–3,17,18]. Theoretically, then, the measurement results from either interface, or both, can be considered [1–3,13–19,21] on the remote side.

However, the following statements are critical for explaining the reasons this research is conducted recording only the vibration data (and not the noise data) on the 'remote side' of the CMS and utilising the collected datasets on the 'server side'.

1. Using the datasets recorded from an interface is sufficient. According to the above review summary, vibration and noise are the two interfaces that can be used to understand a wind turbine's physical deterioration. However, it has been proven that most of the noises generated by a turbine are caused by vibrations.
2. The criteria referenced above do not support judging the faulty conditions of the blades as addressed by this study. The experimental results from our extensive laboratory tests performed prior to this study showed that the criteria f above or judging the faulty conditions of power generators were ineffective for those of blades (i.e., they usually resulted in more incorrect judgments than correct ones) on the remote side. We referred to the usual criteria for power generators, and thus they might not be suitable for other components of turbines (even though a turbine always includes at least one generator) (see Section 2.4).

From the above discussion, as using the vibration datasets is sufficient (and using these is better than using the noise datasets due to the causal relationship), and as the traditional criteria for diagnosing a power generator do not apply in diagnosing the component of turbine blades on the remote side, we decided to seek clues to establish the judgement rules from the vibration datasets received and gathered on the server side.

*2.4. Developing Wind Turbines: A Briefing*

To demonstrate the role of wind turbine blades in designing and maintaining a wind turbine, as a supplemental review, Figure 1 shows the life cycle of wind turbine development.

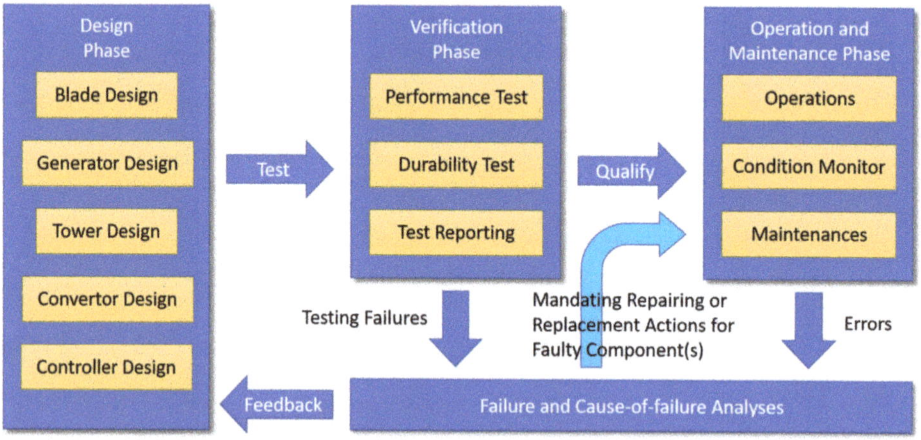


**Figure 1.** The life cycle of wind turbine development.

After a wind turbine is set up, if it becomes inoperable, most wind turbine manufacturers will assert that it was because the turbine was not properly maintained. Then, to argue with the manufacturer, the engineers in a green energy operator company would like to design a simulation program based on a given turbine manufacturer's model and run it to determine if the design was flawed. However, it is questionable whether or not the operator can successfully request the design drawings and related data for subsequent operations, maintenance, monitoring and analysis from the manufacturer in practice, let alone build up a simulated turbine. Wind turbine manufacturers believe that wind turbine design drawings and related technologies are trade secrets, so technology transfers are impossible for them; a chance of negotiation exists only if the business deal involves a large-scale wind farm. In reality, however, most cases do not contain this possibility, even in a large-scale wind farm construction project (to the authors' knowledge).

Therefore, based on the operation and maintenance phase in Figure 1, we must rely on the condition monitoring approach using a CMS (both remote side and server side) and perform the repair or replacement tasks (i.e., the unplanned maintenance outside of the regular, periodic maintenance tasks) when this system detects a sudden faulty status of the turbine while it is operating.

Blade design is the most commonly addressed topic in the design phase for wind turbines because it is usually used to distinguish the brand and type of the turbine directly (visually) and because the technologies for the design of all other components (generator, tower, convertor and controller) were mature long before wind turbines appeared. Thus, this study focuses on the detection of blades' sudden faulty status. Before this, it is necessary to seek clues for establishing the effective rules for the detection process based on the vibration data transmitted to the server side in near real time from the remote side.

An effective CMS providing accurate predictive detections and posing precise warnings for subsequent maintenance by prediction actions cannot proceed without these rules. Strictly speaking, the methods to establish these rules also fall within the scope of the failure and cause-of-failure analyses phase in Figure 1, in addition to the traditional tasks usually defined based upon the testing failures. The next section presents the results obtained from the data analysis, which forms the basis for establishing the rules.

## 3. Results

In this section, we summarise the process for establishing effective rules and present the key results obtained from the data analysis (omitting the results from the extensive trial analysis). The process follows the methods introduced in Sections 4.3 and 4.4, while the flows and methods for the experimental conditions, data collection, data preprocessing and data curation processes are detailed in Sections 4.1 and 4.2.

### *3.1. Source Datasets and Visualisation*

After the collected datasets were preprocessed and curated, the first phase involved visualising the source time-domain datasets, so that observations could be made in order to perform some initial comparisons. For this purpose, the datasets of the vibration data updated on the server-side CMS were sampled and rendered for 3.0-blade (full blades), 2.5-blade, and 2.0-blade wind turbine settings. In Appendix C, a full coverage for these is shown in Figure A1, A2 and A3, respectively. For space reasons, only several subfigures are preserved here in Figures 2–4 as example cases, which are plotted for the data being recorded along the X, Y and Z axes subject to the 'no wind imposed' setting.

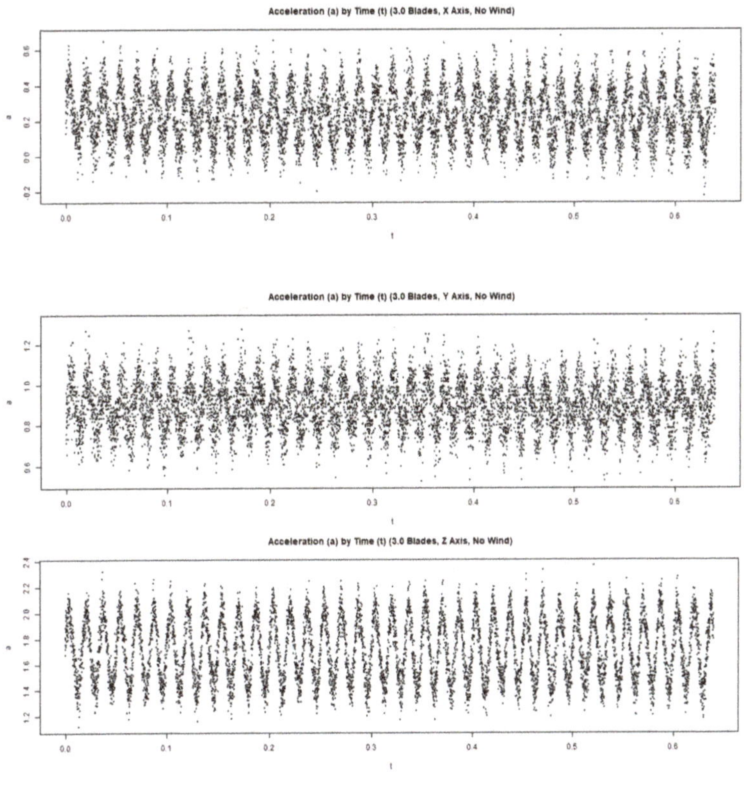


**Figure 2.** Acceleration signals recorded for the full (normal) 3.0-blade turbine.

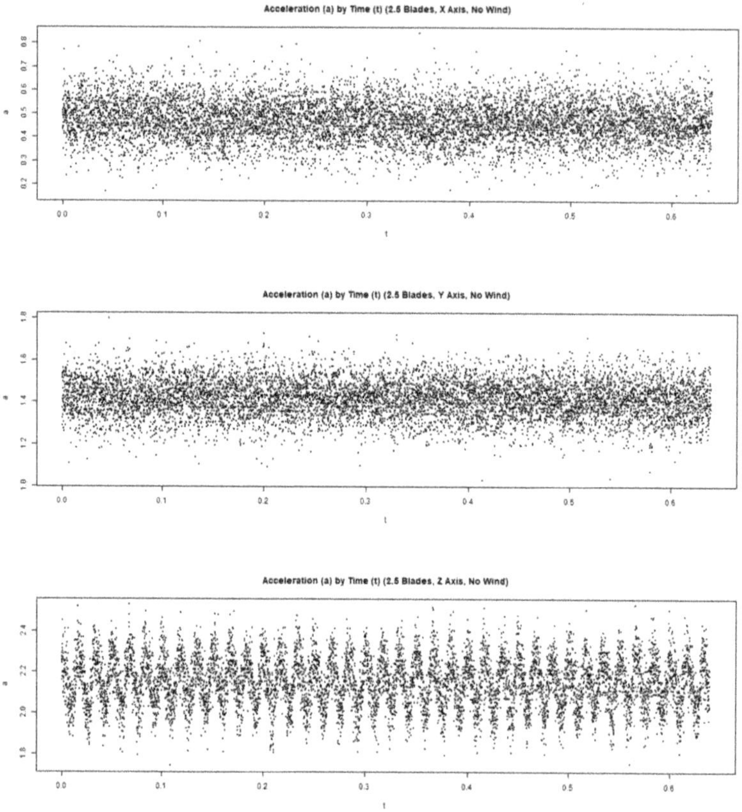


**Figure 3.** Acceleration signals recorded for the 2.5-blade (partially broken) turbine.

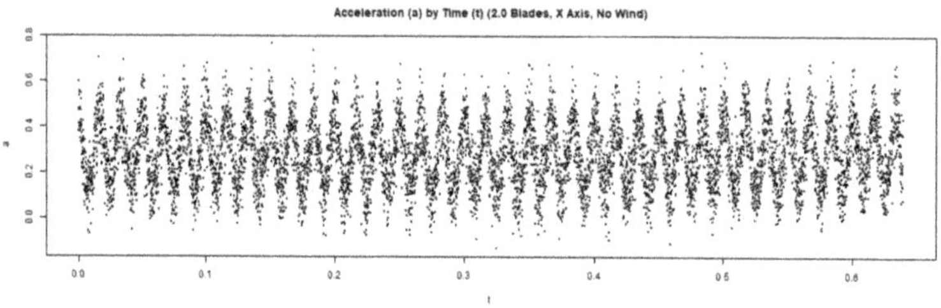


**Figure 4.** *Cont.*

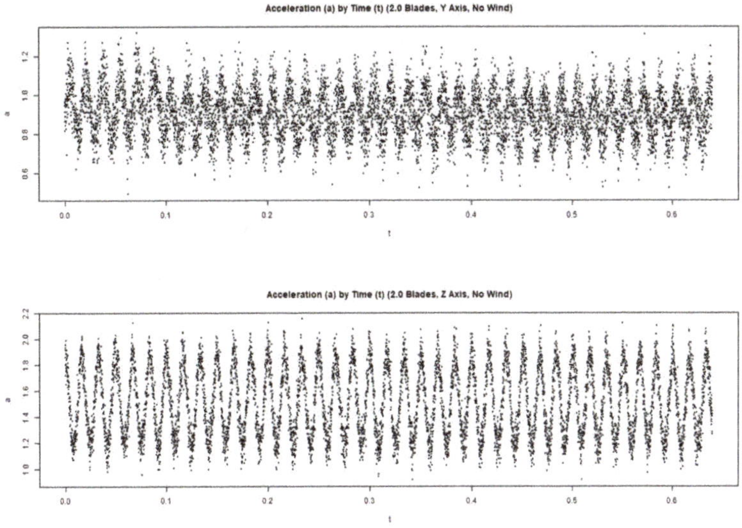


**Figure 4.** Acceleration signals recorded for the 2.0-blade (one blade missing) turbine.

### 3.2. *Using the Local Regression Method (LRM) and the Smooth-Line Approach*

Next, based on the first 1024 data entries in each dataset (data sequence), we found the smooth line to represent the trend of the data points using the local regression method (LRM). In Appendix D, we plotted it as a bold blue line in Figures A4–A6 for different wind-speeds subject to the wind turbine settings of 3.0-blade (full blades), 2.5-blade and 2.0-blade configurations, respectively. For space reason, only several subfigures used for subsequent discussions are preserved here in Figures 5–7.

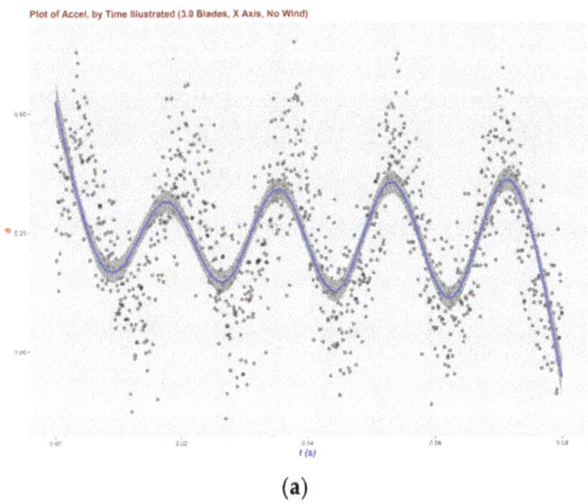


(**a**)

**Figure 5.** *Cont.*

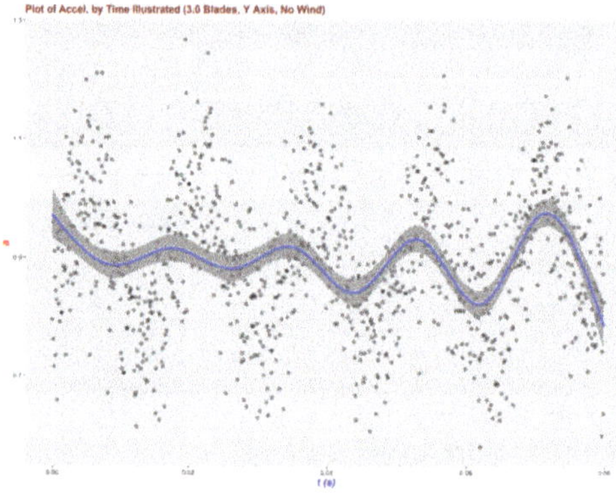


(**b**)

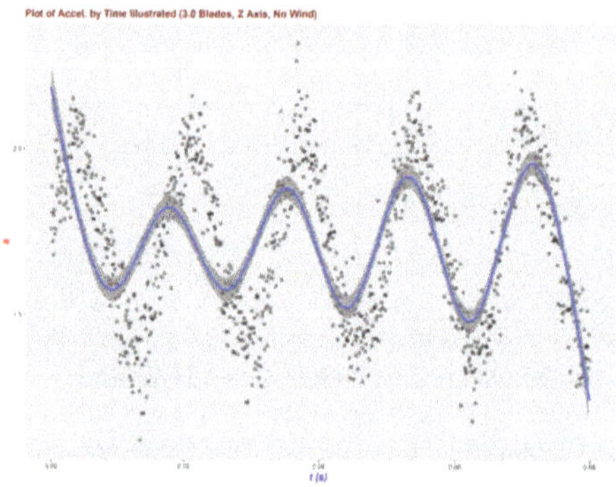


(**c**)

**Figure 5.** Magnitude and trend of acceleration for the full (normal) 3.0-blade turbine. (**a**) Based on datasets recorded along the X axis with no wind imposed. (**b**) Based on datasets recorded along the Y axis with no wind imposed. (**c**) Based on datasets recorded along the Z axis with no wind imposed.

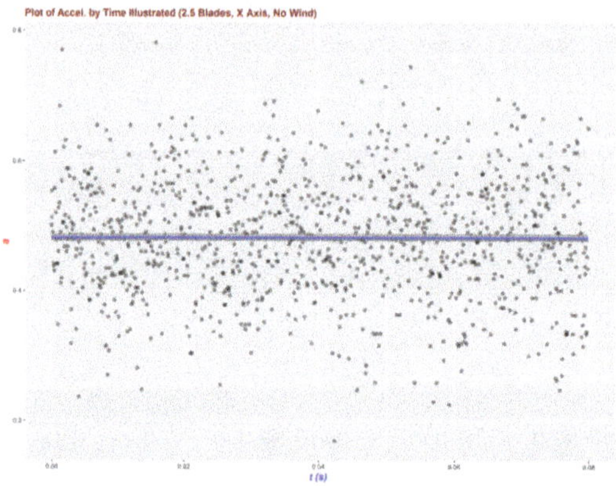


(**a**)

**Figure 6.** *Cont.*

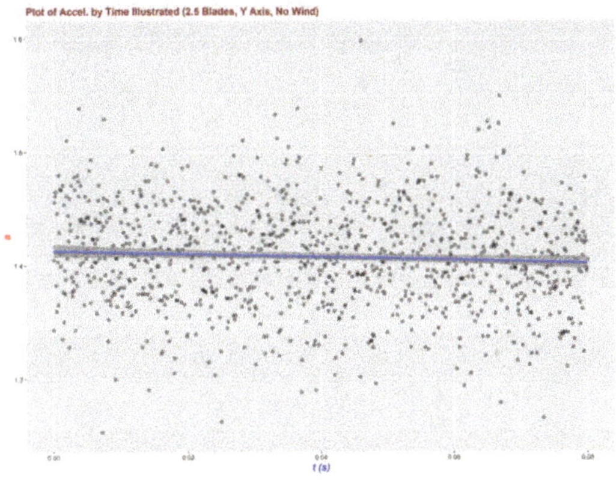


(b)

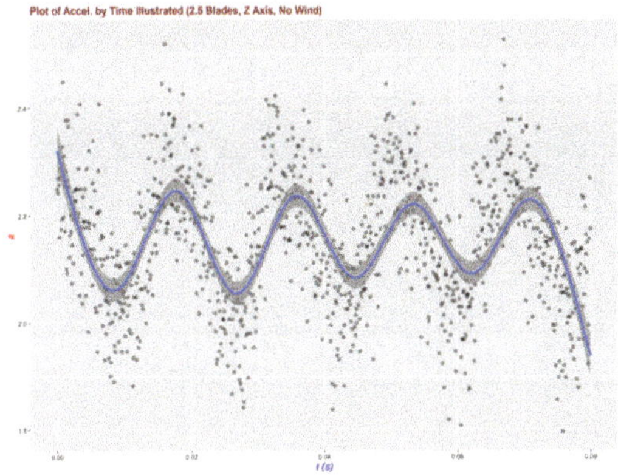


(c)

**Figure 6.** Magnitude and trend of acceleration for the 2.5-blade (partially broken) turbine. (**a**) Based on datasets recorded along the X axis with no wind imposed. (**b**) Based on datasets recorded along the Y axis with no wind imposed. (**c**) Based on datasets recorded along the Z axis with no wind imposed.

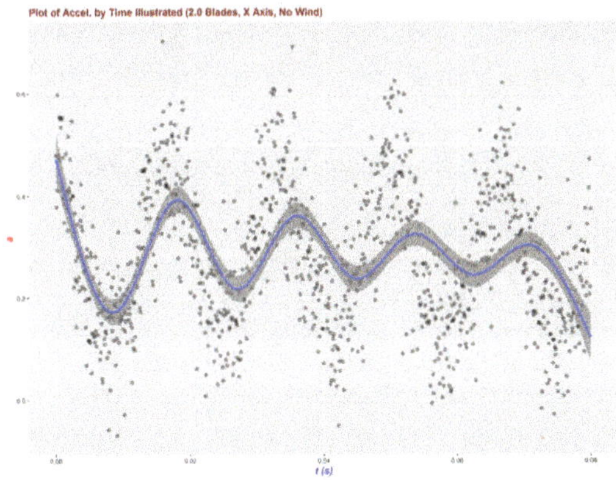


(a)

**Figure 7.** *Cont.*

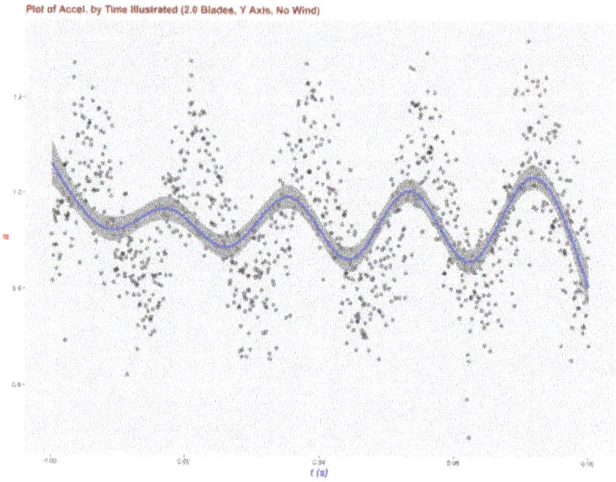


**(b)**

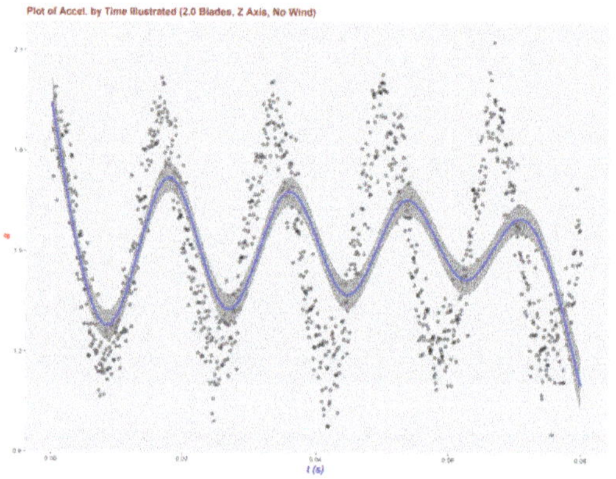


**(c)**

**Figure 7.** Magnitude and trend of acceleration for the 2.0-blade (one blade missing) turbine. (**a**) Based on datasets recorded along the X axis with no wind imposed. (**b**) Based on datasets recorded along the Y axis with no wind imposed. (**c**) Based on datasets recorded along the Z axis with no wind imposed.

In these visualisations, we used a small grey circle, instead of points, to denote each original data point, so that the intensity of the data points could be observed in terms of their overlaps. Through the presentation of these points, we might easily find that the cycle of a waveform roughly concurs with the interval between the 'peaks' or the interval between the 'valleys' of the original acceleration data. As seen in most of these figures (except for the acceleration signals recorded along the X-axis and the Y-axis for the 2.5-blade turbine with no wind, i.e., Figure 6a,b), a waveform containing several repeated waves with a highly static time cycle appeared.

However, despite some 2.5-blade cases being distinguished easily, the 3.0-blade cases and the 2.0-blade cases are not dissimilar visually. Therefore, no rule can be established to differentiate these three cases through the eyes till now, let alone this vision-based process lacks a mathematical foundation. Here, only the observed static time cycle is worthwhile, i.e., in each subfigure, each pair of two nearby peaks, or each pair of two nearby valleys, was almost equally spaced, so the interval for nearby peaks or that for nearby valleys was consistent as well.

However, from these graphs we also saw that (1) the number of data points outside the waveform, (2) the degree to which they were outside the waveform, (3) the concentration of these data points and (4) the proportion of data outliers were all different (i.e., they varied from case to case). For example, in general, on the same axis, there were more acceleration data outliers in Figure A5 for the 2.5-blade setting than in Figure A4 for the 3.0-blade setting and in Figure A6 for the 2.0-blade setting.

Take the settings with 0.6 m/s wind-speed arbitrarily as an example. In Figure A5a (upper-right subfigure), there are 14 outliers (out of the first 1024 data entries), according to the normal outlier equations below:

$$\begin{cases} Outliers(VD) = U(VD) \backslash \{vd \in VD | vd = [Q_1 - 1.5 \times IQR, Q_3 + 1.5 \times IQR]\} \\ IQR = Q_3 - Q_1 \end{cases}$$

At the same time, in the upper-right subfigure in Figure A4a (no blade broken), imposing the same wind-speed, there is no (0) outlier. And surprisingly, in the upper-right subfigure in Figure A6a (a full blade broken) with the same wind-speed being imposed, there is no (0) outlier either.

In addition, in Appendix D in general, on the X axis and the Y axis, apart from the acceleration magnitude ranges, it seems that accelerations with the 2.0-blade setting (e.g., Figure 7a,b) were more concentrated than those with the 3.0-blade setting (e.g., Figure 5a,b). Combined with the previous observations, these provided further clues for obtaining the effective information for constructing the judgement rules in the next subsection.

### 3.3. Further Transformation: Resampling and Resmoothing

In this part of the analysis, we found that the initial 1024 data points in each data sample (i.e., the 'data digest') were sufficient to establish the final rules to identify the different (faulty) situations of turbine blade malfunction after further data transformations using resampling and resmoothing. After conducting experiments, we also found that resampling eight consecutive data points as a representative one (i.e., eight original sampling time units as a 'clock') was appropriate. Therefore, three processes were run as follows.

First, each shortened dataset was resampled using the new clock. We let the original dataset be $a(t)$, where $t = t_j, j = 0, 1, ..., 1023$ was the original sampling time sequence, and then defined the resampled dataset as $\bar{a}(\bar{t})$, where $\bar{t}$ was the redefined clock sequence and $\bar{t}_i = \{t_{8i}, t_{8i+1}, t_{8i+2}, \cdots, t_{8i+7}\}, i = 0, 1, ..., 128$. For each clock, the information from the original data was preserved as follows:

$$m(\bar{t}_i) = \sum_{j=0}^{7} \frac{a(t_{8i+j})}{8},$$

$$v(\bar{t}_i) = \frac{1}{8} \sum_{j=0}^{7} \left(a(t_{8i+j}) - m(\bar{t}_i)\right)^2,$$

where $m(\bar{t}_i)$ and $v(\bar{t}_i)$ are, respectively, the mean of the original data sequence on the redefined clock sequence $\bar{t}$.

Next, the curve-like piecewise line was approached using LRM, and it was in fact a predictor that also produced the 'theoretical value' of the acceleration degree at any specified time point $t$, i.e., $P(t)$. Therefore, when the data was re-considered using a 'clock', this clock was also applied to the predictor function. We named this new predictor a function called $\overline{P}(\bar{t})$, where $\bar{t}$ was the redefined clock sequence and $\bar{t}_i = \{t_{8i}, t_{8i+1}, t_{8i+2}, \cdots, t_{8i+7}\}$. Therefore, $\overline{P}(\bar{t})$ could be simply redefined as:

$$\overline{P}(\bar{t}) = \overline{P}(m(\bar{t}_i))$$

However, in the above equation, the function $\overline{P}(\cdot)$ was identical to $P(\cdot)$ because both used local regression as the smoother function.

Third, the computational results are rendered in Figures A7–A9 for the 3.0-blade, 2.5-blade and 2.0 blade settings, respectively. Once again, for these settings, only the no-wind-imposed cases on the 3 axes are presented in Figures 8–10 for simplicity. These simplified plots were qualified to clarify the features among the different faulty blade cases and establish the rules to distinguish them automatically.

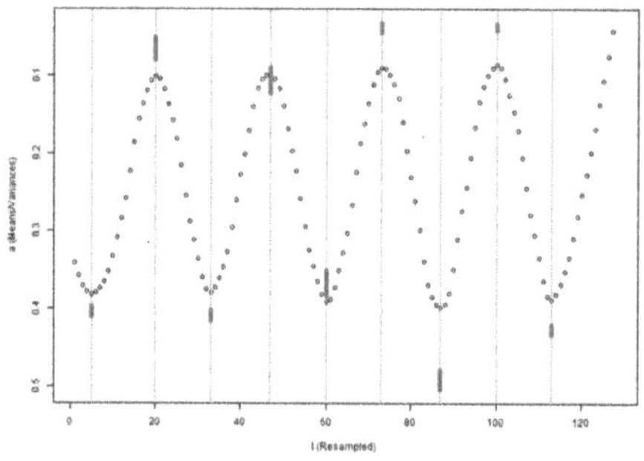

(**a**)

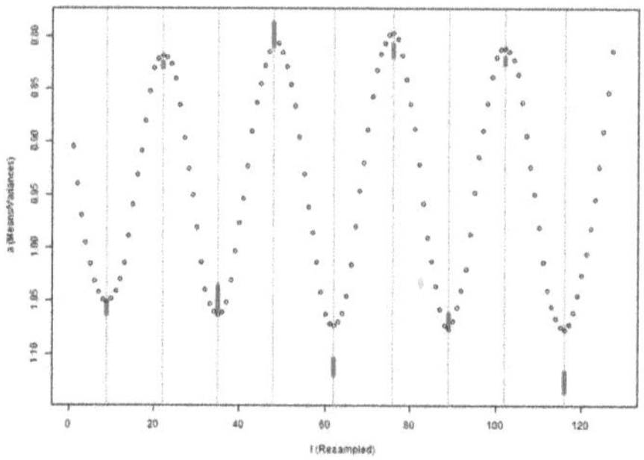

(**b**)

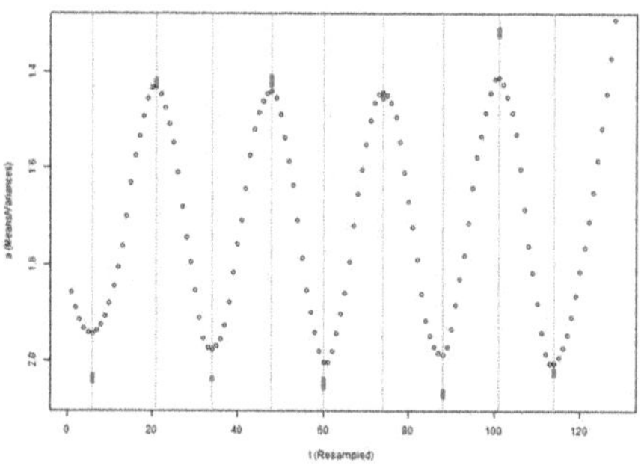

(**c**)

**Figure 8.** Means and peak/valley variances of resampled data for the 3.0-blade turbine. (**a**) Based on datasets recorded along the X axis with no wind imposed. (**b**) Based on datasets recorded along the Y axis with no wind imposed. (**c**) Based on datasets recorded along the Z axis with no wind imposed.

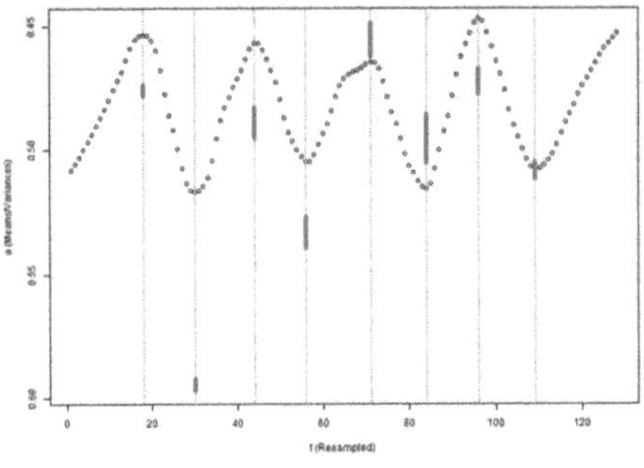

(a)

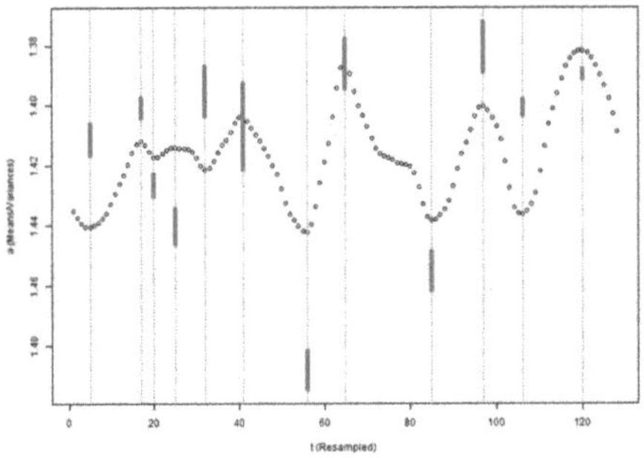

(b)

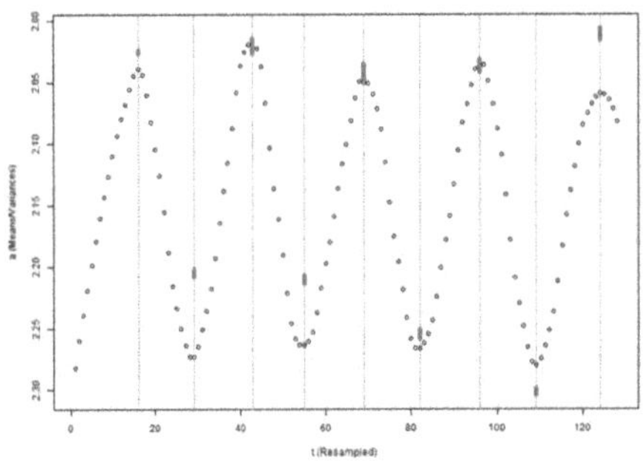

(c)

**Figure 9.** Means and peak/valley variances of resampled data for the 2.5-blade turbine. (**a**) Based on datasets recorded along the X axis with no wind imposed. (**b**) Based on datasets recorded along the Y axis with no wind imposed. (**c**) Based on datasets recorded along the Z axis with no wind imposed.

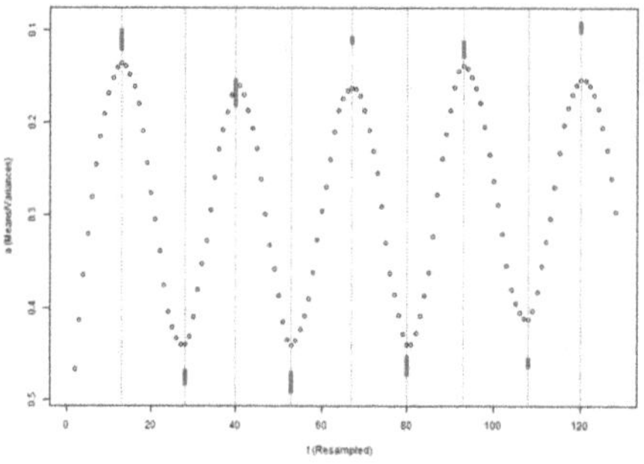

**(a)**

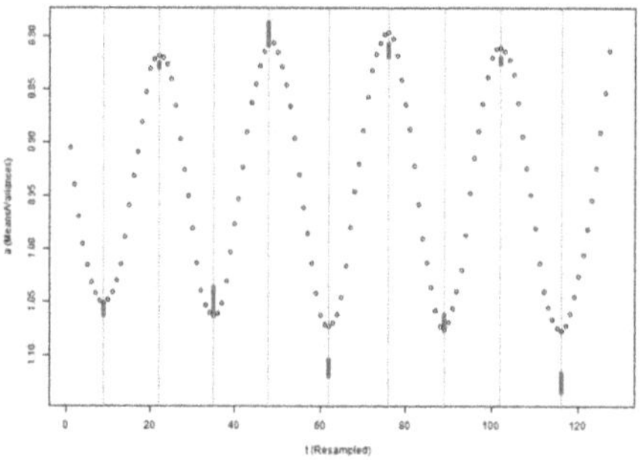

**(b)**

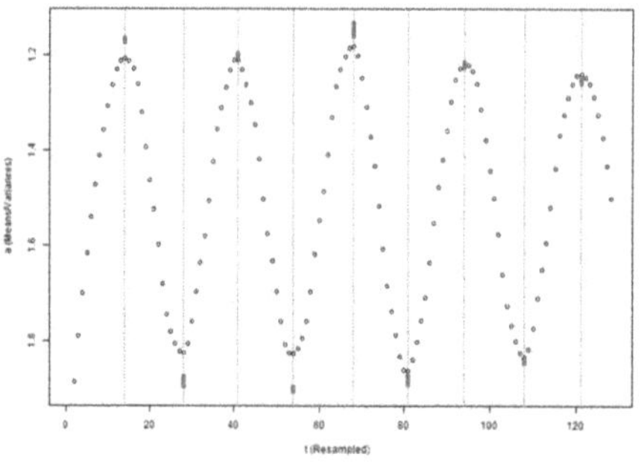

**(c)**

**Figure 10.** Means and peak/valley variances of resampled data for the 2.0-blade turbine. (**a**) Based on datasets recorded along the X axis with no wind imposed. (**b**) Based on datasets recorded along the Y axis with no wind imposed. (**c**) Based on datasets recorded along the Z axis with no wind imposed.

As can be seen, these plots provided clear ways to compare and establish the rules to identify the two malfunctioning blade cases against that with full blades working normally.

## 4. Establishment of Rules and Discussion

Based on these results, the rules to judge whether a blade on the wind turbine was half-broken (i.e., the 2.5-blade case), normal with full-blades running (i.e., the 3.0-blade case), or completely missing a blade (i.e., the 2.0-blade case), could be established.

### 4.1. First Rule to Judge the 2.5-Blade Case

First, certain features of the 2.5-blade cases emerged as salient. From the graphs in Figure A8, e.g., Figure 9b, it was easily observed that many of the 2.5-blade cases had distorted or abnormal waveshapes after the data points ($t$, $a$) were smoothly interpolated using LRM, compared to the 3.0 or 2.0 cases in Figures 8b and 10b. An extreme case of this could be seen in the plot for the 2.5-blade, Y axis, no-wind case, while other trivial cases were the 2.5-blade, X axis, no wind; 2.5-blade, X axis, wind-speed = 12 m/s; 2.5-blade, X axis, wind-speed = 18 m/s; 2.5-blade, Y axis, wind-speed = 12 m/s and 2.5-blade, Y axis, wind-speed = 18 m/s cases, as seen in Figure A8. As seen in the figures, almost all of these cases appeared based on the data series recorded on the X and Y axes, despite the level of the peaks/valleys also being slightly jittered based on the data recorded on the Z axis subject to certain wind-speeds, e.g., '2.5-blade, Z axis, no wind' and '2.5-blade, Z axis, wind-speed = 18 m/s'.

However, despite its simplicity, a rule based on the visualisation process was difficult to implement because it was established through human-based pattern recognition. For example, to what extent could the so-called 'distortion' and 'abnormality' be justified for a wave-like plot? Therefore, a numerical rule needed to be established so that an algorithm could be implemented based on the results for automatic detection in the future. This relied on the true mean and variance values of the acceleration source data that corresponded to a peak or valley in a resampled and smoothed waveform, which were displayed as bold vertical red line segments in the plot. Each such line ranged from (mean − variance) to (mean + variance) of a source data slice associated with some peak or valley.

As can be observed, in general, the red line segments were longer in the figures plotted based on the X-axis and Y-axis vibration data subject to the 2.5-blade setting, no matter how great the wind-speed imposed on the turbine, compared to those plotted based on the X-axis and Y-axis vibration data subject to either the 2.0-blade setting or the 3.0-blade setting. Moreover, no such situation was found for the figures plotted based on the Z-axis vibration data subject to the same 2.5-blade setting.

As such, this feature (Rule 1), i.e., occurrences of the long red line segments around the peaks and valleys of the LRM-smoothed and resampled waveform that appear for the X-axis and Y-axis vibration data under different wind-speed settings, can be used to identify whether half (0.5) a blade on the wind turbine, or a part of a blade, is broken (i.e., the 2.5-blade case).

### 4.2. Second Rule to Judge the 2.0-Blade Case

Next, since the 2.5-blade case could be excluded using the above rule, the remaining problem involved how to distinguish the 2.0-blade (one blade totally missing) case from the 3.0-blade case. A clue to the reasoning was that a turbine having full blades (the 3.0-blade setting) should be heavier than the same turbine with a blade totally missing (the 2.0-blade setting) in the case of no wind (static without other conditions changing). That is, unlike the vibration data recorded along the X axis or Y axis, the Z axis data corresponded to the vertical power (i.e., the weight factor of the turbine) of the turbine interacting with the foundation structure and the land. Therefore, lower accelerations should be detected along the Z axis for a turbine with a full blade missing than for a normal (full-bladed) turbine, and this effect could be clearly compared and displayed when no wind was imposed.

This was reflected in the experimental results. Comparing the acceleration data recorded along the Z axis for the situation with no wind (0 m/s) (see Figures 8c and 10c), the interval of the predicted acceleration values for the 3.0-blade normal case was [−1.4,−2.0] (m/s) (bounded by the peaks and valleys of the waveform). In contrast, the interval of those predictions for the 2.0-blade case was [−1.2,−1.8] (m/s), a lower window.

Therefore, the results supported our theoretical suppositions. These became the second rule to distinguish a turbine with a blade totally missing (i.e., the 2.0-blade case) from a normal turbine (Rule 2): if the case is not 'a part of a blade is broken' (which can be detected based on the first rule), the predictive waveform identified from the Z axis data can be used to see whether or not a blade is missing for any possible reason by checking to see if waveform fluctuation of the resampled data, in terms of the interval delimited by the peaks and valleys of the smoothed function (i.e., $\{\max\overline{P}(\bar{t}), \min\overline{P}(\bar{t})\}$), is narrower than usual.

*4.3. Discussion*

In short, the means and variances of the time-domain data recorded along the X axis or the Y axis around the peaks and valleys of the locally regressed and resampled waveform can be used to determine whether the wind turbine already has a partially broken blade (i.e., the 2.5-blade case). The bold vertical red line segments are measured based on these means and variances, and the extraordinary appearance of these line segments may indicate the faulty 2.5-blade case or perhaps harm to a blade.

Following this rule and excluding the 2.5-blade case, the fluctuation range of the predictive waveform obtained based on the data interval recorded along the Z axis when there is no wind can be used as a measure to distinguish the 2.0-blade case from the normal 3.0-blade case. When this interval rises further upward than usual, the case in which a blade is totally missing (the 2.0-blade case) can be detected.

Combing these two rules, all three of the cases—the normal case and the two faulty cases of wind turbine fan-blade damage—can be explained systematically. As these rules are simple (with limited computational complexity), they can be used to detect these malfunctions almost in real time and to transmit the necessary warning messages to those in charge just in time, given that the vibration datasets for the wind turbines are synchronised routinely (within a short period of time) on the server side of the CMS. This greatly benefits the unplanned maintenance efforts of wind farm operations.

The Circum-Pacific Belt area is prone to super typhoons and strong typhoons, whose wind-speeds may easily reach 150 km/h or above, and thus a situation in which a wind turbine blade breaks apart or falls off completely should not be news to anyone in the green energy industry. Although the proposed set of rules does not fit the case in which a turbine totally collapses (as it is unclear whether the accelerometer and remote-side CMS would still work in this case), in most cases it can serve as a computerised remedy to detect whether a blade is broken or falling off, if the wireless transmission works.

Since another common cause of turbine damage in the studied area involves earthquakes, which are usually as unpredictable as typhoons, the maintenance by prediction mechanism is suggested as a supplement to regular predictive maintenance, even during periods of predictable weather conditions (i.e., to best control that which is controllable). In short, from the above discussions for the Circum-Pacific Belt area and articulating back to the outset of this study (see Section 1.1), it should be clear that the proposed mechanism, even just putting a puzzle piece for detecting the damages on the fan component of a turbine (with respect to the whole integrative perspective of turbine maintenance), can improve the operating efficiency of turbines. This reduces the maintenance costs and benefits the unplanned maintenance of wind farm operations.

Note that in terms of digital signal processing (DSP), we established the set of detection rules based on the converted (original) time-domain data, rather than the data in any other domain, e.g., frequency-domain data. Doing so not only maintains simplicity for future implementation and makes the entire process faithful to the original data but also avoids possible ambiguity. For example, if the upper and lower limits of the predictive waveform window determined by the peaks and valleys are logged, it becomes harder to examine whether there has been an upward shift in the interval from the 3.0-blade setting to the 2.0-blade setting (i.e., for the second rule) just because the window in the logged domain would become narrower.

*4.4. Extensive Materials*

This subsection provides extensive descriptions for how the rules are justified using tabularised quantitative information.

Table 1 digests how the first rule was established in Section 4.1, based on the resampled, resmoothed, and mean–var. transformed data on the X axis (see Section 3.3 and Figures A7a, A8a and A9a), with the setting that wind-speed (WS) = 6 m/s. Table 2 is numerically presented on the Y axis (see Section 3.3 and Figures A7b, A8b and A9b), because rule 1 says that using the information either on the X-axis or on the Y-axis is okay.

**Table 1.** Resampled, resmoothed mean–var. transformed data on the X axis (WS = 6 m/s). (**a**) Min/var. peak/valley Stats for the 3.0-blade turbine. (**b**) Min/var. peak/valley stats for the 2.5-blade turbine. (**c**) Min/var. peak/valley stats for the 2.0-blade turbine.

(a)

| ID | Mean | Var. | Peak | Valley | Peak Top | Peak Bot. | Valley Top | Valley Bot. | Peak Range | Valley Range |
|---|---|---|---|---|---|---|---|---|---|---|
| 4 | −0.23097 | 0.016014 | 0 | 1 | - | - | −0.21496 | −0.24699 | - | 0.032028 |
| 19 | 0.914043 | 0.003893 | 1 | 0 | 0.917935 | 0.91015 | - | - | 0.007785 | - |
| 32 | −0.28885 | 0.008199 | 0 | 1 | - | - | −0.28065 | −0.29705 | - | 0.016398 |
| 45 | 0.66058 | 0.011351 | 1 | 0 | 0.671931 | 0.64923 | - | - | 0.022701 | - |
| 59 | −0.45689 | 0.00657 | 0 | 1 | - | - | −0.45032 | −0.46346 | - | 0.013139 |
| 72 | 0.83469 | 0.005848 | 1 | 0 | 0.840538 | 0.828842 | - | - | 0.011696 | - |
| 85 | −0.37567 | 0.010922 | 0 | 1 | - | - | −0.36475 | −0.3866 | - | 0.021845 |
| 99 | 0.661981 | 0.006781 | 1 | 0 | 0.668762 | 0.6552 | - | - | 0.013562 | - |
| 112 | −0.51664 | 0.018861 | 0 | 1 | - | - | −0.49778 | −0.5355 | - | 0.037722 |

(b)

| ID | Mean | Var. | Peak | Valley | Peak Top | Peak Bot. | Valley Top | Valley Bot. | Peak Range | Valley Range |
|---|---|---|---|---|---|---|---|---|---|---|
| 11 | −0.00738 | 0.112158 | 0 | 1 | - | - | 0.104774 | −0.11954 | - | 0.224315 |
| 24 | 0.879034 | 0.05588 | 1 | 0 | 0.934914 | 0.823154 | - | - | 0.11176 | - |
| 38 | 0.365574 | 0.043997 | 0 | 1 | - | - | 0.409572 | 0.321577 | - | 0.087994 |
| 52 | 0.878567 | 0.015044 | 1 | 0 | 0.893612 | 0.863523 | - | - | 0.030089 | - |
| 64 | −0.05406 | 0.017749 | 0 | 1 | - | - | −0.03631 | −0.07181 | - | 0.035497 |
| 77 | 0.927113 | 0.884839 | 1 | 0 | 1.811952 | 0.042274 | - | - | 1.769678 | - |
| 90 | 0.233009 | 0.022373 | 0 | 1 | - | - | 0.255381 | 0.210636 | - | 0.044746 |
| 104 | 0.990595 | 0.006383 | 1 | 0 | 0.996978 | 0.984212 | - | - | 0.012765 | - |
| 117 | 0.004286 | 0.00893 | 0 | 1 | - | - | 0.013216 | −0.00464 | - | 0.017861 |

(c)

| ID | Mean | Var. | Peak | Valley | Peak Top | Peak Bot. | Valley Top | Valley Bot. | Peak Range | Valley Range |
|---|---|---|---|---|---|---|---|---|---|---|
| 15 | 0.846359 | 0.002904 | 1 | 0 | 0.849264 | 0.843455 | - | - | 0.005808 | - |
| 28 | −0.23564 | 0.005518 | 0 | 1 | - | - | −0.23012 | −0.24116 | - | 0.011036 |
| 41 | 0.851494 | 0.001186 | 1 | 0 | 0.85268 | 0.850308 | - | - | 0.002373 | - |
| 55 | −0.23797 | 0.008646 | 0 | 1 | - | - | −0.22933 | −0.24662 | - | 0.017291 |
| 68 | 0.837491 | 0.00561 | 1 | 0 | 0.8431 | 0.831881 | - | - | 0.011219 | - |
| 81 | −0.27158 | 0.003622 | 0 | 1 | - | - | −0.26796 | −0.2752 | - | 0.007243 |
| 95 | 0.774475 | 0.002352 | 1 | 0 | 0.776827 | 0.772123 | - | - | 0.004705 | - |
| 108 | −0.30332 | 0.008789 | 0 | 1 | - | - | −0.29453 | −0.31211 | - | 0.017578 |
| 122 | 0.760005 | 0.004668 | 1 | 0 | 0.764673 | 0.755336 | - | - | 0.009337 | - |

**Table 2.** Resampled, resmoothed mean–var. transformed data on the Y axis (WS = 6 m/s). (**a**) Min/var. peak/valley stats for the 3.0-blade turbine. (**b**) Min/var. peak/valley stats for the 2.5-blade turbine. (**c**) Min/var. peak/valley stats for the 2.0-blade turbine.

(**a**)

| ID | Mean | Var. | Peak | Valley | Peak Top | Peak Bot. | Valley Top | Valley Bot. | Peak Range | Valley Range |
|---|---|---|---|---|---|---|---|---|---|---|
| 7 | 0.553172 | 0.008447 | 0 | 1 | 0 | 0 | 0.561619 | 0.544724 | 0 | 0.016895 |
| 20 | 1.113486 | 0.009612 | 1 | 0 | 1.123098 | 1.103874 | 0 | 0 | 0.019224 | 0 |
| 34 | 0.597753 | 0.010292 | 0 | 1 | 0 | 0 | 0.608045 | 0.58746 | 0 | 0.020585 |
| 48 | 0.91686 | 0.008372 | 1 | 0 | 0.925231 | 0.908488 | 0 | 0 | 0.016743 | 0 |
| 61 | 0.746513 | 0.019163 | 0 | 1 | 0 | 0 | 0.765676 | 0.72735 | 0 | 0.038326 |
| 75 | 1.28571 | 0.002974 | 1 | 0 | 1.288683 | 1.282736 | 0 | 0 | 0.005947 | 0 |
| 88 | 0.474803 | 0.007407 | 0 | 1 | 0 | 0 | 0.48221 | 0.467396 | 0 | 0.014814 |
| 101 | 1.31762 | 0.01647 | 1 | 0 | 1.33409 | 1.30115 | 0 | 0 | 0.032939 | 0 |
| 115 | 0.595876 | 0.002217 | 0 | 1 | 0 | 0 | 0.598092 | 0.593659 | 0 | 0.004433 |

(**b**)

| ID | Mean | Var. | Peak | Valley | Peak Top | Peak Bot. | Valley Top | Valley Bot. | Peak Range | Valley Range |
|---|---|---|---|---|---|---|---|---|---|---|
| 12 | 1.145866 | 0.154981 | 0 | 1 | 0 | 0 | 1.300847 | 0.990884 | 0 | 0.309963 |
| 25 | 1.584168 | 0.018942 | 1 | 0 | 1.60311 | 1.565226 | 0 | 0 | 0.037884 | 0 |
| 38 | 1.204525 | 0.18757 | 0 | 1 | 0 | 0 | 1.392095 | 1.016955 | 0 | 0.37514 |
| 52 | 1.810359 | 0.021382 | 1 | 0 | 1.831741 | 1.788976 | 0 | 0 | 0.042765 | 0 |
| 65 | 1.112547 | 0.109336 | 0 | 1 | 0 | 0 | 1.221883 | 1.003211 | 0 | 0.218672 |
| 78 | 2.019655 | 0.094675 | 1 | 0 | 2.11433 | 1.92498 | 0 | 0 | 0.18935 | 0 |
| 92 | 1.097061 | 0.010047 | 0 | 1 | 0 | 0 | 1.107108 | 1.087014 | 0 | 0.020093 |
| 105 | 1.745129 | 0.010731 | 1 | 0 | 1.75586 | 1.734399 | 0 | 0 | 0.021461 | 0 |
| 118 | 1.113955 | 0.005631 | 0 | 1 | 0 | 0 | 1.119586 | 1.108324 | 0 | 0.011263 |

(**c**)

| ID | Mean | Var. | Peak | Valley | Peak Top | Peak Bot. | Valley Top | Valley Bot. | Peak Range | Valley Range |
|---|---|---|---|---|---|---|---|---|---|---|
| 7 | 0.532054 | 0.010882 | 0 | 1 | 0 | 0 | 0.542937 | 0.521172 | 0 | 0.021765 |
| 20 | 1.098469 | 0.00871 | 1 | 0 | 1.107179 | 1.089759 | 0 | 0 | 0.01742 | 0 |
| 34 | 0.570066 | 0.011347 | 0 | 1 | 0 | 0 | 0.581412 | 0.558719 | 0 | 0.022693 |
| 48 | 0.934223 | 0.007859 | 1 | 0 | 0.942082 | 0.926363 | 0 | 0 | 0.015719 | 0 |
| 61 | 0.798602 | 0.016174 | 0 | 1 | 0 | 0 | 0.814776 | 0.782429 | 0 | 0.032347 |
| 75 | 1.308704 | 0.004071 | 1 | 0 | 1.312775 | 1.304633 | 0 | 0 | 0.008142 | 0 |
| 88 | 0.445239 | 0.002191 | 0 | 1 | 0 | 0 | 0.447429 | 0.443048 | 0 | 0.004381 |
| 101 | 1.283832 | 0.013534 | 1 | 0 | 1.297366 | 1.270298 | 0 | 0 | 0.027068 | 0 |
| 115 | 0.610893 | 0.001298 | 0 | 1 | 0 | 0 | 0.612191 | 0.609594 | 0 | 0.002597 |

In these tables, ID is the number of the data entry in the resmoothed and resampled dataset in 3.3, based on the retrieval of the initial 1024 signals of each data sequence in 3.2. For example, in Table 2b, the 'Mean' and 'Var.' for 'ID' = 12 means the data entries with clocks $[(12-1)\times 8, (12)\times 8-1]$ in the source accelerometer data sequence (which is 0-started) have a mean value of 1.145866 and a variance of 0.154981. 'Peak' = 0 and 'Valley' = 1 means that a valley appears here (as can be seen, in this table, entries that are neither a peak nor a valley in the figure are not shown here), and this valley is having a value delimited by ['Valley Top', 'Valley Bot.'] = [0.990884, 1.300847], while the 'Valley Range' of it is 0.309963. For this identified extreme, 'Peak Top', 'Peak Bot.' and 'Peak Range' do not receive any value because they are not peaks (see also in the corresponding subfigure in Figure A8b).

Rule 1 is obvious from these tables. Along the X or Y axis, either the peak range or the valley range of the 2.5-blade turbine is far greater than the peak range or the valley range of a 3.0-blade turbine or a 2.0-blade turbine. This is more salient when a total summary for all cases and the average is given in Table 3 (e.g., when wind-speed = 6 m/s, on Y axis, 0.072865

(2.5-blade) >> 0.018713 (3.0-blade) > 0.017087 (2.0-blade); 0.187026 (2.5-blade) >> 0.01901 (3.0-blade) > 0.016757 (2.0-blade)). That is, for a 2.5-blade turbine, the variance at the peak or valley along the X or Y axis is at least triple or more than the 3.0-blade or 2.0-blade cases.

**Table 3.** Summary for peak/valley information for all cases.

| Turbine Case | Accel. Axis | WindSpeed | #Peaks/Valleys | Avg. All Peaks | Avg. Peaks' Variances | Avg. All Valleys | Avg. Valley Variances |
|---|---|---|---|---|---|---|---|
| 3.0 | X | 00 | 9 | 0.419068 | 0.019037 | 0.010037 | 0.020074 |
| 3.0 | X | 06 | 9 | 0.767823 | 0.013936 | 0.012113 | 0.024227 |
| 3.0 | X | 12 | 9 | 0.862113 | 0.016336 | 0.010408 | 0.020817 |
| 3.0 | X | 18 | 9 | 0.948771 | 0.014 | 0.010141 | 0.020282 |
| 3.0 | Y | 00 | 9 | 1.027233 | 0.018741 | 0.005394 | 0.010787 |
| 3.0 | Y | 06 | 9 | 1.158419 | 0.018713 | 0.009505 | 0.01901 |
| 3.0 | Y | 12 | 8 | 1.212972 | 0.013009 | 0.014553 | 0.029105 |
| 3.0 | Y | 18 | 9 | 1.364735 | 0.025744 | 0.015587 | 0.031175 |
| 3.0 | Z | 00 | 9 | 2.041538 | 0.013117 | 0.006551 | 0.013102 |
| 3.0 | Z | 06 | 9 | 3.094599 | 0.026482 | 0.018394 | 0.036788 |
| 3.0 | Z | 12 | 9 | 3.197712 | 0.029044 | 0.012246 | 0.024493 |
| 3.0 | Z | 18 | 9 | 3.446215 | 0.050402 | 0.021442 | 0.042884 |
| 2.5 | X | 00 | 8 | 0.532799 | 0.010725 | 0.004935 | 0.00987 |
| 2.5 | X | 06 | 9 | 0.918827 | 0.481073 | 0.041041 | 0.082083 |
| 2.5 | X | 12 | 8 | 0.852311 | 0.113416 | 0.036543 | 0.073086 |
| 2.5 | X | 18 | 9 | 0.873083 | 0.231489 | 0.032187 | 0.064373 |
| 2.5 | Y | 00 | 12 | 1.429386 | 0.010936 | 0.006929 | 0.013859 |
| 2.5 | Y | 06 | 9 | 1.789828 | 0.072865 | 0.093513 | 0.187026 |
| 2.5 | Y | 12 | 9 | 1.696677 | 0.032836 | 0.043012 | 0.086024 |
| 2.5 | Y | 18 | 8 | 1.819861 | 0.125277 | 0.040118 | 0.080236 |
| 2.5 | Z | 00 | 9 | 2.242601 | 0.005715 | 0.004855 | 0.00971 |
| 2.5 | Z | 06 | 9 | 3.405228 | 0.036168 | 0.017841 | 0.035681 |
| 2.5 | Z | 12 | 9 | 3.288273 | 0.068447 | 0.048333 | 0.096666 |
| 2.5 | Z | 18 | 9 | 3.69345 | 0.042324 | 0.032733 | 0.065466 |
| 2.0 | X | 00 | 9 | 0.46955 | 0.014728 | 0.007318 | 0.014636 |
| 2.0 | X | 06 | 9 | 0.813965 | 0.006688 | 0.006644 | 0.013287 |
| 2.0 | X | 12 | 9 | 0.769434 | 0.025687 | 0.00883 | 0.017661 |
| 2.0 | X | 18 | 9 | 0.904054 | 0.012814 | 0.00654 | 0.01308 |
| 2.0 | Y | 00 | 9 | 1.082138 | 0.015552 | 0.005682 | 0.011363 |
| 2.0 | Y | 06 | 9 | 1.156307 | 0.017087 | 0.008378 | 0.016757 |
| 2.0 | Y | 12 | 8 | 1.213207 | 0.014997 | 0.013445 | 0.026891 |
| 2.0 | Y | 18 | 9 | 3.088382 | 0.023286 | 0.013972 | 0.027944 |
| 2.0 | Z | 00 | 9 | 1.876956 | 0.013995 | 0.007343 | 0.014686 |
| 2.0 | Z | 06 | 9 | 2.940388 | 0.020016 | 0.011664 | 0.023329 |
| 2.0 | Z | 12 | 9 | 3.114471 | 0.014417 | 0.003698 | 0.007396 |
| 2.0 | Z | 18 | 9 | 3.088382 | 0.023286 | 0.013972 | 0.027944 |

Finally, Rule 2 should also be obvious from these tables. See in Table 3. For the 3.0-blade settings, on the Z axis, from the average of all valley values to the average of all peak values (i.e., the fluctuation), the numbers are 2.034987, 3.076206, 3.185465 and 3.424772, respectively, imposing the four wind-speeds. However, for the 2.0-blade settings, on the same axis, these numbers are 1.869613, 2.928724, 3.110773 and 3.074410, respectively. All of these numbers are below those of the 3.0-blade settings (while other settings are identical) by 7% (8.8%, 5%, 2.4% and 11.4%) in average. This quantified rule can be applied, and the reason for this is related to electromechanical conversion: more electricity is produced by the wind turbine when no blade is missing.

## 5. Conclusions

Green energy has become a major power source over the past two decades. Wind does not pollute and is currently one of the most promising clean and inexhaustible energy sources for power generation [24]. Recent advances in wind energy production have helped to solve practical problems [25] and improve quality of life [26]. Due to the torsional and flexural coupling of the pre-bent blades, the dynamic characteristics of blades made from orthotropic composite materials with conical pre-torsion are much more complex than those of isotropic blades. On the other hand, this means that—compared to other components of a wind turbine whose designs have had time to mature—the relevant technologies and designs for the fan blades are relatively new.

Condition monitoring and fault diagnosis of wind turbines have gained more and more practical value for reducing maintenance costs and improving wind farms' operational efficiency [27], because more and more wind farms are being constructed and operated globally. Thus, the market has become competitive, and wind farm operators usually need to reduce their operating and maintenance costs in order to make their operations more profitable, and to maintain the sustainable competitive advantage (SCA) of the company, the maintenance strategies must be effective as operations continue. Due to these drivers, the condition monitoring and early fault diagnosis of wind turbines have become required industrial practices because they help improve the reliability and productivity of wind farms [28].

Due to the high maintenance costs (and efforts) incurred, the failure of wind turbine blades during the operation and maintenance phases has become a major problem for the wind power industry. Therefore, the utilisation of quality real-time data [29] and the development of methods to monitor the turbine blades' integrity is critical [30] because of the novelty of the blades' designs (which also determine the feature(s) of some turbine types of certain turbine brands).

In order to detect blade damage, after a review, we found that vibrations and noises are the two interfaces through which to determine mechanical faults, and the signals these carry can be analysed to determine the cause of the event. However, after long laboratory trials, we also found that taking these signals and using the existing justification criteria from research on power generators is infeasible and ineffective for diagnosing fan-blade faults using CMS with the wind turbine on the remote side. The rules (criteria) for detecting faulty situations therefore had to be reconsidered, and the idea of detecting faults using remote-side CMS with limited computation power had to be abandoned.

This led to the idea to establish effective new rules for the server-side CMS to detect faulty situations based on the vibration data transmitted and updated from the remote-side CMS, because using the datasets recorded from an interface were sufficient and using the interface recording vibrations would be better. This resulted in the creation of a new plug-in for the failure and cause-of-failure analysis module on the server-side CMS, which detects sudden faulty events and types of fan-blade damage in almost real time (see Figure 1). It can then send an alert message via SMS or email to the engineers on duty, so that they can make repairs immediately.

In this sense, it may help to establish a maintenance by prediction mechanism for unplanned maintenance when any fan blade is out of order, which supplements the common planned tasks carried out for the preventive maintenance of the fan blades. Despite the suggested mechanism playing a supplementary role to regular preventive maintenance being the main possible contribution, it should be particularly noted that this point is also exactly the boundary of this study, i.e., no work related to traditional preventive maintenance is presented. Another boundary of the study should be that the mechanism suggested by this study is for turbines having blades made of isotropic materials (because of current experimental limitations), so there is still room for exploring whether or not it still holds (or is there any other more effective mechanism) for those turbines having orthotropic blades.

All of this relies on the effective rules established to identify the 3.0-blade (full blade or normal), 2.5-blade (half or part of the blade broken) and 2.0-blade (one blade totally missing) cases while the turbines are operating. Fortunately, with the help of a contemporary data-driven approach and the adoption of suitable data processing/analysis methods in both DSP and statistics, we found that watching a continuous (but short, in terms of sampling time) sequence of the vibration dataset was sufficient to establish these rules. Fortunately, despite the considerable time and effort put into the data experiments to determine the rules, these rules are not difficult to implement on any given CMS, because:

(1) No real domain transformation is required, and only data conversions happen: the rules work with the data in the source domain, with just a few conversions (rather than transformations) of the data required for the computation;

(2) The rules are data-pattern-based, rather than learning-based: fixed judgements are made on the observable patterns or characteristics in the converted data, so no other processes, such as training, verifying, or parametric tuning efforts, are needed; and

(3) The rules are simple: only two rules are included (and required) to make the judgements, and they are simple, so they can be designed as additional functions of the CMS without utilising significant run-time computational resources.

The real use of these rules on the server-side CMS is expected in the future, and their true value will be shown when the unplanned maintenance by prediction tasks is someday carried out for the turbine blades. We sincerely hope that in the future, these rules can benefit not only the company operating the wind farms but also the entire wind turbine industry. They have the potential to change the ways people approach CMS design for wind turbines, as the effectiveness of data-driven server-side fault diagnostics has been made evident in this study. Finally, while the proposed mechanism works for shoreside wind turbines, a similar logic can be generalised to other wind turbines (e.g., offshore or inland) as well.

**Author Contributions:** Conceptualization, funding acquisition, methodology, writing—original draft preparation, M.-H.H. and Z.-Y.Z.; data curation, software (data experiments), writing—review and editing, visualization, Z.-Y.Z.; formal analysis, investigation (physical experiments), validation, M.-H.H. All authors have read and agreed to the published version of the manuscript.

**Funding:** This research was funded by National Science and Technology Council, Taiwan, ROC ('Ministry of Science and Technology' or 'MOST' before July 2022), under grants MOST-109-2221-E-346-001, MOST-110-2410-H-992-020 and MOST-111-2410-H-992-011.

**Data Availability Statement:** This research did not use publicly archived datasets.

**Conflicts of Interest:** The authors declare no conflict of interest.

## Appendix A

(C1) The vibration generated by the generator is caused by an uneven magnetic force acting on the rotor or the stator. In the case of vibration caused by the impulse torque, the induction motor essentially generates a pulsed torque and causes vibration, and impulse torque is thus generated. It is the result of rotating the magnetic field to actuate the poles of the stator. Since each stator pole is actuated twice when the AC power source completes a cycle, the resulting vibration frequency is twice that of the AC power source. That is if the frequency of the AC power used is 60 Hz, the vibration frequency of the impulse torque is 120 Hz. This type of vibration due to pulsed torque is often not significant unless the motor needs to operate in a relatively low vibration environment or because the pulsed torque vibration frequency matches the tower's natural frequency or blade body frame. When an abnormal phenomenon occurs, however, it must be dealt with. The magnitude of the vibration caused by the impulse torque depends on the evenness of the rotor winding. The more uniform the rotor winding, the lower the vibration value of the impulse torque. Therefore, for more demanding applications, good uniformity of windings should be required during a motor's manufacturing. Moreover, when the rotor winding rod is loose, or the stator winding is open, it will generate 2×, 3×, or even a higher multiple of the vibration frequency at the pulse torque frequency, i.e., 120 Hz and 180 Hz will be generated in the frequency spectrum analysis.

(C2) Vibrations generated by the generator or vibrations caused by an abnormal rotor or rotor coils. The causes of faults within the generator related to the rotor coil can consist of issues such as a rotor winding bar break, a rotor winding open or short circuit, rotor deflection, rotor eccentricity, and so on. The vibrations caused by the causes mentioned above of the fault will all be unstable. In other words, if the amplitude is measured with a vibrating instrument, the pointer or digit on the instrument will oscillate continuously. It will be a single-frequency vibration due to the causes of malfunction; the amplitude will be modulated. This modulation phenomenon is based on the reasons mentioned above of failure, which causes the imbalance of the magnetic forces between the upper and lower magnetic poles and between the rotor coil and the permanent magnet. Suppose a two-pole induction rotor is winding breaks when the generator's stator winding is actuated by AC power to generate a magnetic field. In that case, the broken part of it will be precisely aligned with the magnetic pole in a specific direction. At this moment, the current will reach a maximum value. That is, the magnetic force of the magnetic pole will be at its maximum; the magnetic pole of the other pole of the stator will act on the rotor with the maximum magnetic force. However, due to the induced magnetic poles, the rotor corresponding to one pole will be intact. The rotor corresponding to the other bar will be fractured, so the two poles' magnetic force will not be equal, resulting in an unbalanced force being generated between the rotor and the stator. Meanwhile, the broken wire rod gradually moves away from the magnetic field, which no longer precisely aligns with the magnetic pole in either direction. Then, the magnetic force difference between the two pole-operated rotors will be reduced, and the vibration force will be reduced. As a result of this cyclic operation, the vibration amplitude generated by the rotor will form a periodic vibration, and the vibration frequency will be the rotational frequency of the magnetic field, not the rotational frequency of the rotor.

(C3) Vibrations generated by the generator or vibrations caused by an abnormal stator or stator coils. If the motor problem occurs in the winding of the stator of the generator, the frequency of the vibration generated by the generator will be the frequency of the magnetic field, but its amplitude will not be. In other words, there is a modulation phenomenon. However, abnormal vibrations appearing on a stator or stator coils often have pulsating amplitudes. This pulsation is not the result of amplitude modulation, but rather the result of a slap phenomenon caused by two very similar vibration waveforms. The stator-related faults are usually caused by stator windings or short circuits, gaps of different sizes, and phase imbalances. If the vibration problem associated with the stator produces a pulsation amplitude, it needs two vibration amplitudes near each other in their frequency spectrum. One of the vibrations of the two near-frequency frequencies may be the vibration of the rotation shaft frequency caused by an imbalance of the shaft or poor centring of the post, while the motor factor will cause the pulse of the other frequency. Since the frequency of the shaft factor is very close to the electrical equipment's frequency, the two's vibration amplitude will be alternately added or subtracted at the rate of the difference between the two frequencies. It causes the motor to produce a significant slap phenomenon. As a result, amplitude ripples are formed. Suppose you want to determine the pulsation amplitude. In that case, regulated by the amplitude of a single frequency or by the result of adding two very similar vibration frequencies, you can use the internal key amplification function of the frequency analyser. First, amplify the frequency coordinate axis. Then, analyse the frequency spectrum near the shaft rotation frequency. If the frequency-axis amplification analysis shows that there are two closely related vibration frequencies stacked together, the cause of the pulsation in the amplitude can be determined to be caused by a slap phenomenon. If the frequency is amplified, there is still only a single frequency. A significant up-and-down variation in the spectrum analyser's amplitude causes the pulsation amplitude, which is the result of the modulation.

(C4) Unbalanced weight of the shaft of the driveshaft. When the shaft quality is not uniform due to the centrifugal force after the rotation, the wind turbine drive mechanism generates vibration. This unbalanced phenomenon is caused by factors such as non-uniformity in the structure of the shaft during the manufacturing process or shape asymmetry during the processing. Because of the unbalanced vibrations, radial vibrations occur. Therefore, the amount of vibration in each direction is measured. The amplitude ratio in the horizontal, vertical, and axial directions is 5:4:1, and the amplitude is proportional to the unbalanced mass. The vibration frequency based on the imbalance is one time the rotating frequency of the rotating shaft. Therefore, for the frequency spectrum analysis, the primary vibration frequency and speed are the same for determining the unbalanced force's vibration. However, in addition to the imbalance, many vibration frequencies are the same as the vibration speed. Therefore, pure spectrum analysis often cannot determine the vibration of this one-time rotational speed, which is purely due to the imbalance phenomenon. If supplemented by phase analysis, however, the imbalance phenomenon is apparent. When the phase analysis is applied, the radial phase on the bearing seats at both ends of the shaft can be measured. The vibration value and phase can be measured at a position separated by 90°, and the results of corresponding measurements of the two bearing blocks can be compared. If the stage measured by the 90° position is different by about 90° and the vibration amount calculated by the two bearing housings is a ratio, it can be determined that this vibration frequency is generated by the unbalanced force.

(C5) The shaft of the driveshaft is bent. Due to a lack of precision in its manufacturing or external force being applied during transportation and installation, the transmission shaft may cause the shaft to bend. This bending may occur at the midpoint of the distance between the two bearings or at one of the paths. This will cause the shaft to have higher vibration in the axial direction. If the beam's bending occurs at the centre of the distance between the two bearing seats, the axial vibrations derived from it will arise in both the free end and the bearing's load end. If the shaft's bending occurs at the load end of the post passing through the path, the load end's bearing will measure higher axial vibration than the free end. The spectrum analyser's pulse is generated by the beam bending phenomenon to find that the primary vibration frequency is double the rotation speed. Sometimes, it is accompanied by a slight dual rotation frequency. If we want to perform further analysis, we need to supplement the above analysis with a phase analysis to confirm its results. In general, when the shaft bending position is at the centre of the two bearings, the corresponding axial phases measured by both approaches are 180° out of step with each other, and the shaft is twisted so that it is measured at different positions on the bearing housing. The phases are different. When the shaft bending position is at the bearing or the bearing's outer end, the stage measured by the direction at the load end is different from the degrees measured in other paths, but the step is measured at the free end is the same at each position.

(C6) Improper installation of the shaft of the driveshaft. Poor installation of the shaft can be attributed to material-processing factors and installation techniques. When the rotary shaft is processed, the journal's roundness may be unsatisfactory due to a lack of precision, and ellipses or triangles may be generated, resulting in a subpar installation. The vibration caused by such a lack of roundness mainly occurs in the radial direction, horizontal or vertical. For an elliptical journal, measuring its phase angle will produce a phase difference of 180° between the horizontal and vertical directions. Still, below the vertical direction, the phase of the measured phase and the other directions will be 180° out of phase. If the frequency spectrum analyses the vibration frequency, it will be found that the main vibration frequency at the elliptical ellipse is twice the rotational frequency. The primary vibration frequency when the journal is triangular will be triple the rotational frequency. When the shaft is installed, if the shaft's centreline cannot be installed parallel to the centreline of the two bearing seats, it will result in poor centring. When the post has a poor centring, it will produce a higher vibration in the axial direction.

When the axial vibration is more generous than 1/2 of the radial (horizontal or vertical) vibration, various problems may result. If a spectrum analysis is performed, it can be found that the vibration frequency of the shaft centre is poor. In addition to indicating 1.0 times the speed, there will be 2.0 times or 3.0 times the rotation frequency, and the phases of the two bearings' corresponding points will be 180° out of phase. For the bearing measurement horizontal and vertical phases, the phase angles will also have a difference of 180°.

(C7) When the transmission shaft is misaligned with the coupling connecting the machine during installation, it will cause the machine to vibrate violently. Generally, there are two types of misalignment: misaligned misalignment and low centripetal misalignment. Decentralization and angling do not occur at the same time. When a machine has a poor centring, it will produce high axial vibration values, and the frequency spectrum shown in the analysis of the vibration frequency will, in addition to being double the speed, have twice or three times the speed frequency. If the form of the poor centring is poor, the phase angle of the corresponding point of the bearing on both sides of the coupling is measured at the phase angle, and the horizontal phase is 180° out of phase. If the bad machine core part's pattern is a purely eccentric type, the vertical phase of the corresponding point will also differ by 180°. When the axial vibration is more than half the horizontal vibration, it is necessary to pay attention to whether the image's adverse effect is bad or the coupling is damaged. When the coupling is damaged, loosening occurs in the coupling portion due to the excessive clearance, and in the frequency spectrum analysis, there will be an abnormal vibration at a multiple of the rotational speed frequency that is the same as the number of coupling claws. If the cause of the diagnosis's vibration is caused by poor coring, the axis centring correction must be done appropriately. If the coupling is damaged, the coupling needs to be replaced.

(C8) When the bearing is installed, the shaft diameter may be too small, resulting in the bearing's relative movement and the inner ring of the bearing. The bearing seat's diameter may be too little, causing the path to slip during operation. When this phenomenon occurs, the horizontal and vertical directions will produce more severe vibrations. For spectrum analysis, harmonic vibration frequencies with an integer multiple speeds will be generated in the low-frequency range. This harmonic frequency is based on the mains frequency or the pulsed torque frequency. The resulting harmonic vibration of the magnetic field frequency is caused by the air gap.

(C9) When the bearing is installed, the shoulder's angle may not be good, resulting in the centre of the inner and outer rings no longer being on the same straight line. Rotating causes the motor-transmission mechanism to generate a large axial force due to this twisting. A significantly higher level of noise will accompany the amount of vibration during operation. When spectrum analysis is performed, it will show that the vibration amplitude of the spindle-rotation speed is increased. If a phase analysis is performed, the axial phase will be 90° apart every 90°.

(C10) When the bearing is mounted, the ball bearing can cause temperature rise and vibration due to insufficient lubrication or improper lubrication viscosity. In the diagnostic analysis, if the fan temperature is found to be above 30 °C above the ambient temperature, it suggests that there is poor lubrication. Sleeve bearings, on the other hand, tend to wear with the shaft and cause loosening. This loosening phenomenon will cause the wind turbine to produce eccentricity and poor centring, and different vibration patterns. Frequency spectrum analysis of the vibration frequency will be accompanied by a harmonic frequency of twice or a high multiple of its radial and axis, in addition to a doubling of the number of revolutions. There will also be no relation to the phase. As for gasification bearings, abnormal vibration may occur due to the intrusion of foreign matter. Since a magnetic field's action forms the rotation of a gasification bearing, if foreign matter invades, the rotation speed of the wind turbine will be unstable, and it may even stop due to failure. In the spectrum analysis, it can be found that its central frequency will be intricate and cluttered, such that it is almost impossible to stabilise. Moreover, if the average signal function is applied, noise cannot be eliminated. This phenomenon is caused by the intrusion of foreign material into the gasification bearing, which results in varying rotation speed.

(C11) Gearbox. The wind turbine drives the rotating shaft by rotating the blades through the winds provided by nature. The initial low speed accelerates the rotating shaft via the speed-change gearbox, and the output is driven by the high-speed rotating shaft to drive the generator. Gearboxes play an essential role in wind power transmission systems, usually using a third-order parallel shaft gear drive. They also have a transmission mechanism designed as a hybrid of a third-order planetary gear and a helical gear. A wind turbine generator gearbox is affected by wind power generation and is subjected to extreme loads. It is also deployed in harsh environments. Its frequent faults and breakdowns are time-consuming to repair, resulting in a heavy operational burden. In general, wind turbine gearbox failures, the causes of which are complex, will not damage a single part, but rather a chain effect of damage across multiple parts. For example, when the gearbox's oil fails, its bearings will be damaged by wear and tear, which will gradually affect the gear, causing it to crack, eventually leading to gearbox failure. The gearbox structure and components are large and complex, easily causing the deformation of the gearbox housing and the main driveshaft. When starting up, it is easy to cause wear on the gearbox bearings and cause the post's eccentricity to run in an unbalanced manner, causing excessive vibration and noise.

The vibration frequency caused by gearbox meshing during transmission is much higher than the vibration frequency generated by an unbalanced shaft and poor centring of the gear. When the gearbox fault causes vibration or noise, the analysis of its frequency occurs at the spindle speed and is based on the shaft's number of teeth. Besides, when the two corresponding motion gears mesh, the contact point collision is rigid and rapid, and it efficiently induces the gear vibration's harmonic frequency. Measuring the vibration signal of the gearbox requires two sets of probe sensors. One set of probes is mounted on the bearing to measure the gear transmission shaft's vibration signal and the bearing, and the other set of probes is fixed on the gear section near the gear mesh. The meshing frequency is the number of teeth multiplied by the number of revolutions. Usually, the gears' meshing frequency is not much larger than the frequency of one frequency, and a large amount of vibration often occurs at twice the frequency. When the gear is engaged, the clearance between the two teeth is too large, or the rotating shaft is too loose in the bearing, there will be a loosening phenomenon during the transmission. Under these circumstances, the vibration frequency will be 0.5 times the meshing frequency. Analysing the spectrum of the complicated gearbox and studying the damage and cracking of gears, inverse Fourier analysis can be used to invert the Fourier function of the frequency domain into the time domain to calculate the defect position. The weight of a wind turbine's gearbox can be as much as 15 tons. If the structure is deformed during installation, the base vibration will cause the gearbox to cause abnormal wear, gear eccentricity, and shaft bending. The cause of vibration due to gear deterioration is typically improper lubrication or insufficient lubrication or even the oil's infiltration of metal impurities, all of which will allow the gearbox is operating temperature to increase, causing tooth surface wear or the intrusion of foreign matter or dust particles in gear. Relatedly, the excessive, long-term intrusion of salt into the gearbox will cause pitting on the tooth surface and even cracking of the teeth.

(C12) Bearing damage. The bearings mounted on a wind turbine drive mechanism consist of ball bearings, sleeve bearings, and gasification bearings. These bearings are often inaccessible due to handling, installation, or miscellaneous infiltration. They operate at high temperatures or high loads for a long time and have insufficient lubrication or wear. The resulting flaws can cause other defects to appear on the inner and outer ring raceways or rolling elements and retainers in ball bearings. According to the relative speed calculation, when any kind of bearing assembly is damaged, there is a specific rate. This particular frequency is the frequency at which the ball rolls over the inner and outer rings. This relative rolling frequency often occurs when the bearing is subjected to a local load. Therefore, in a frequency spectrum analysis, the passing frequency of the low frequency is often filtered out, and the frequency band of the high frequency generated by the crucible is used. Then the absolute value and envelope processing are applied. In other words, when there is a ball-rolling frequency through the frequency spectrum, it means that the bearing has a high-frequency vibration due to impact, and the specific frequency generated when the ball bearing components are damaged.

## Appendix B

Noise measurement of wind turbines. The noise generated by wind turbines is itself a kind of sound, so it has all the characteristics of a sound. The vibration of an object causes a sound. Energy is then transmitted through a given elastic medium in the form of sound waves. Therefore, all the factors that cause the vibration of a rotating machine body or cause the pressure field of a transmission medium to change are all mechanical equipment noise sources. When a wind generator is in regular operation, the blades generate periodic noise due to the pulsating force of the outlet's uneven flow. On the other hand, due to the uneven distribution of the pressure on the blades themselves, the surrounding gas and components' disturbance during rotation also constitutes rotational noise. The turbulent pressure distribution on the blades causes turbulence noise due to the turbulent flow, airflow, and vortex shedding generated when the gas flows through the blade. The low-frequency sound is generated by the wind turbine results from this. Large-scale wind turbines have a small number of diaphragms. Under regular operation, there are not many turns, so low-frequency sounds occur. If one measures the fundamental frequency $f$ (Hz) of the noise of the wind turbine, then the following relationship exists between the speed of the wind fan $R$ (rpm) and the wind turbine blade $Z$ (pieces): $f = \frac{RZ}{60}$ (Hz).

## Appendix C

The sampled datasets of the vibration data updated on the server-side CMS were rendered for 3.0-blade (full blades), 2.5-blade and 2.0-blade wind turbine settings.

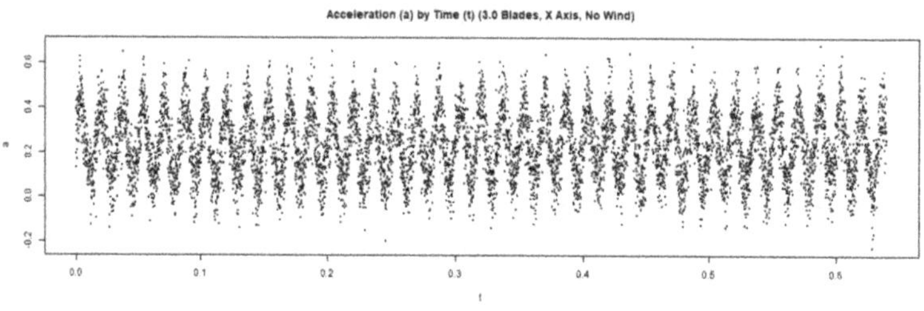


**Figure A1.** *Cont.*

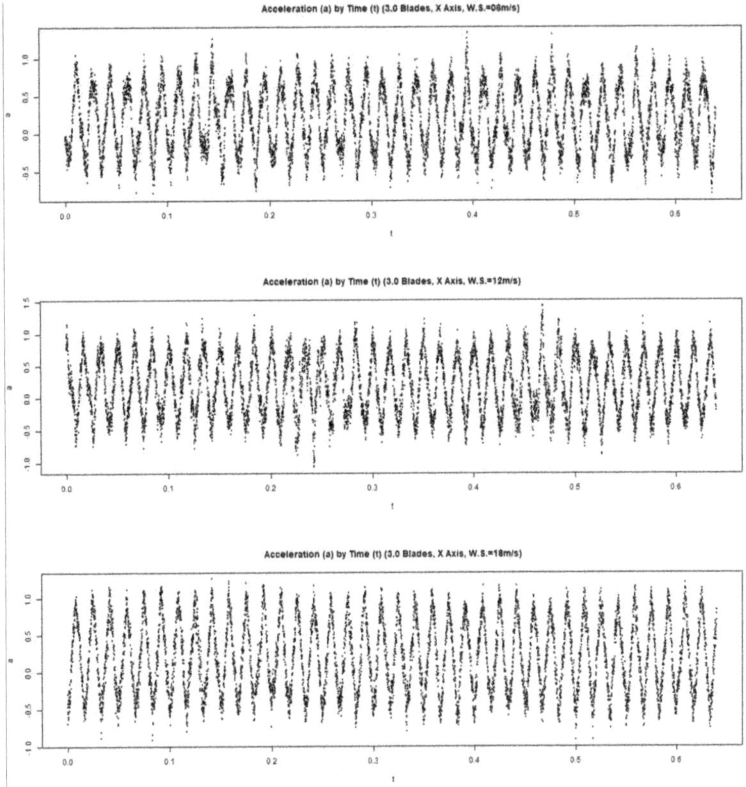


(a)

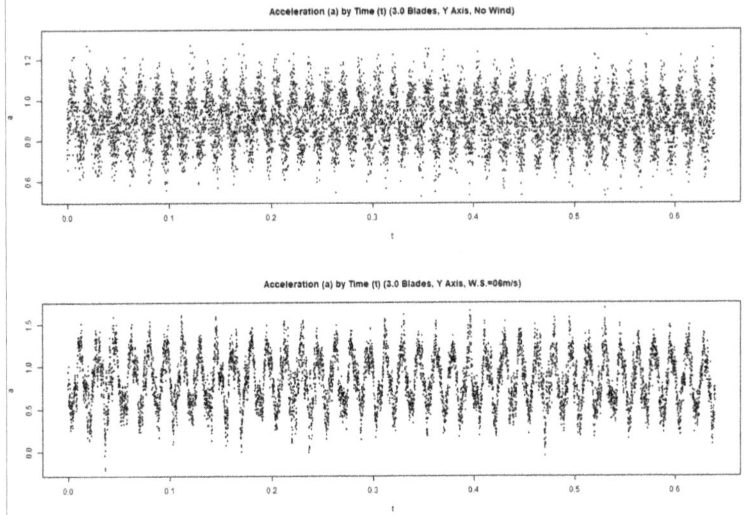


**Figure A1.** *Cont.*

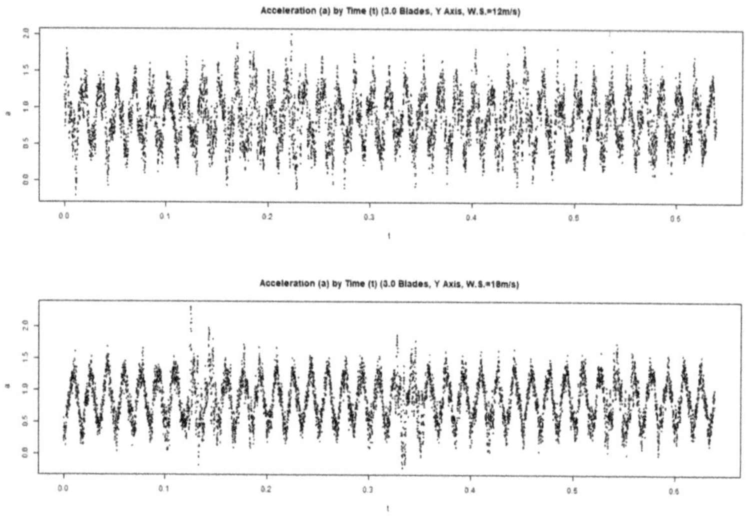


(**b**)

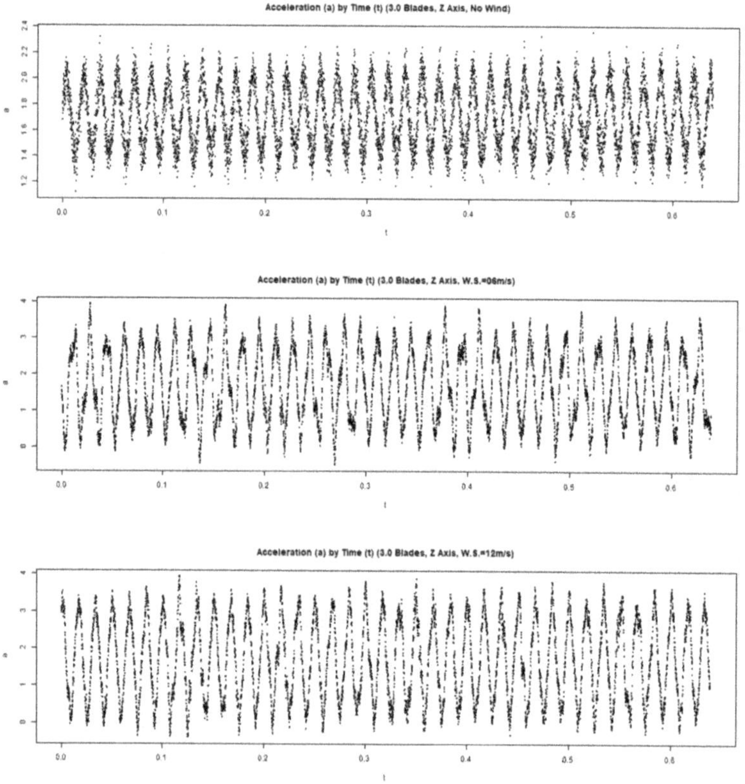


**Figure A1.** *Cont.*

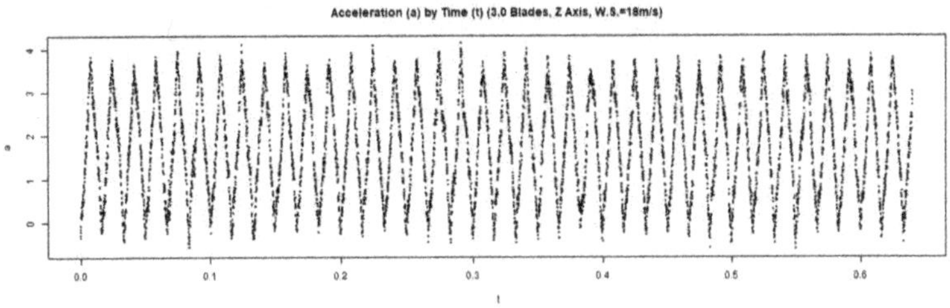


(c)

**Figure A1.** Acceleration signals recorded for the full (normal) 3.0-blade turbine. (**a**) Data recorded along the X axis with different wind-speeds imposed. (**b**) Data recorded along the Y axis with different wind-speeds imposed. (**c**) Data recorded along the Z axis with different wind-speeds imposed.

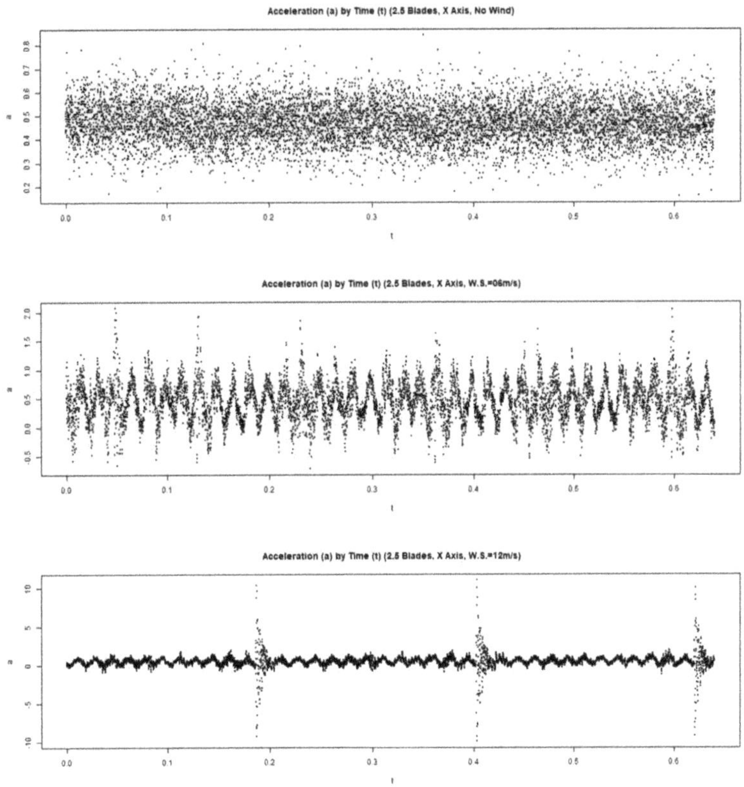


**Figure A2.** *Cont.*

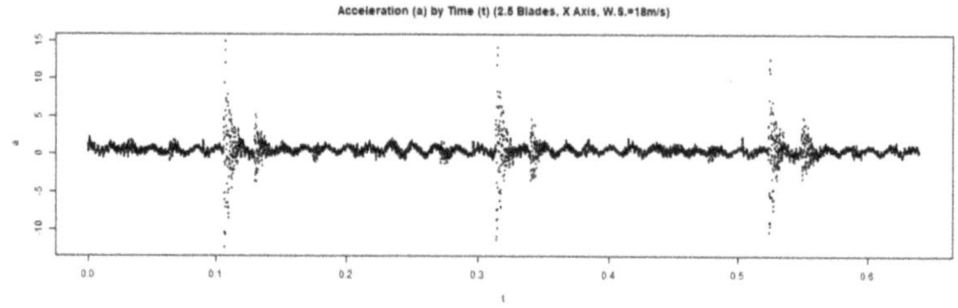


**(a)**

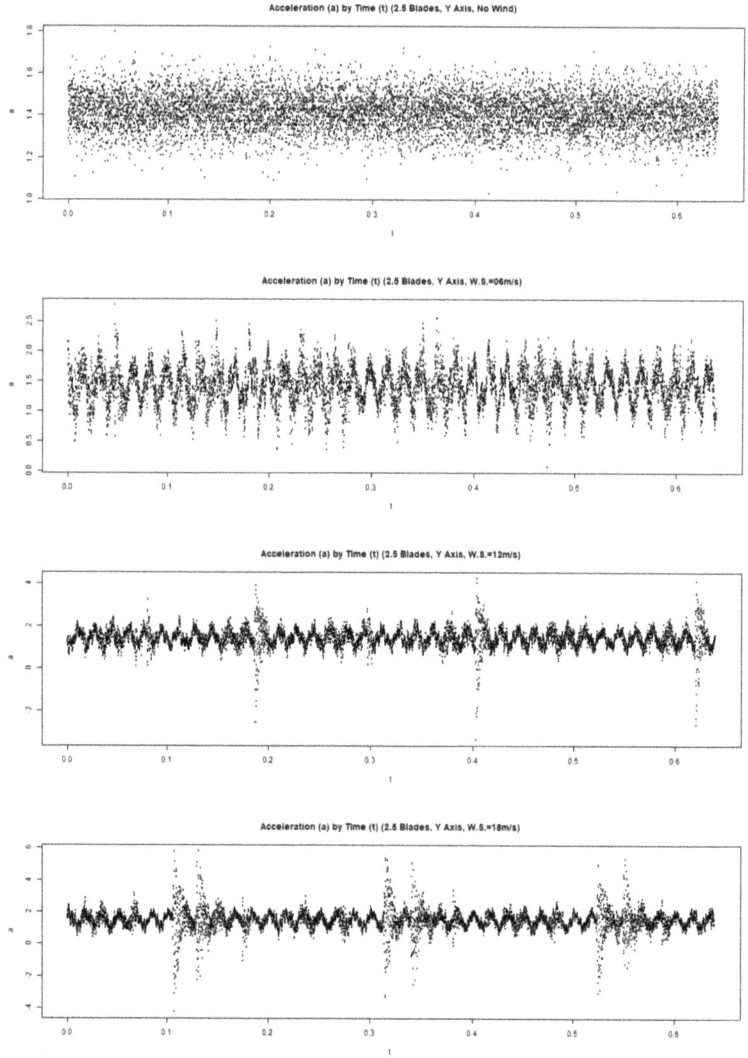


**(b)**

**Figure A2.** *Cont.*

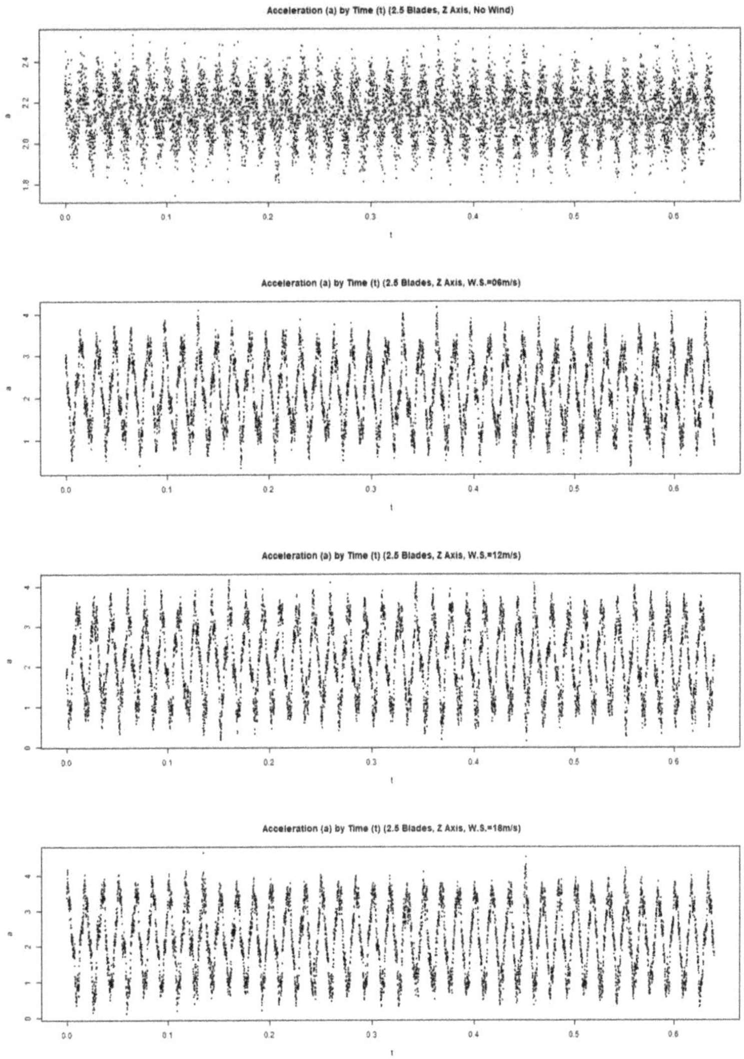


(c)

**Figure A2.** Acceleration signals recorded for the 2.5-blade (partially broken) turbine. (**a**) Data recorded along the X axis with different wind-speeds imposed. (**b**) Data recorded along the Y axis with different wind-speeds imposed. (**c**) Data recorded along the Z axis with different wind-speeds imposed.

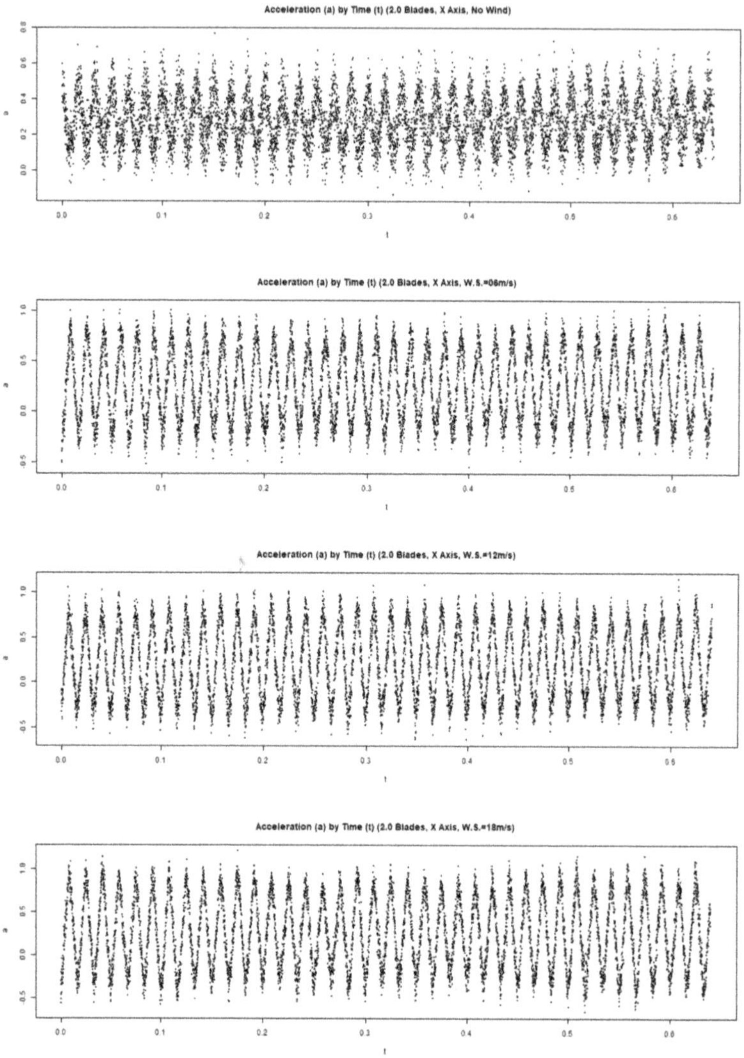


**(a)**

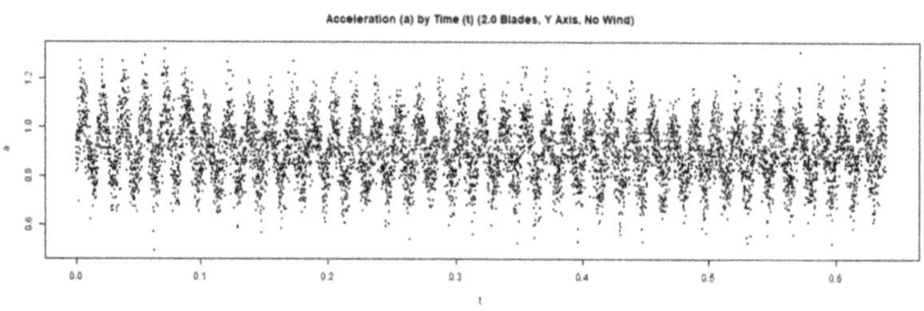


**Figure A3.** *Cont.*

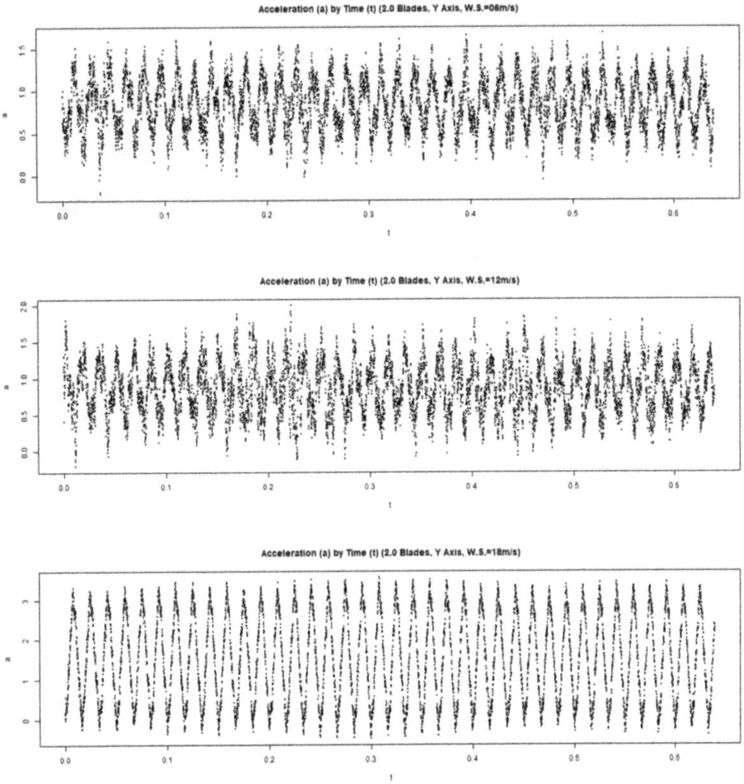


**(b)**

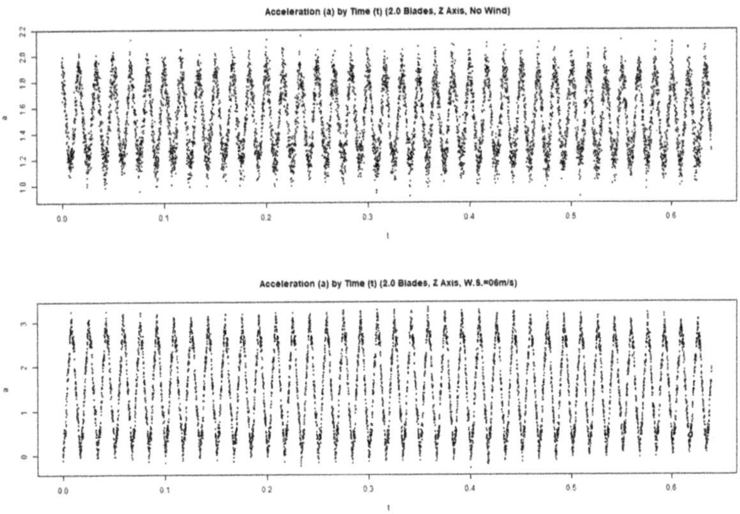


**Figure A3.** *Cont.*

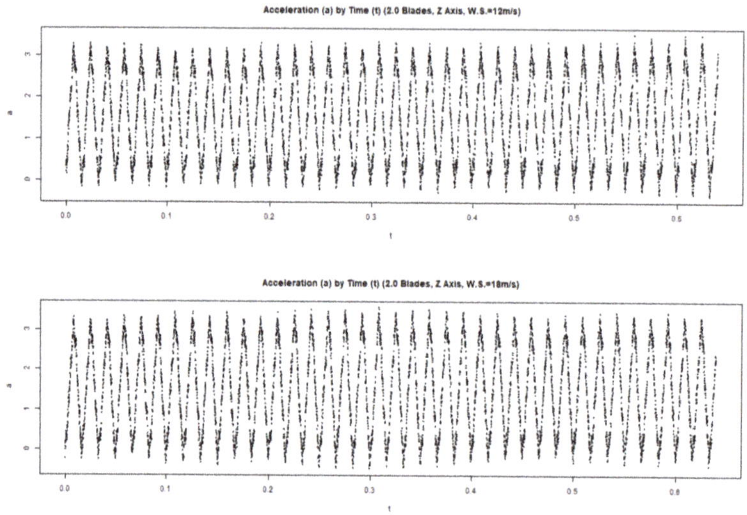


(c)

**Figure A3.** Acceleration signals recorded for the 2.0-blade (one blade missing) turbine. (**a**) Data recorded along the X axis with different wind-speeds imposed. (**b**) Data recorded along the Y axis with different wind-speeds imposed. (**c**) Data recorded along the Z axis with different wind-speeds imposed.

## Appendix D

We plotted each result (obtained from LRM) as a series of piecewise linear line segments using the 'connect the dots' method as a bold blue line in Figures A4–A6 for different wind-speeds subject to the wind turbine settings of 3.0-blade (full blades), 2.5-blade and 2.0-blade configurations, respectively. In each figure, the plots were divided into three groups, i.e., the X-axis, Y-axis and Z-axis groups, and each group includes subfigures plotted for the no-wind (0 m/s), 6 m/s, 12 m/s and 18 m/s wind-speed settings.

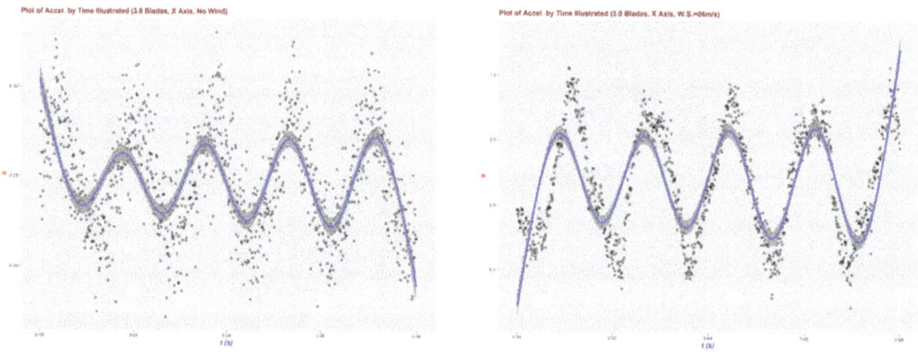


**Figure A4.** *Cont.*

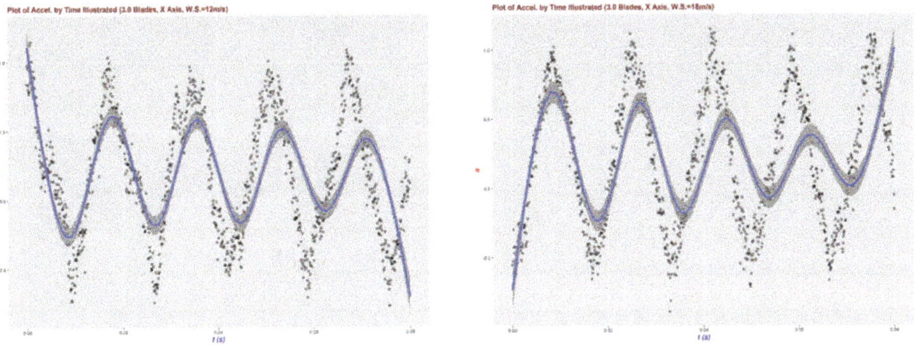


(**a**)

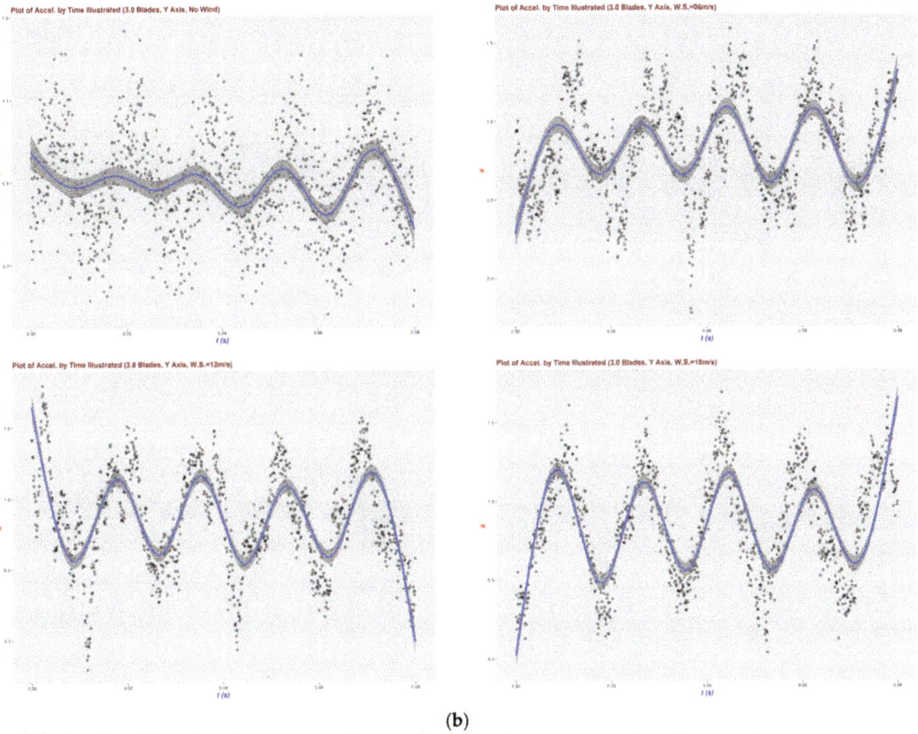


(**b**)

**Figure A4.** *Cont.*

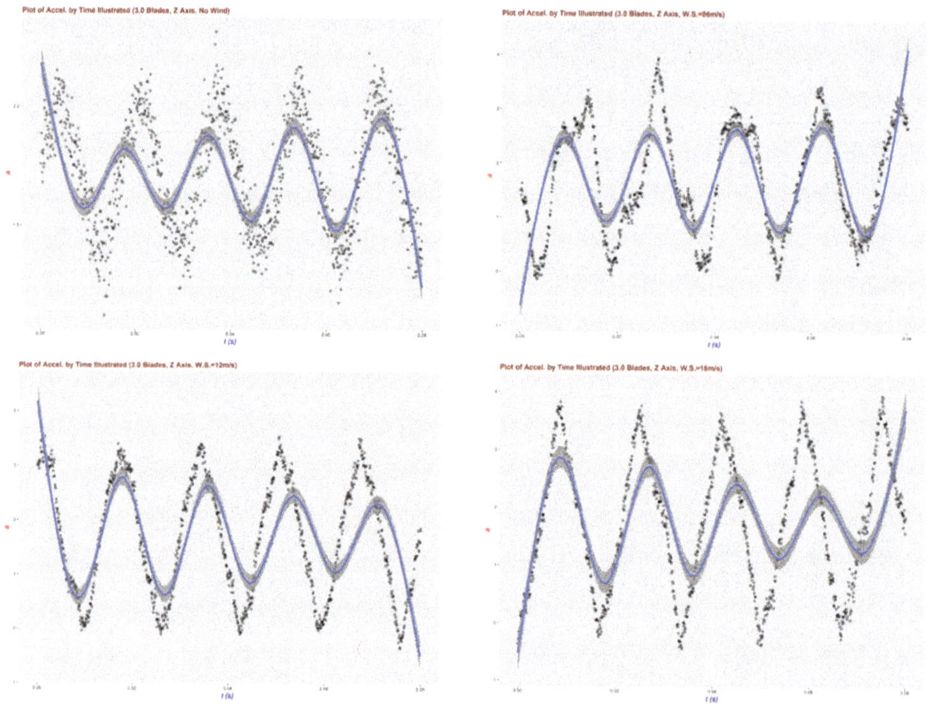


(**c**)

**Figure A4.** Magnitude and trend of acceleration for the full (normal) 3.0-blade turbine. (**a**) Based on datasets recorded along the X axis with different wind-speeds imposed. (**b**) Based on datasets recorded along the Y axis with different wind-speeds imposed. (**c**) Based on datasets recorded along the Z axis with different wind-speeds imposed.

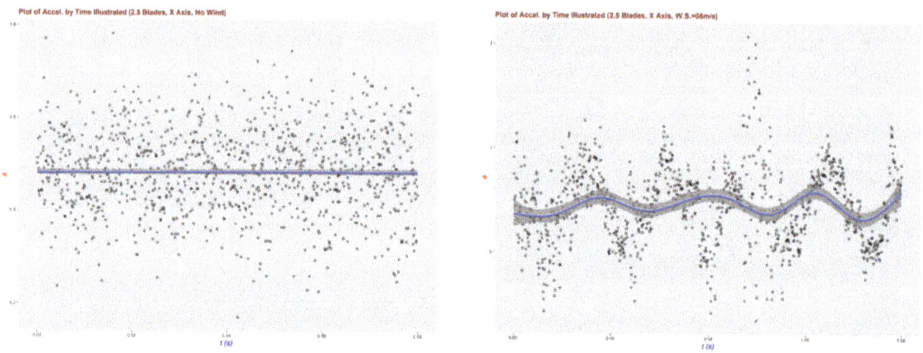


**Figure A5.** *Cont.*

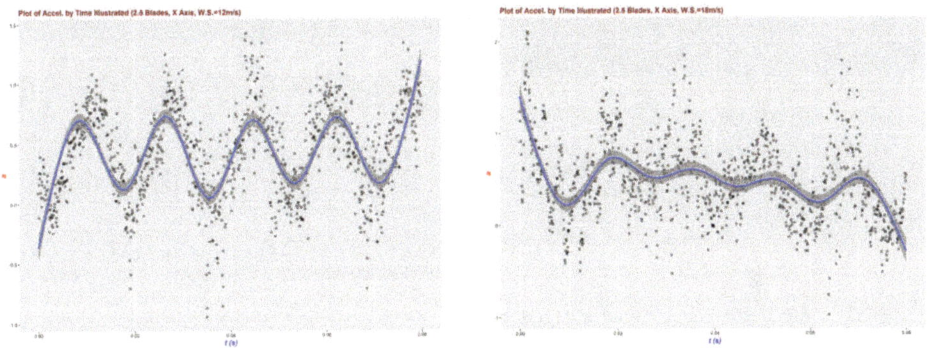


(**a**)

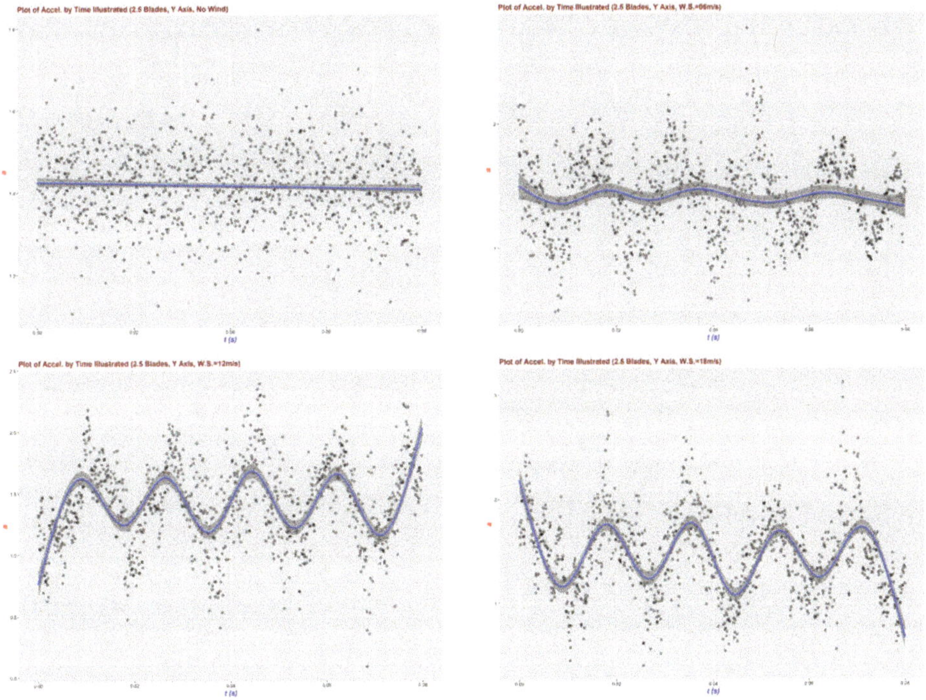


(**b**)

**Figure A5.** *Cont.*

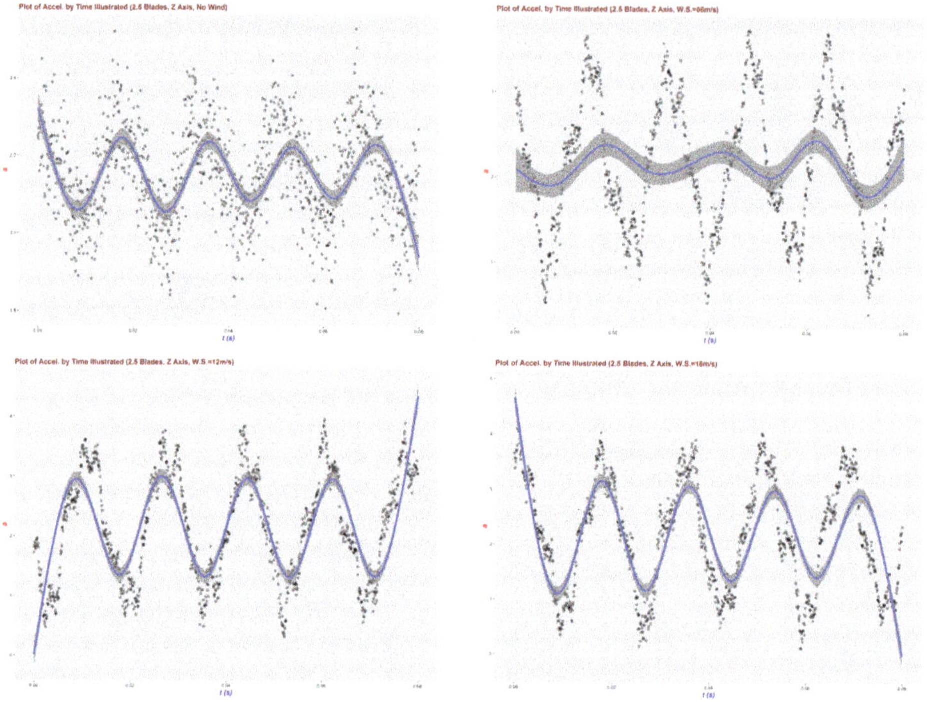


(**c**)

**Figure A5.** Magnitude and trend of acceleration for the 2.5-blade (partially broken) turbine. (**a**) Based on datasets recorded along the X axis with different wind-speeds imposed. (**b**) Based on datasets recorded along the Y axis with different wind-speeds imposed. (**c**) Based on datasets recorded along the Z axis with different wind-speeds imposed.

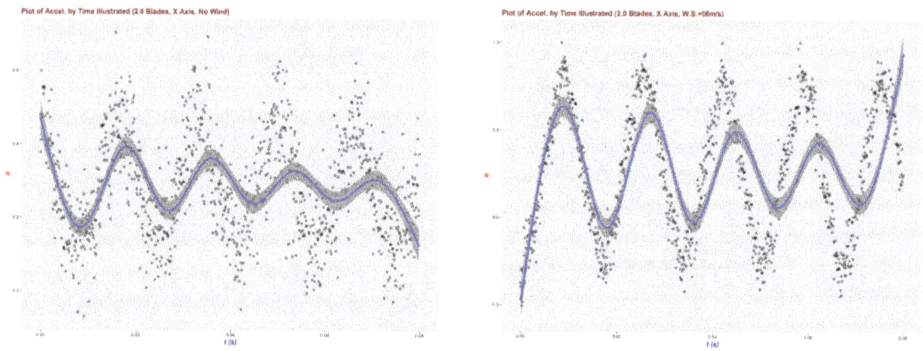


**Figure A6.** *Cont.*

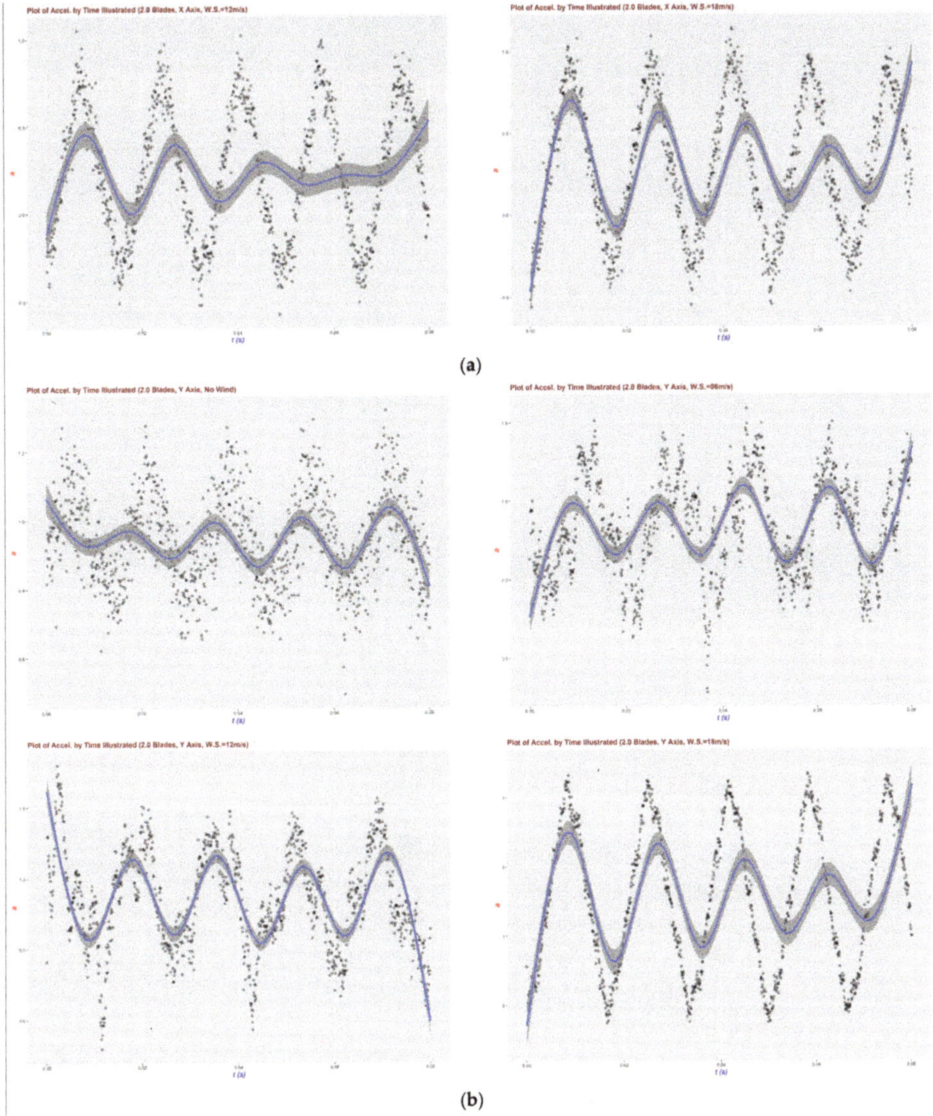


**Figure A6.** *Cont.*

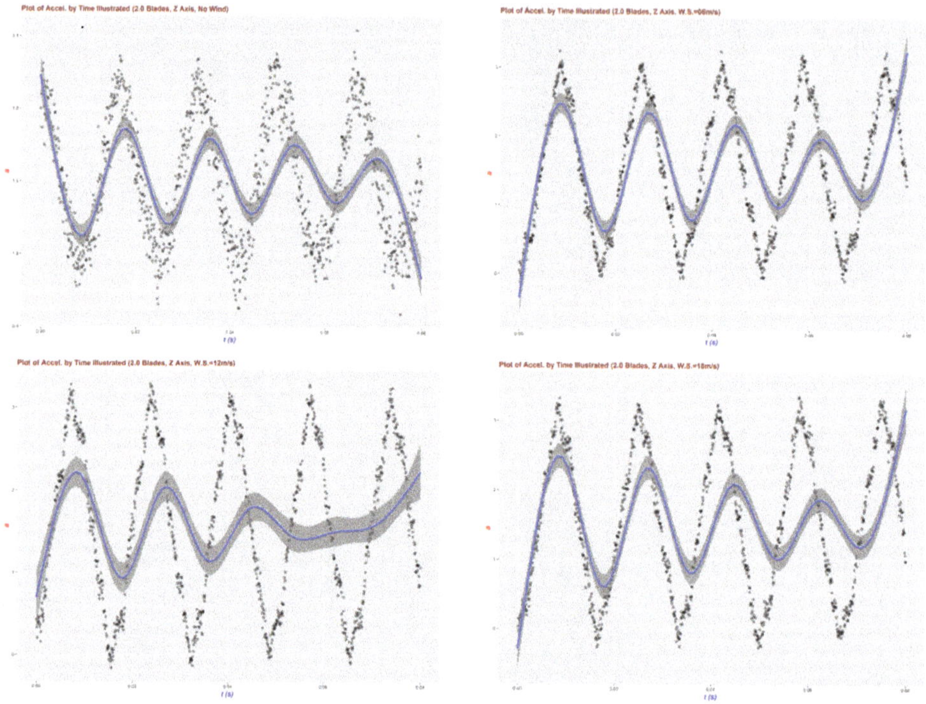


(c)

**Figure A6.** Magnitude and trend of acceleration for the 2.0-blade (one blade missing) turbine. (**a**) Based on datasets recorded along the X axis with different wind-speeds imposed. (**b**) Based on datasets recorded along the Y axis with different wind-speeds imposed. (**c**) Based on datasets recorded along the Z axis with different wind-speeds imposed.

## Appendix E

The computational results after resampling and resmoothing for all wind-speed cases on all axes are rendered in Figures A7–A9 for the 3.0-blade, 2.5-blade and 2.0 blade settings, respectively.

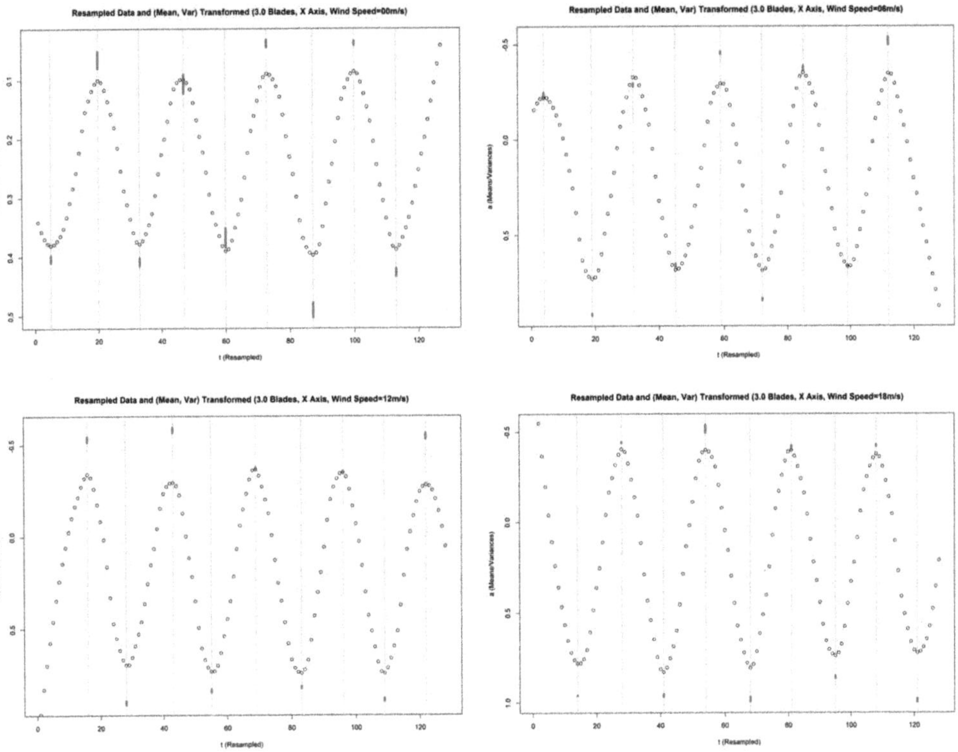


**(a)**

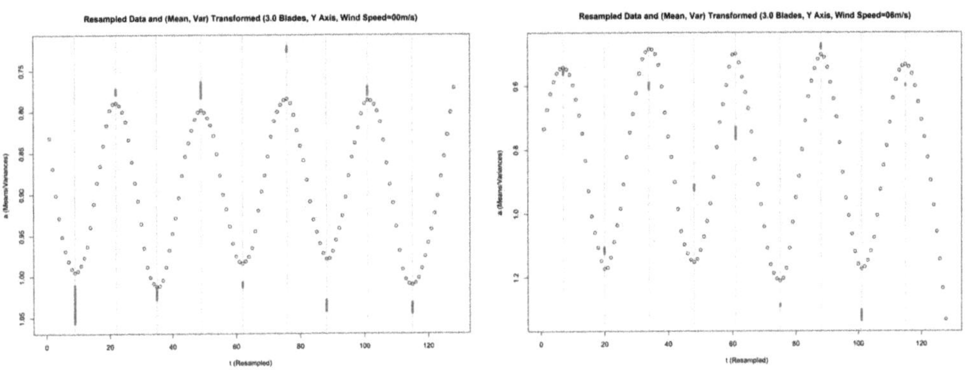


**Figure A7.** *Cont.*

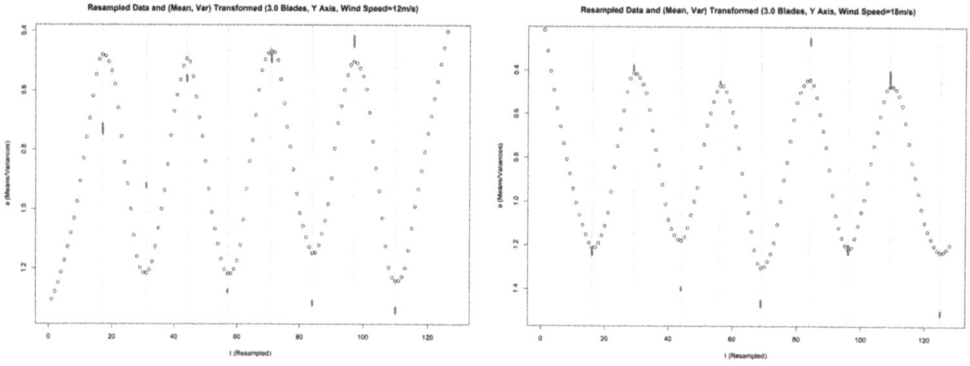


(b)

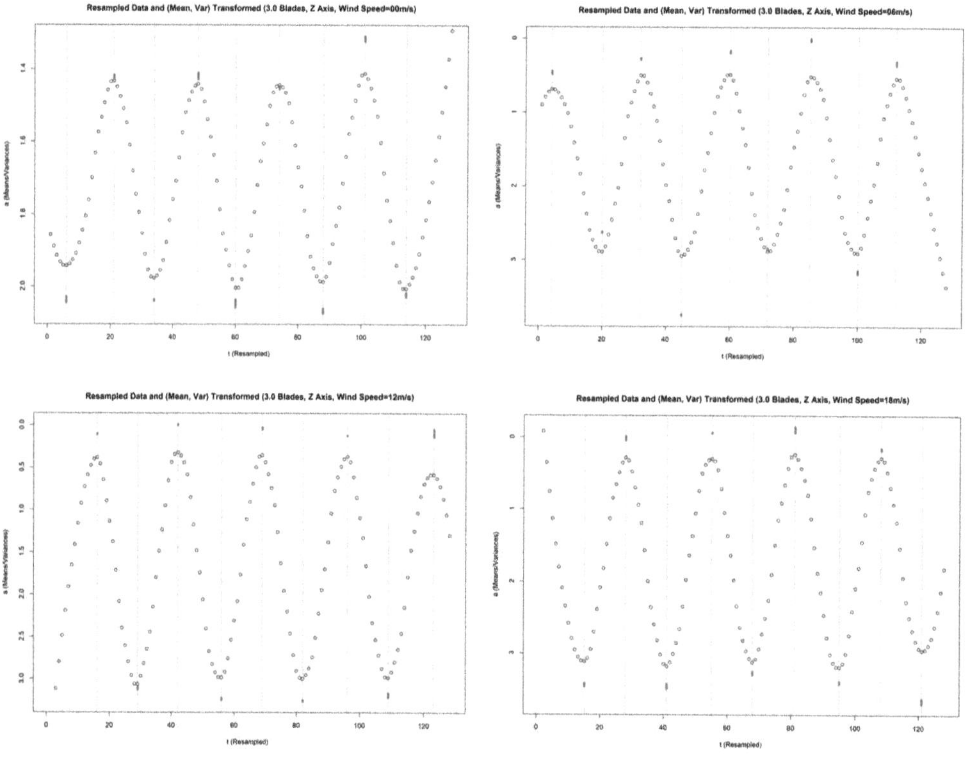


(c)

**Figure A7.** Means and peak/valley variances of resampled data for the 3.0-blade turbine. (**a**) Based on datasets recorded along the X axis with different wind-speeds imposed. (**b**) Based on datasets recorded along the Y axis with different wind-speeds imposed. (**c**) Based on datasets recorded along the Z axis with different wind-speeds imposed.

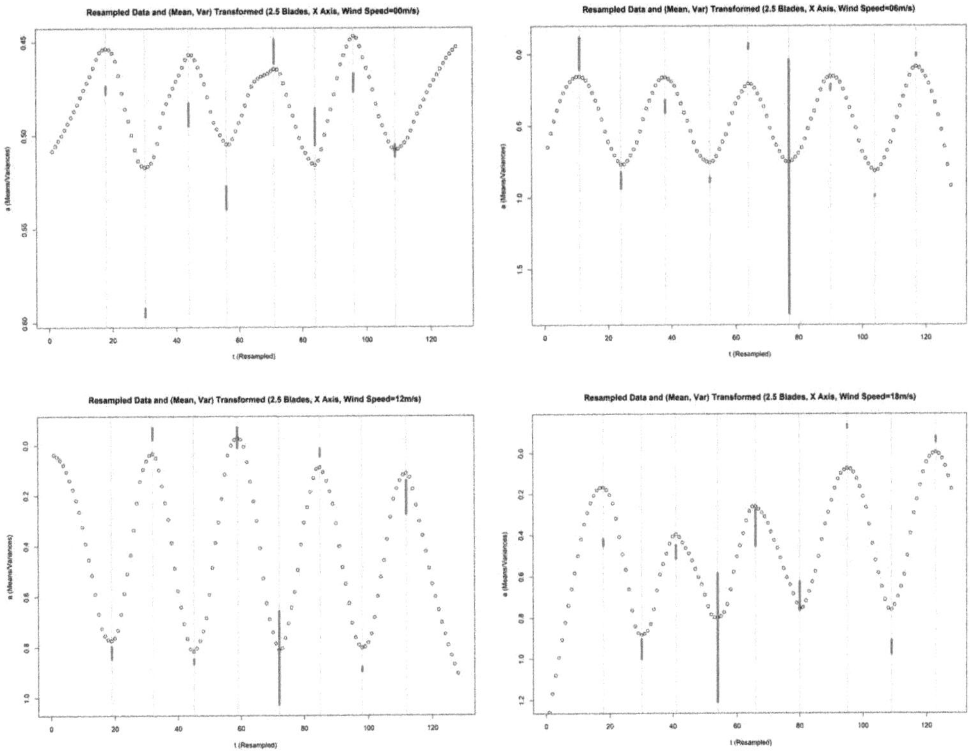


**(a)**

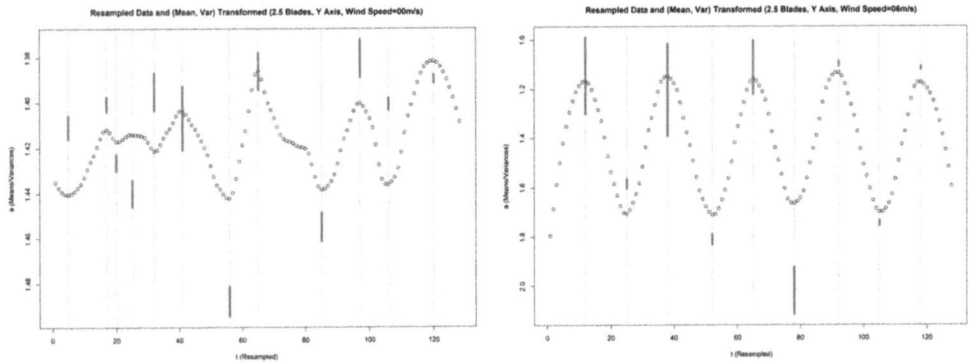


**Figure A8.** *Cont.*

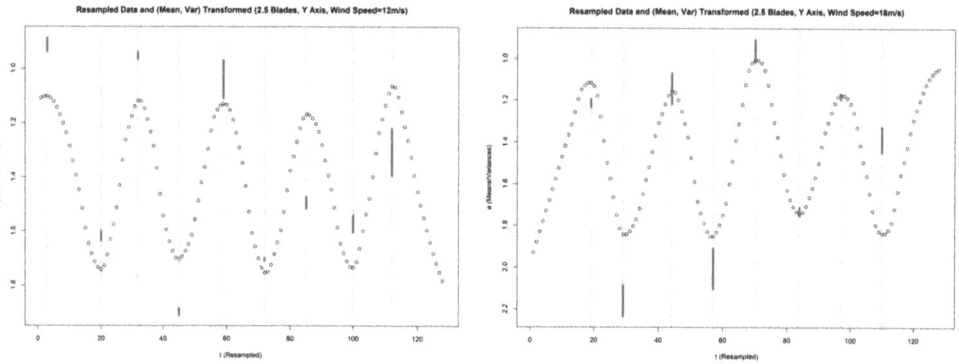


(**b**)

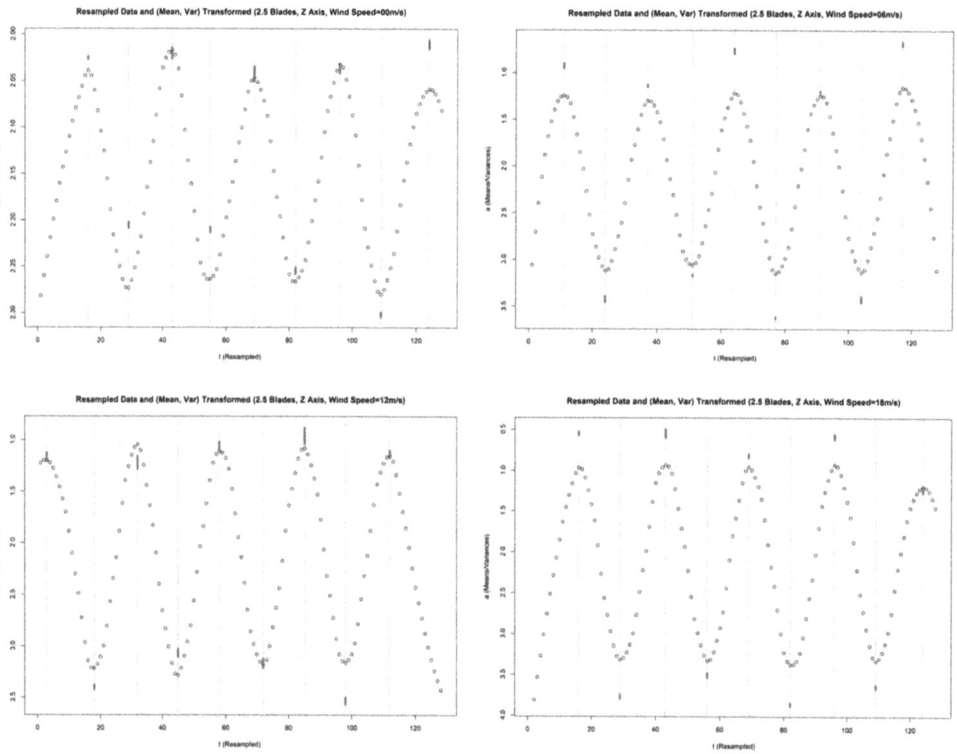


(**c**)

**Figure A8.** Means and peak/valley variances of resampled data for the 2.5-blade turbine. (**a**) Based on datasets recorded along the X axis with different wind-speeds imposed. (**b**) Based on datasets recorded along the Y axis with different wind-speeds imposed. (**c**) Based on datasets recorded along the Z axis with different wind-speeds imposed.

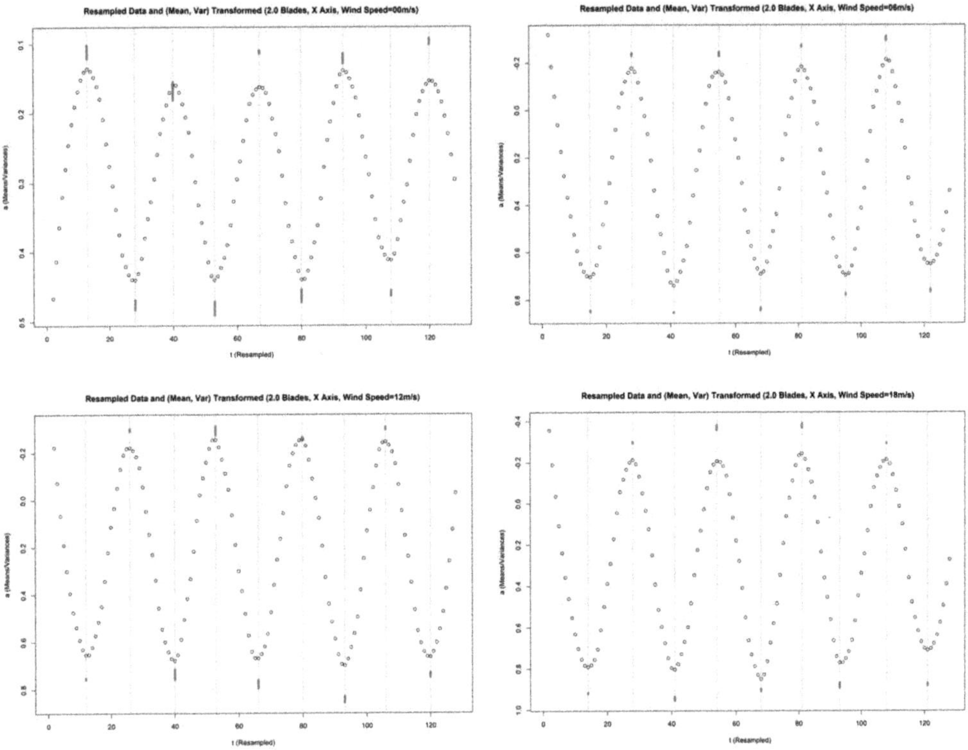


**(a)**

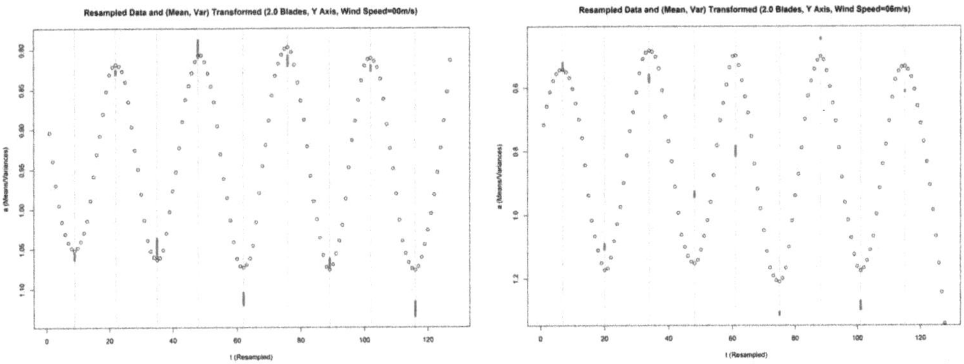


**Figure A9.** *Cont.*

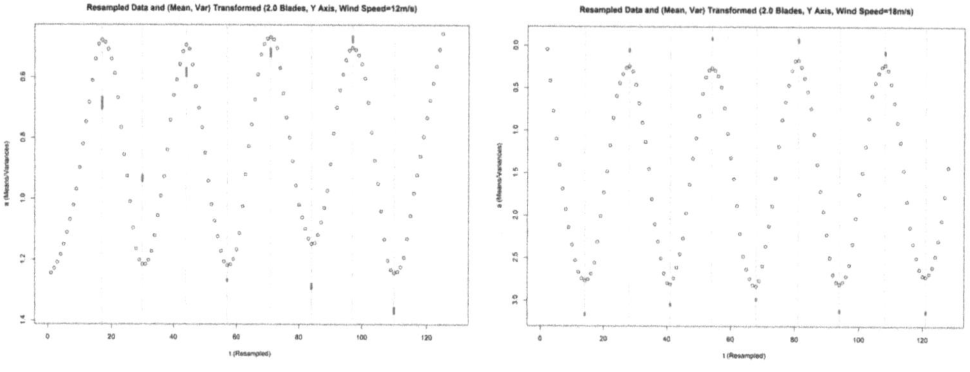


(b)

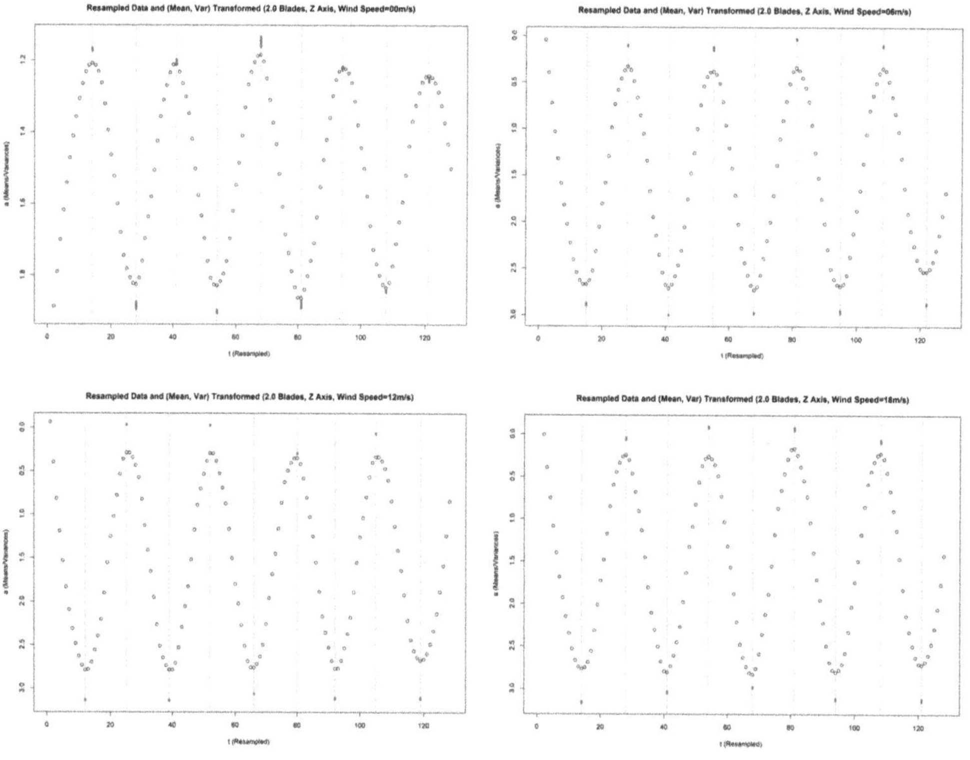


(c)

**Figure A9.** Means and peak/valley variances of resampled data for the 2.0-blade turbine. (**a**) Based on datasets recorded along the X axis with different wind-speeds imposed. (**b**) Based on datasets recorded along the Y axis with different wind-speeds imposed. (**c**) Based on datasets recorded along the Z axis with different wind-speeds imposed.

## References

1. Hsu, M.H.; Hsia, S.Y.; Chou, Y.T.; Chu, H.M.; Cheng, J.W. *Fault Diagnosis of Offshore Wind Power Generation Systems*; Project Proposal; Ministry of Science and Technology: Taiwan, China, 2014.
2. Tan, J.B.; Hsu, M.H. Diagnosis of Faults in Wind Power Generation Systems. In Proceedings of the IEEE Conference on Industrial Electronics and Applications (ICIEA), Hefei, China, 5–7 June 2016; pp. 1459–1462.

3. Tan, J.B.; Chao, C.C.; Lin, M.C.H.; Hsu, M.H. Wind Turbine Monitoring Warning Device. In Proceedings of the IEEE Conference on Industrial Electronics and Applications, Siem Reap, Cambodia, 18–20 June 2017; pp. 510–515.
4. Benbouzid, M.; Berghout, T.; Sarma, N.; Djurović, S.; Wu, Y.; Ma, X. Intelligent Condition Monitoring of Wind Power Systems: State of the Art Review. *Energies* **2021**, *14*, 5967. [CrossRef]
5. Natili, F.; Daga, A.P.; Castellani, F.; Garibaldi, L. Multi-Scale Wind Turbine Bearings Supervision Techniques Using Industrial SCADA and Vibration Data. *Appl. Sci.* **2021**, *11*, 6785. [CrossRef]
6. Papi, F.; Cappugi, L.; Salvadori, S.; Carnevale, M.; Bianchini, A. Uncertainty quantification of the effects of blade damage on the actual energy production of modern wind turbines. *Energies* **2020**, *13*, 3785. [CrossRef]
7. Santolamazza, A.; Dadi, D.; Introna, V. A data-mining approach for wind turbine fault detection based on SCADA data analysis using artificial neural networks. *Energies* **2021**, *14*, 1845. [CrossRef]
8. Gonzalez, E.; Nanos, E.M.; Seyr, H.; Valldecabres, L.; Yürüşen, N.Y.; Smolka, U.; Muskulus, M.; Melero, J.J. Key performance indicators for wind farm operation and maintenance. *Energy Procedia* **2017**, *137*, 559–570. [CrossRef]
9. Ackermann, T.; Söder, L. Wind energy technology and current status: A review. *Renew. Sustain. Energy Rev.* **2000**, *4*, 315–374. [CrossRef]
10. Manwell, J.F.; McGowan, J.G.; Rogers, A.L. *Wind Energy Explained: Theory, Design and Application*; John Wiley & Sons: Hoboken, NJ, USA, 2010.
11. Crawford, C.A. An Integrated CAD Methodology Applied to Wind Turbine Optimisation. Ph.D. Thesis, Massachusetts Institute of Technology, Cambridge, MA, USA, 2003.
12. Zhong, L.; Gu, D.; Lai, Y.; Wu, L. Optimal damper design for tuning vibration of the mass damper to the base. *Struct. Eng.* **2012**, *27*, 70–90.
13. Brune, C.S.; Spee, R.; Wallace, A.K. Experimental evaluation of a variable-speed, doubly-fed wind-power generation system. *IEEE Trans. Ind. Appl.* **1994**, *30*, 648–655. [CrossRef]
14. Hilloowala, R.M.; Sharaf, A.M. A rule-based fuzzy logic controller for a PWM inverter in a stand-alone wind energy conversion scheme. *IEEE Trans. Ind. Appl.* **1996**, *32*, 57–65. [CrossRef]
15. Simoes, M.G.; Bose, B.K.; Spiegel, R.J. Design and performance evaluation of a fuzzy-logic-based variable-speed wind generation system. *IEEE Trans. Ind. Appl.* **1997**, *33*, 956–965. [CrossRef]
16. Cory, W.T.W. Overview of condition monitoring methods with emphasis on industrial fans. *Proc. Inst. Mech. Eng. Part A: J. Power Energy* **1991**, *205*, 225–240. [CrossRef]
17. Yan, R.; Gao, R.X. Hilbert–Huang transform-based vibration signal analysis for machine health monitoring. *IEEE Trans. Instrum. Meas.* **2006**, *55*, 2320–2329. [CrossRef]
18. Bruel and Kjær. *Machine Health Monitoring Using Vibration Analysis*; Bruel and Kjær: Nærum, Denmark, 1988.
19. Chang, H.C.; Lin, S.C.; Kuo, C.C.; Lin, C.Y.; Hsieh, C.F. Using neural network based on the shaft orbit feature for online rotating machinery fault diagnosis. In Proceedings of the 2016 International Conference on System Science and Engineering (ICSSE), Puli, Taiwan, 7–9 July 2016; IEEE: Piscataway, NJ, USA, 2016; pp. 1–4.
20. Lin, S.C.; Chang, H.C.; Kuo, C.C.; Hsu, T.C.; Shen, W.C. Assessing Motor Conditions by Using Electrical and Vibrational Detection Methods. In Proceedings of the 2015 Conference on Engineering and Natural Sciences Conference, Tokyo, Japan, 3–5 February 2015; pp. 41–58.
21. Tandon, N.; Choudhury, A. A review of vibration and acoustic measurement methods for the detection of defects in rolling element bearings. *Tribol. Int.* **1999**, *32*, 469–480. [CrossRef]
22. Kıral, Z.; Karagülle, H. Vibration analysis of rolling element bearings with various defects under the action of an unbalanced force. *Mech. Syst. Signal Process.* **2006**, *20*, 1967–1991. [CrossRef]
23. Guo, P.; Fu, J.; Yang, X. Condition Monitoring and Fault Diagnosis of Wind Turbines Gearbox Bearing Temperature Based on Kolmogorov-Smirnov Test and Convolutional Neural Network Model. *Energies* **2018**, *11*, 2248. [CrossRef]
24. Zhuang, Z.Y.; Hocine, A. Meta goal programming approach for solving multi-criteria de Novo programming problem. *Eur. J. Oper. Res.* **2018**, *265*, 228–238. [CrossRef]
25. Zhuang, Z.-Y.; Hocine, A.; Kouaissah, N.; Kiker, G.A. Optimising sustainable renewable energy portfolios using a multi-tolerance fuzzy goal programming approach. *Int. J. Green Energy* **2022**, 1–16. [CrossRef]
26. Ligęza, P. Basic, Advanced, and Sophisticated Approaches to the Current and Forecast Challenges of Wind Energy. *Energies* **2021**, *14*, 8147. [CrossRef]
27. Pandit, R.; Infield, D. Gaussian Process Operational Curves for Wind Turbine Condition Monitoring. *Energies* **2018**, *11*, 1631. [CrossRef]
28. Hossain, M.L.; Abu-Siada, A.; Muyeen, S.M. Methods for Advanced Wind Turbine Condition Monitoring and Early Diagnosis: A Literature Review. *Energies* **2018**, *11*, 1309. [CrossRef]
29. Leahy, K.; Gallagher, C.; O'Donovan, P.; O'Sullivan, D.T.J. Issues with Data Quality for Wind Turbine Condition Monitoring and Reliability Analyses. *Energies* **2019**, *12*, 201. [CrossRef]
30. Sanati, H.; Wood, D.; Sun, Q. Condition Monitoring of Wind Turbine Blades Using Active and Passive Thermography. *Appl. Sci.* **2018**, *8*, 2004. [CrossRef]

buildings

MDPI

Article

# Reliability Analysis of RC Slab-Column Joints under Punching Shear Load Using a Machine Learning-Based Surrogate Model

**Lulu Shen, Yuanxie Shen and Shixue Liang ***

School of Civil Engineering and Architecture, Zhejiang Sci-Tech University, Hangzhou 310018, China
* Correspondence: liangsx@zstu.edu.cn

**Abstract:** Reinforced concrete slab-column structures, despite their advantages such as architectural flexibility and easy construction, are susceptible to punching shear failure. In addition, punching shear failure is a typical brittle failure, which introduces difficulties in assessing the functionality and failure probability of slab-column structures. Therefore, the prediction of punching shear resistance and corresponding reliability analysis are critical issues in the design of reinforced RC slab-column structures. In order to enhance the computational efficiency of the reliability analysis of reinforced concrete (RC) slab-column joints, a database containing 610 experimental data is used for machine learning (ML) modelling. According to the nonlinear mapping between the selected seven input variables and the punching shear resistance of slab-column joints, four ML models, such as artificial neural network (ANN), decision tree (DT), random forest (RF), and extreme gradient boosting (XGBoost) are established. With the assistance of three performance measures, such as root mean squared error (RMSE), mean absolute error (MAE), and coefficient of determination ($R^2$), XGBoost is selected as the best prediction model; its RMSE, MAE, and $R^2$ are 32.43, 19.51, and 0.99, respectively. Such advantages are also reflected in the comparison with the five empirical models introduced in this paper. The prediction process of XGBoost is visualized by SHapley Additive exPlanation (SHAP); the importance sorting and feature dependency plots of the input variables explain the prediction process globally. Furthermore, this paper adopts Monte Carlo simulation with a machine learning-based surrogate model (ML-MCS) to calibrate the reliability of slab-column joints in a real engineering example. A total of 1,000,000 samples were obtained through random sampling, and the reliability index $\beta$ of this practical building was calculated by Monte Carlo simulation. Results demonstrate that the target reliability index requirements under design provisions can be achieved. The sensitivity analysis of stochastic variables was then conducted, and the impact of that analysis on structural reliability was deeply examined.

**Keywords:** reliability analysis; RC slab-column structure; machine learning; Monte Carlo simulation; shapley additive explanation

**Citation:** Shen, L.; Shen, Y.; Liang, S. Reliability Analysis of RC Slab-Column Joints under Punching Shear Load Using a Machine Learning-Based Surrogate Model. *Buildings* **2022**, *12*, 1750. https://doi.org/10.3390/buildings12101750

Academic Editors: Ming-Hung Hsu, Zheng-Yun Zhuang and Ying-Wu Yang

Received: 12 September 2022
Accepted: 18 October 2022
Published: 20 October 2022



## 1. Introduction

Reinforced concrete (RC) slab-column structures comprised of slabs and columns are susceptible to punching shear, because the beams are not arranged for the considerations of structural layout under slabs [1]. Under excessive punching shear loads, the interior slab-column joint is usually destroyed first, the rest of the joints are destroyed in succession, and the progressive collapse of overall structure takes place [2]. Accidents (Figure 1), such as the collapse of a 16-storey apartment building [3] in Boston, US and Skyline Plaza [4] in Virginia, US, have caused severe damage, which arouse the researchers' attention regarding the reliability analysis of RC slab-column joints.

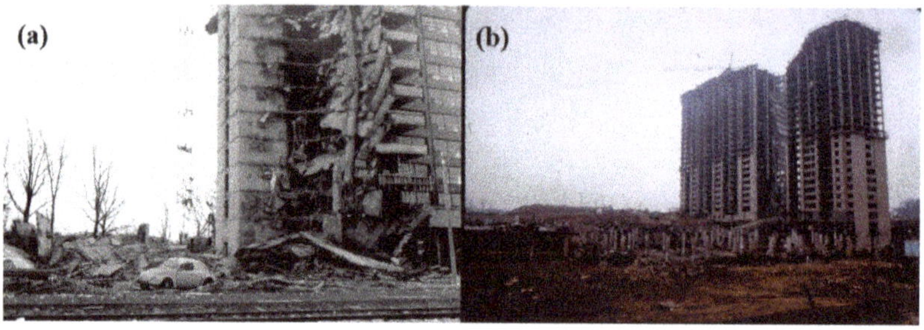


**Figure 1.** Collapse of slab-column structures: (**a**) a 16-storey apartment building, Boston [3]; (**b**) Skyline Plaza, Virginia [4].

To assess the performance of slab-column structures, especially the slab-column joints, a number of experimental studies have been conducted on the punching shear resistance. With the experimental results, some empirical models [5–15] have been proposed based on a variety of mechanical theories. Kinnunen and Nylander [5] analyzed the experimental data of circle slab-circle columns, and created the sector model. Based on this, Broms [6,7] proposed a modified model considering the impact of size effect, which obtained the solution of the ultimate angle of the slabs. Tian et al. [8] proposed a prediction model considering the impact of reinforcement strength ($\rho f_y$). According to the eccentric shear stress model proposed by Stasio et al. [9], an improved model with stronger applicability was proposed by Moe [10], which became the theory basis of both GB 50010-2010 [11] and ACI 318-19 [12]. After analyzing the critical cracks of slab-column joints and considering the impact of aggregate size, the critical shear crack theory (CSCT) was proposed by Muttoni [13]. Based on the modified compression field theory (MCFT), Wu et al. [14] developed a prediction model; its prediction performance was validated by many experimental data. According to the regression analysis of the experimental data, a prediction model was proposed by Chetchotisak et al. [15].

However, the aforementioned mechanical or empirical models possess the problem of prediction precision [16,17]. As a typical data-driven model with advantages such as superior prediction performance and high computational efficiency, machine learning (ML) is applied to many engineering fields successfully [18–25]. In the resistance prediction of slab-column joints, Nguyen et al. [16] established a prediction model using extreme gradient boosting (XGBoost), the performance of which was validated by empirical models and other two ML models. Mangalathu et al. [17] also constructed XGBoost models, and used SHapley Additive exPlanation (SHAP) to illustrate the prediction process of XGBoost. Shen et al. [23] established an ML model to predict the punching shear resistance of fiber-reinforced polymer (FRP) -reinforced concrete slabs, the performance of which was better than that of the compared empirical models. Truong et al. [24] studied the punching shear strength of FRP-RC slab column connections with the assistance of ML models.

The objective of reliability analysis is to evaluate the safety of structures by considering how their performances are affected by the uncertainties, which are introduced by random material properties or stochastic loads [26]. There are two types of methods for reliability analysis, namely the gradient-based method and the simulation-based method [27]. The first method contains the first-order reliability method (FORM), and the second-order reliability method (SORM) aims to find the most likely failure point through the limit state function estimation. Such a method has a high computational efficiency, but it introduces approximations that are sometimes unacceptable from a precision point of view [28]. As the main simulation-based method, the Monte Carlo sampling method is conventional, clear, and easy to use, but such a method requires numerous samples [29,30]. Nassim et al. [31] studied the reliability of two cases by using the response surface method (RSM) as well as Monte Carlo simulation (MCS). Olmati et al. [32] proposed a simplified analysis framework and used MCS to analyze the reliability of an office building. Chetchotisak et al. [15]

studied the structural reliability within two kinds of concrete (normal-strength concrete and high-strength concrete) by using MCS. Ricker et al. [33] utilized three reliability analysis techniques, such as the mean-value first-order second moment method (MVFOSM), the first-order second moment method (FOSM), and MCS, to assess the safety levels of the punching shear resistance of flat slabs without shear reinforcement. However, the relatively low prediction accuracy of the aforementioned mechanical or empirical models led to unsatisfying results of the reliability analysis. To obtain more accurate reliability analysis results, the finite element method (FEM) is popularly applied as the surrogate model of structural response under stochastic material properties or loading conditions [34]. The complexity and nonlinearity existing in structures, as well as the randomness produced by influential factors of a structure itself, prove that FEM becomes a fine choice. However, the mechanical property-based analysis restricts the computational efficiency of FEM, which is inapplicable to practical projects [35]. Furthermore, as the most commonly used parallel analysis method in a stochastic context, MCS has a problem of inadequate computational efficiency, because the number of samples needed for analysis is considerably large [36]. The ML model is a prospective solution for the contradiction between computational efficiency and accuracy, and has been applied in the reliability analyses of RC structures in the latest studies [37].

To the best knowledge of the authors, there is no available example combining reliability analysis of RC slab-column joints and ML; thus, this paper establishes an ML-MCS model for reliability analysis to meet the requirements of practical projects. The candidate ML models selected in this paper are artificial neural network (ANN), decision tree (DT), random forest (RF), and XGBoost. The final prediction model is screened from these four ML models, and the performance comparison between them is implemented through three performance measures: root mean squared error (RMSE), mean absolute error (MAE), and coefficient of determination ($R^2$). To display the advantages of the ML models, two design provisions (GB 50010-2010 [11] and ACI 318-19 [12]), as well as three prediction models proposed by Tian et al. [8], Wu et al. [14], and Chetchotisak et al. [15], are used for prediction performance comparison with ML models. Furthermore, SHAP is introduced for model explanation and analysis of influential factors; the prediction process can be visualized to facilitate the understanding [22]. Based on the established ML model, a slab-column structure in an actual engineering application is used for reliability analysis through MCS. Moreover, the safety assessment of the structure is discussed through sensitivity analysis.

## 2. Punching Shear Resistance Database of RC Slab-Column Joints

The high-fidelity data is the basis of the construction of ML models, so that the compilation of the experimental database is required. The punching shear resistance database containing 610 experimental data is shown in Appendix A, and the statistic information of input variables is listed in Table 1. Some relevant studies [8,14,38] report that there are seven main influential factors affecting slab-column joints: cross-section shape of column ($s$), cross-section area of column ($A$), slab's effective depth ($d$), compressive strength of concrete ($f'_c$), yield strength of reinforcement ($f_y$), reinforcement ratio ($\rho$), and span-depth ratio ($\lambda$). Their distributions are described in four measures: minimum, maximum, standard deviation, and average. The cross-section of each column has three shapes: square ($s$ = 1), circle ($s$ = 2), and rectangle ($s$ = 3). The prediction target of the ML models is the punching shear resistance ($V$) of slab-column joints.

**Table 1.** Statistic information of input variables.

| Parameter | Minimum | Maximum | Standard Deviation | Average | Type |
|---|---|---|---|---|---|
| $s$ | 1.00 | 3.00 | 0.58 | 1.40 | Input |
| $A$ (cm$^2$) | 20.43 | 6375.87 | 596.68 | 489.31 | Input |
| $d$ (mm) | 29.97 | 668.50 | 58.52 | 113.74 | Input |
| $f'_c$ (Mpa) | 9.40 | 130.10 | 18.56 | 35.39 | Input |
| $f_y$ (Mpa) | 234.70 | 749.00 | 115.83 | 456.60 | Input |
| $\rho$ (%) | 0.25 | 7.31 | 0.70 | 1.26 | Input |
| $\lambda$ | 0.61 | 32.51 | 4.83 | 6.59 | Input |
| $V$ (kN) | 24.00 | 4915.00 | 406.56 | 403.25 | Output |

The histograms displayed in Figure 2 show the relative frequency distributions of the input variables and the output, and the red lines represent the cumulative distribution functions (CDF) of the parameters. To further understand the correlations between the input variables, they are quantified as a Pearson correlation coefficient matrixand shown in Figure 3, where coefficients represent the degree of linear correlation between input variables [39]. The coefficients close to −1 or 1 represent the obvious negative or positive linear correlation, and the degree of linear correlation between $A$ and $d$ is highest.

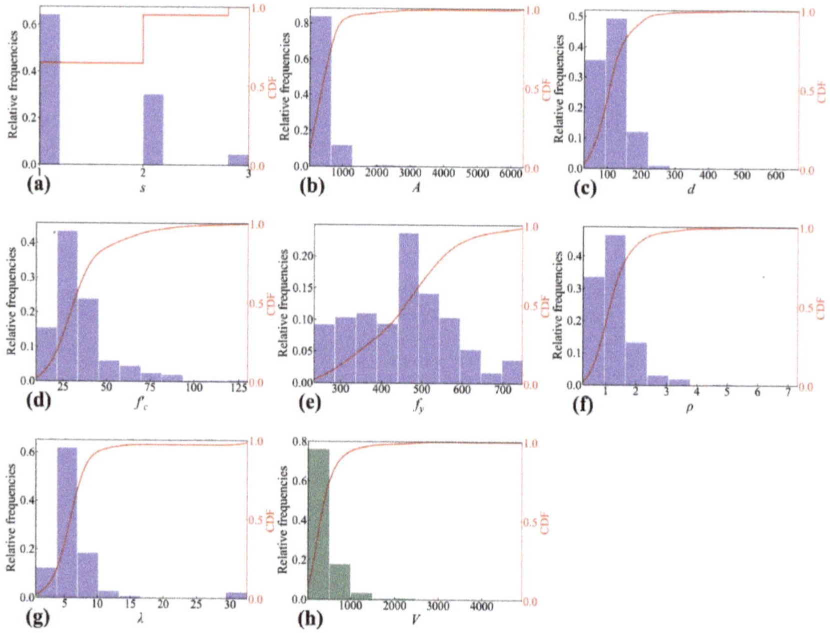

**Figure 2.** Distributions of the parameters in the database: **(a)** s; **(b)** $A$; **(c)** $d$; **(d)** $f'_c$; **(e)** $f_y$; **(f)** $\rho$; **(g)** $\lambda$; **(h)** $V$.

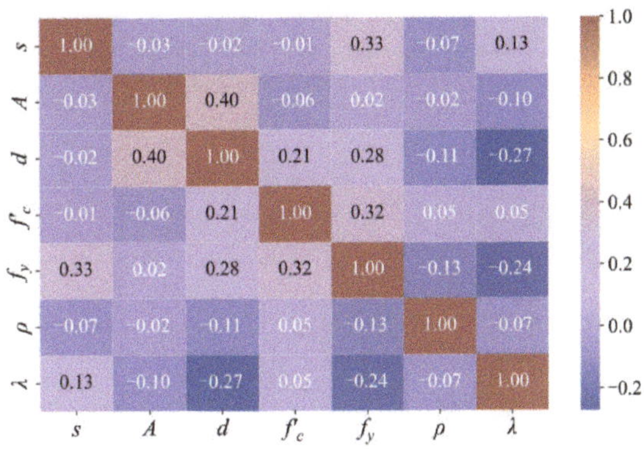

Figure 3. Correlation coefficient matrix of input variables.

## 3. Machine Learning Model for Punching Shear Resistance Prediction

The flow of establishment of an ML model is shown in Figure 4, which can be generalized as the following steps [40]: (1) Divide the compiled database as a training set (containing 500 data) and test set (containing 110 data) based on the ratio of 80% and 20%; (2) obtain the optimal hyperparameters by model training; (3) examine the generalization ability of the candidate model by the test set; (4) output the final prediction model. The four ML models selected in this paper are all established following this procedure, and the related introductions for models are displayed in Section 3.1.

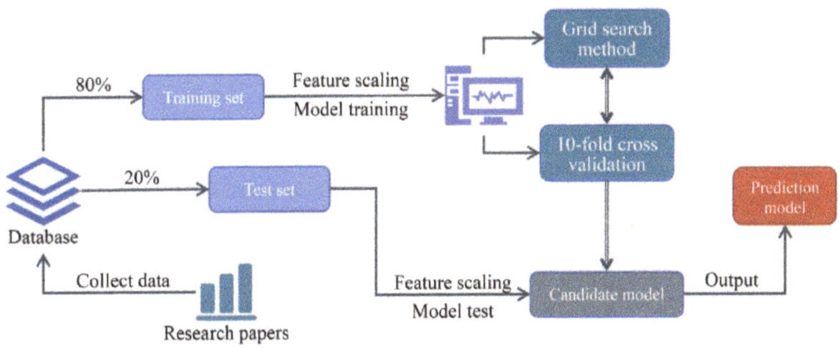


Figure 4. Flowchart of ML modelling.

### 3.1. Overview of Machine Learning Models

As the basic ML algorithms, ANN and DT have been widely studied and thus become the beginning of two types of artificial intelligence algorithms: deep learning and ensemble learning [41]. Among ensemble learning algorithms, RF and XGBoost are two representative algorithms constructed by different ensemble tactics such as bagging and boosting [40]. Due to the four typical ML models possessing different fitting techniques, the comparison of them enhances the credibility of the final prediction model.

#### Extreme Gradient Boosting

Extreme gradient boosting (XGBoost) inherits the combination strategy of gradient boosting decision tree (GBDT) and becomes the advanced implementation of the latter [42]. The employ of two regularization coefficients and the optimization of the second-order Taylor approximation guarantees not only the generalization ability, but also the prediction accuracy. The complexity of each base learner can be defined as:

$$\Omega(f_t) = \gamma T + \frac{1}{2}\lambda' \|w'\|^2 \quad (1)$$

where $\gamma$ and $\lambda'$ are the L1 and L2 regularization coefficients; $T$ is the number of base learners; $w'$ is the score of the node. Based on the fitting of the residuum of the prediction result, the prediction error of XGBoost can be further decreased. The fitting objective of each base learner can be formulated as:

$$\mathcal{L}(\varphi) = \sum_i l\left(y_{\text{pred}}^{(i)}, y^{(i)}\right) + \sum_k \Omega(f_k) \tag{2}$$

where $l$ is the loss function; $y_{\text{pred}}$ is the prediction value of the sample; $y$ is the true value of the sample. Based on these, a prediction value generated by XGBoost can be expressed as the linear addition of the prediction values of all the base learners:

$$y_{\text{pred}} = \sum_{k=1}^{K} f_k \tag{3}$$

where $K$ is the number of base learners.

### 3.2. Prediction Results of Machine Learning Models

The optimal hyperparameters of each ML model are obtained through the grid search method and through 10-fold cross-validation [43], which are listed in Table 2. To compare the prediction performances of different ML models, three performance measures, root mean squared error (RMSE), mean absolute error (MAE), and coefficient of determination ($R^2$), are adopted and expressed as:

$$\text{RMSE} = \sqrt{\frac{1}{m}\sum_{i=1}^{m}\left(y_{\text{pred}}^{(i)} - y^{(i)}\right)^2} \tag{4}$$

$$\text{MAE} = \frac{1}{m}\sum_{i=1}^{m}\left|y_{\text{pred}}^{(i)} - y^{(i)}\right| \tag{5}$$

$$R^2 = 1 - \frac{\sum_{i=1}^{m}\left(y_{\text{pred}}^{(i)} - y^{(i)}\right)^2}{\sum_{i=1}^{m}\left(y^{(i)} - \frac{1}{m}\sum_{i=1}^{m} y^{(i)}\right)^2} \tag{6}$$

where $m$ is the number of samples.

**Table 2.** Optimal hyperparameters of ML models.

| ML Model | Optimal Hyper-Parameter |
|---|---|
| ANN | Learning rate = 0.1, neurons number of hidden layer = 17, maximum iteration = 2000 |
| DT | Maximum depth = 8 |
| RF | Number of weak learner = 100, maximum depth = 14 |
| XGBoost | Number of weak learners = 100, learning rate = 0.5, maximum depth = 3, $\gamma$ = 0.9, $\lambda'$ = 1.4 |

After the determination of the optimal hyper-parameters, four ML models are all established. To examine the prediction performance of ML models, five empirical models containing two design provisions [11,12], two mechanical models [8,14], and a regression analysis-based model [15] are introduced and listed in Table 3. Their prediction results are shown in Figure 5, where gray-green and blue-pink represent the prediction results of empirical models and ML models in the training set and the test set, respectively. XGBoost has the highest prediction accuracy, which indicates that XGBoost has been well-trained and possesses the best generalization ability. Such a conclusion is also in line with that of some studies [17,44]. RF and DT also have good prediction performance; their prediction tactics are suitable for the regression analysis of the punching shear resistance of RC slab-column joints [45]. However, the prediction performance of ANN must be improved; its characteristic of a nonconvex function suggests that the obtained optimal solution is often local rather than global [18]. Utilizing the good fitting ability of the regression analysis method, the prediction model proposed by Chetchotisak et al. [15] has the best prediction result, but its credibility is low due to its lack of theoretical derivation. The prediction values of mechanical models proposed by Tian et al. [8] and Wu et al. [14] have a large deviation with true punching shear resistance, where coefficients reflected the relationships between influential factors and where punching shear resistance must be further modified. Furthermore, the prediction results of design provisions such as GB 50010-2010 [11] and ACI 318-19 [12] skew conservative; the prediction accuracy must be improved.

**Table 3.** Empirical models used for prediction performance comparison.

| Empirical Model | Punching Shear Resistance Calculation Equation |
|---|---|
| GB 50010-2010 [11] | $V_1 = 0.7\beta_h f_t \eta b_{0,0.5d} d; \eta = \min\begin{cases} \eta_1 = 0.4 + \frac{1.2}{\beta_s} \\ \eta_2 = 0.5 + \frac{\alpha_s d}{4b_{0,0.5d}} \end{cases}$ |
| ACI 318-19 [12] | $V_2 = \min\left[\frac{1}{3}, \frac{1}{6}\left(1 + \frac{2}{\beta_s}\right), \frac{1}{12}\left(2 + \frac{\alpha_s d}{b_{0,0.5d}}\right)\right]\lambda_s \sqrt{f'_c} b_{0,0.5d} d; \lambda_s = \sqrt{\frac{2}{1+0.004d}} \leq 1$ |
| Tian et al. [8] | $V_3 = 0.65\xi A_c \left(\rho f_y \sqrt{f'_c}\right)^{\frac{1}{2}}; \xi = \sqrt{\frac{d}{c}}; A_c = 4(c+d)d$ |
| Wu et al. [14] | $V_4 = 0.00040(\rho)^{\frac{1}{5}} b_{0,2d} dL\sqrt{f'_c} / \left(0.31 + \frac{24\omega}{a_d+16}\right); \omega = 0.0005\frac{0.9d}{\sin\theta}; a_d = 20; \theta = 45^\circ$ |
| Chetchotisak et al. [15] | $V_5 = 92.43(f'_c)^{1.21}\left(\frac{1}{100\rho}\right)^{1.47}(b_{0,0.5d})^{0.42} d^{1.35} k^{4.66}; k = \sqrt{(n\rho)^2 + 2(n\rho)} - (n\rho);$ <br> $n = E_s/E_c = 2 \times 10^5/4700\sqrt{f'_c}$ |

$\beta_h$ is the sectional depth influence coefficient; $f_t$ is the design value of the tensile strength of concrete; $b_{0,0.5d}$ is the critical section perimeter at a distance of 0.5$d$ away from the column; $\beta_s$ is the ratio of the long side to the short side of the column; $\alpha_s$ is the influential coefficient of the column type (40 for interior columns); $c$ is the column size; $b_{0,2d}$ is the critical section perimeter at a distance of 2$d$ away from the column; $L$ is the perimeter of the column.

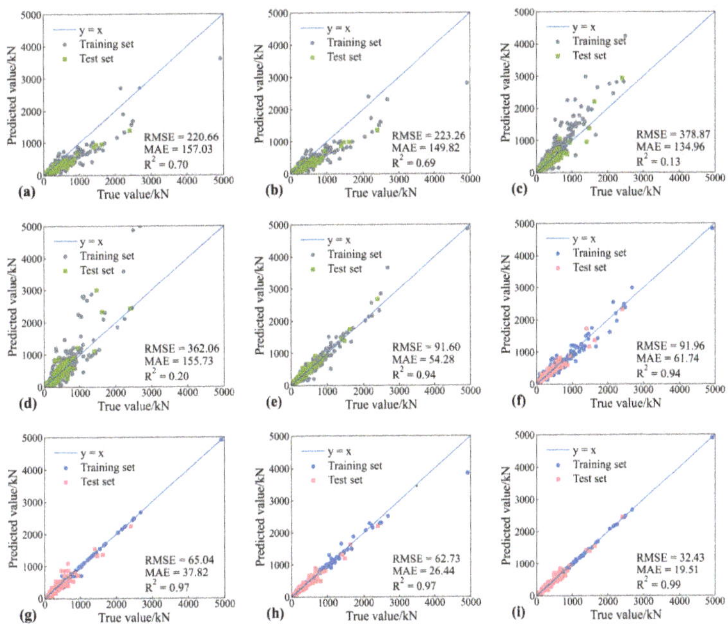


**Figure 5.** Scatter plots of the prediction results of the empirical models and the ML models: (**a**) GB 50010-2010; (**b**) ACI 318-19; (**c**) Tian et al.; (**d**) Wu et al.; (**e**) Chetchotisak et al.; (**f**) ANN; (**g**) DT; (**h**) RF; (**i**) XGBoost.

### 3.3. Interpretation of the ML Prediction Model

According to the performance comparison of the ML models in Section 3.2, XGBoost can be regarded as the final prediction model with the best prediction performance. The feature importance sorting produced by the built-in method of XGBoost [46], as shown in Figure 6, $d$ has the greatest influence on punching shear resistance. However, this method can only provide the importance of influential factors; the effect tendency is unknown yet. Therefore, SHAP is introduced in this paper and utilized for model interpretation.

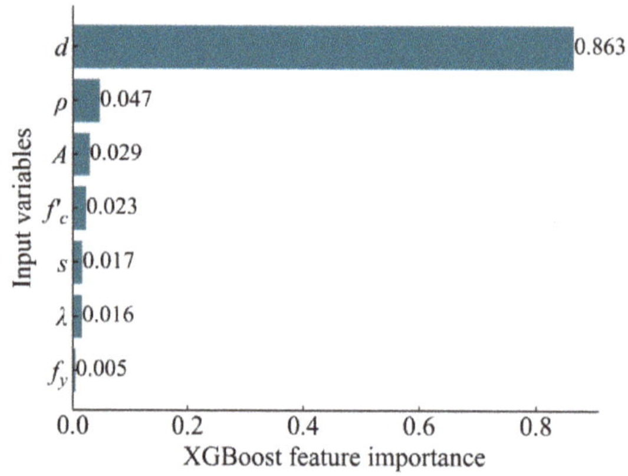


**Figure 6.** Importance sorting using XGBoost feature importance.

#### 3.3.1. Overview of Shapley Additive Explanation

SHapley Additive exPlanation (SHAP) is useful for illustrating the prediction process of any ML model; it originates from the game theory and was proposed by Lundberg et al. [47,48]. For each prediction value, it can be formulated as the linear addition of the baseline value $y_{\text{base}}$ and the SHAP value of each feature $f(x)$:

$$y_{\text{pred}}^{(i)} = y_{\text{base}} + \sum_{j=1}^{n} f(x_{ij}) \tag{7}$$

where $n$ is the number of features. The quantified contribution of the feature is calculated through:

$$f(x_{ij}) = \sum_{S \subseteq N \setminus \{j\}} \frac{|S|!(M-|S|-1)!}{M!}[f_x(S \cup \{j\}) - f_x(S)] \tag{8}$$

where $N$ is the $M$-dimensional set containing all of the features; $S$ is the $|S|$-dimensional subset extracted from $N$; $f_x(S\cup\{j\})$ is the prediction calculated through set $S$ and feature $j$; $f_x(S)$ is the prediction calculated through set $S$.

#### 3.3.2. Model Interpretation Using Shapley Additive Explanation

The importance sorting provided by SHAP is shown in Figure 7, which is calculated by sum of the SHAP values of each sample. The feature importance sorting provided by SHAP is similar to that provided by XGBoost, but they conflict on the impact of $s$. Figure 7b shows the impact of each feature on punching shear resistance as positive or negative, and a feature can be regarded as the positive influential factor if the color of dot transforms from blue to red with the increase of the SHAP value. It can be seen that $d$, $\rho$, $A$, $f'_c$, $f_y$, and $s$ have positive impacts on resistance, and $\lambda$ has a negative impact on resistance, which is consistent with some experimental studies [49–52]. Based on the importance sorting shown in Figure 7, the global impact of each influential factor is revealed, i.e., SHAP explains the global prediction process of XGBoost.

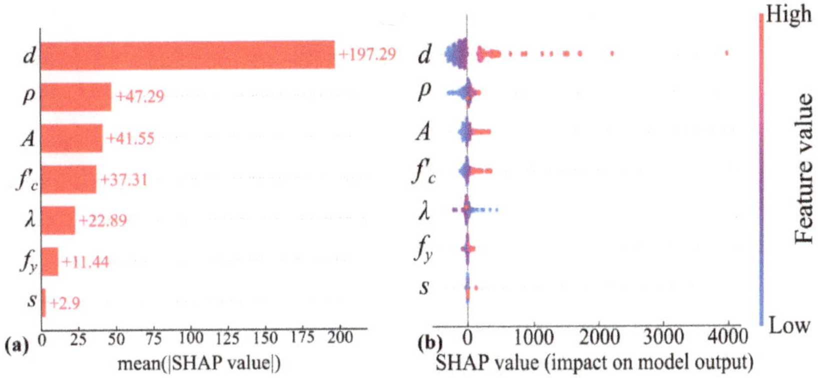

**Figure 7.** Global interpretation of punching shear resistance: (**a**) feature importance sorting; (**b**) SHAP value summary plot.

Figure 8 provides further insight for the impact of influential factors in the form of dependency plots, where the secondary axis represents the input variable interacting most frequently with the variable displayed in the x-axis. According to the variation range of the SHAP values, $d$ and $s$ have the greatest and least impacts, respectively, on punching shear resistance, which is consistent with the findings expressed by Figure 7a. Furthermore, the interaction between input variables is too complicated, such that the simple linear relationship cannot be used to represent it.

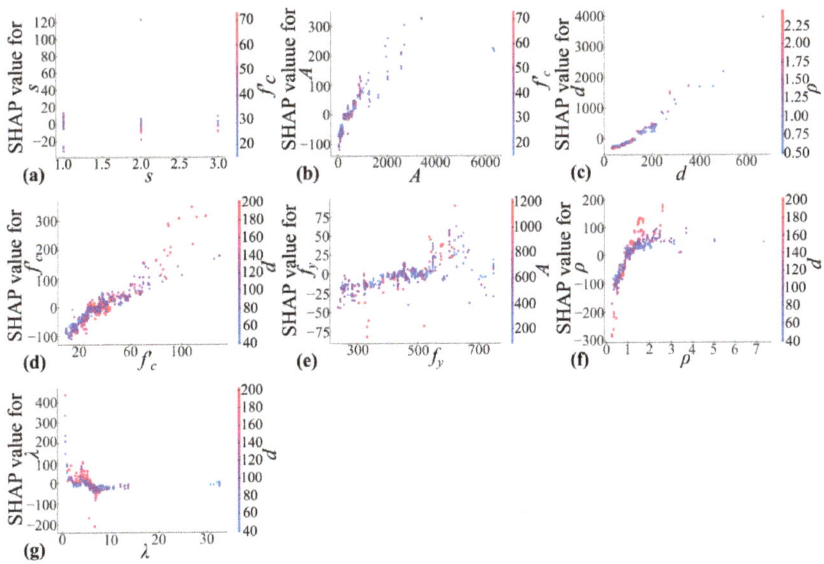

**Figure 8.** Feature dependencies of influential factors: (**a**) $s$; (**b**) $A$; (**c**) $d$; (**d**) $f'_c$; (**e**) $f_y$; (**f**) $\rho$; (**g**) $\lambda$.

## 4. Reliability Analysis: RC Slab-Column Joint of an Office Building

The prototype building used for reliability analysis is a 7-story, 5-span RC slab-column shear wall office building [53], and it was designed using GB 50010-2010 [11] and GB 50011-2010 [54]. The building itself contains 3 m storey height and is supported by a 7.5 m × 7.5 m column grid, and the interior joint shown in Figure 9 is selected as the study object. The selected joint consists of a slab with an effective depth of 209 mm and a square column with side length of 530 mm, which is subjected to the specified dead load of 7.0 $kN/m^2$ and live load of 2.0 $kN/m^2$. According to the requirement of GB 50068-2018 [55], the dead load and live load used for limit state design must be adjusted by multiplying the partial safety factors for the load, such as 1.3 and 1.5. Therefore, the limit state function Z of structure can be defined as:

$$Z = R - 1.3S_G - 1.5S_Q \tag{9}$$

where $R$ is the punching shear resistance; $S_G$ is the dead load; $S_Q$ is the live load. Furthermore, the measured compressive strength of C50 concrete in the slab is 39.31 Mpa, and the measured yield strength of HRB400 reinforcement is 421 Mpa. The reinforcement ratio of the joint is 0.81%, and the main influential factors are listed in Table 4.

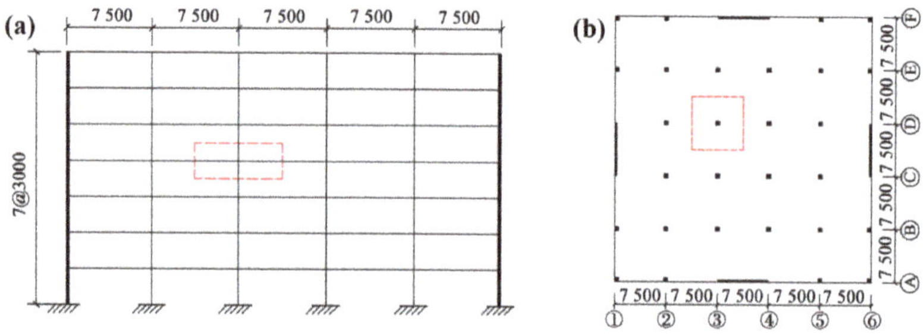

**Figure 9.** Prototype building [53]: (**a**) elevation; (**b**) plan.

**Table 4.** Main influential factors of the selected interior joint.

| $s$ | $A/cm^2$ | $d$/mm | $f'_c$/Mpa | $f_y$/Mpa | $\rho$/% | $\lambda$ |
|---|---|---|---|---|---|---|
| 1 | 2809 | 209 | 39.31 | 421 | 0.81 | 16.67 |

The statistic information and suitable probability density functions of the stochastic variables used for reliability analysis are listed in Table 5 [33,56], where COV is the coefficient of variance. According to the study conducted by Chojaczyk et al. [27], the COV of failure probability $P_f$ calculated by MCS is accepted when its value is around 0.1; then the $P_f$ around $10^{-4}$ (the normal failure probability of an existing structure) can be calculated through the simulation based on $N$ samples [57]:

$$\text{COV}\left(P_f\right) = \frac{1}{P_f}\sqrt{\frac{\left(1-P_f\right)P_f}{N}} \tag{10}$$

where $N$ signifying 1,000,000 can be determined according to the aforementioned conditions. Another method used in the study conducted by Hadianfard et al. [58] stipulates that the number of samples needed for MCS can be determined through:

$$N > \frac{-\ln(1-C)}{P_f} \tag{11}$$

where C is the confidence level, with values of 0.95 in this paper. Equation (11) suggests that the number of samples should not be less than 30,000, so that the value range calculated by Equations (10) and (11) is determined between 30,000 and 1,000,000. In this range, the variation of COV of failure probability $P_f$ within 10 simulations is shown in Figure 10. The COV of failure probability decreases with the increase of the sample size from 30,000 to 1,000,000, which means that the result of the reliability analysis increasingly stabilizes.

Based on this knowledge, 1,000,000 samples are produced randomly and conduced for reliability analysis by XGBoost and MCS.

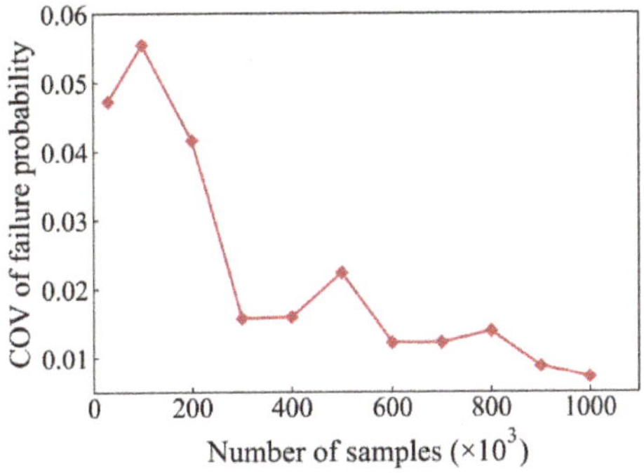


**Figure 10.** Effect of sample size on COV of failure probability.

**Table 5.** Stochastic variables used for Monte Carlo simulation.

| Parameter | Average | Standard Deviation | COV | Distribution |
|---|---|---|---|---|
| $d$: slab's effective depth (mm) | 209 | 6.27 | 0.03 | Gaussian |
| $f'_c$: compressive strength of concrete (Mpa) | 39.31 | 4.32 | 0.11 | Gaussian |
| $f_y$: yield strength of reinforcement (Mpa) | 421 | 33.68 | 0.08 | Gaussian |
| $S_G$: dead load (kN) | 393.75 | 27.56 | 0.07 | Gaussian |
| $S_Q$: live load (kN) | 112.5 | 32.4 | 0.288 | Gumbel |

### *4.1. Results of Structural Reliability Analysis*

The efficient implementation of Monte Carlo simulation (MCS) is restricted by the sample size and the computational efficiency of the surrogate model [59], but this can be solved by XGBoost. The average computation time for 1,000,000 samples and the reliability analysis of the slab-column joint is 30 s. This is done by a laptop with four-core CPU and 8 GB memory, which demonstrates the efficiency of ML-MCS. Based on the regression prediction of punching shear resistance, the distribution and CDF of structural resistance are shown in Figure 11. The average and standard deviations of the distribution of structural resistance are 955.96 kN and 52.42 kN, respectively. MCS can estimate the failure probability of a structure effectively by calculating the probability of $Z < 0$ in Equation (9), and the related reliability analysis can also be realized. Table 6 displays the result of reliability analysis, where $P_f$ is the failure probability of the structure; $\beta$ is the reliability index; $\alpha_R$ and $\alpha_S$ are the sensitivity coefficients of resistance and load; $r^*$ and $s^*$ are the coordinates of the design point. The reliability index $\beta$ indicates that the reliability and safety of the selected interior joint are good and meet the requirement of GB 50068-2018 [55].

**Table 6.** Results of reliability analysis.

| $P_f$ | $\beta$ | $\alpha_R$ | $\alpha_S$ | $r^*$ | $s^*$ |
|---|---|---|---|---|---|
| 0.00546 | 3.443 | −0.655 | 0.755 | 837.625 | 837.625 |

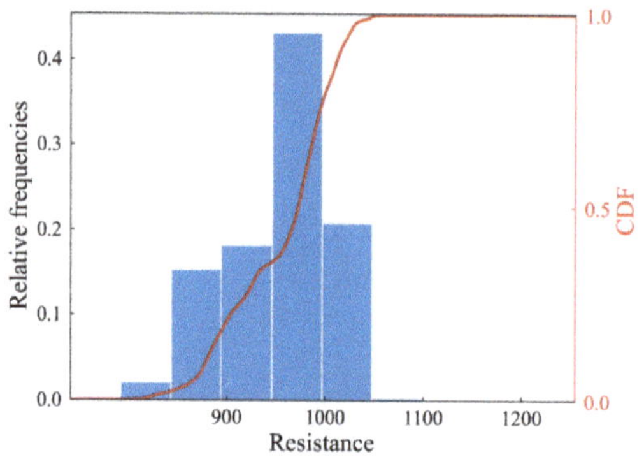


**Figure 11.** Distribution of structural resistance.

### 4.2. Sensitivity Analysis

The relationship between structural reliability and stochastic variables can be examined through sensitivity analysis [60]. The reliability index of the structure in the other two stochastic contexts (the stochastic structural resistance and stochastic loading condition) is shown in Figure 12. The reduction of the randomness of structural resistance or loading conditions can improve the reliability index, and the safety and stability of the structure also can be enhanced.

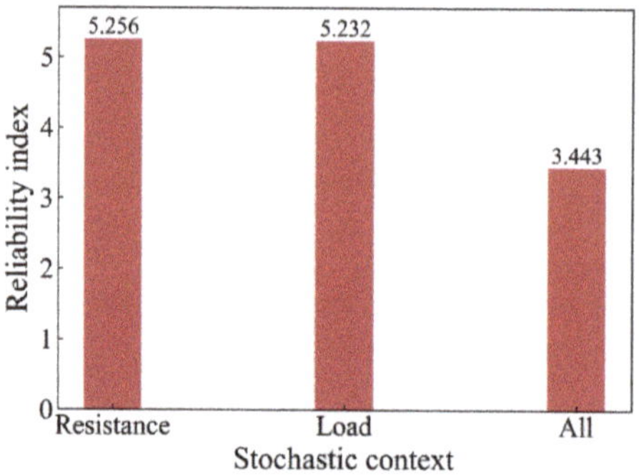


**Figure 12.** The reliability index in different stochastic contexts.

To study the impact of stochastic variables such as $d, f'_c, f_y$, and $S_Q$, their multiples are changed and their relationship with the reliability index is shown in Figure 13. The reliability index can be improved effectively with the increase of $f'_c$ and $f_y$ or the decrease of $S_Q$. However, there exists a complex relationship between slab depth $d$ and reliability index; the reliability index is reduced when the multiple of $d$ is between 1.15 and 1.35. The distribution of structural resistance with 1.3$d$ is shown in Figure 14, which can be used to understand the reason for the reduction of the reliability index. The discontinuous distribution of structural resistance is existed, and the transition of failure modes from flexure to punching shear may exist, both through experimental and theoretical observations [22,61–63]. Therefore, the standard deviation of structural resistance is large, and the reliability index calculated by that is small.

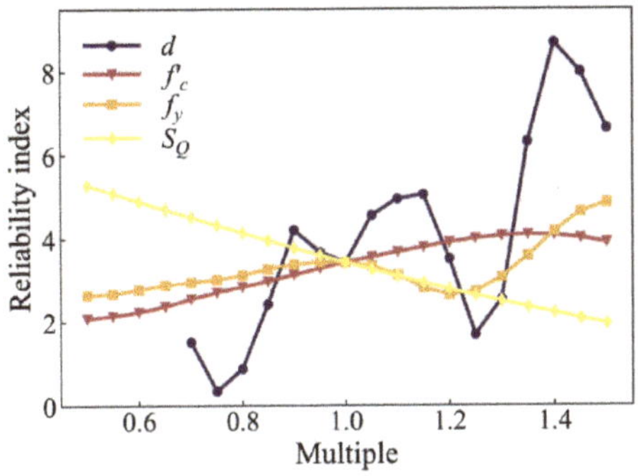


**Figure 13.** Impact of stochastic variables with different multiples on reliability index.

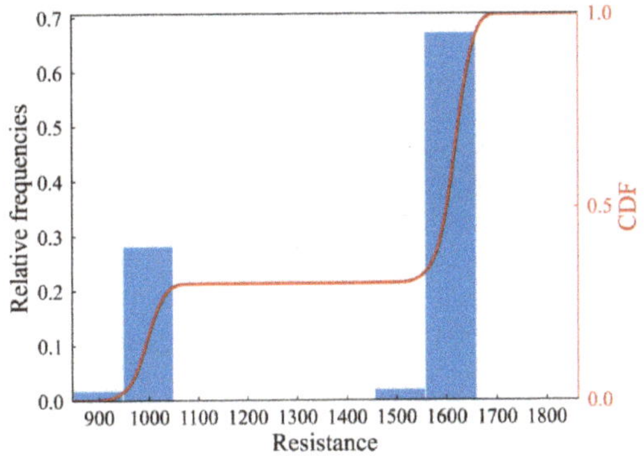


**Figure 14.** Distribution of structural resistance with 1.3d.

## 5. Conclusions

Structural reliability reflects the safety and stability of the entire practical structure subjected to permanent action and variable action [36], the calculation of which, through MCS, is restricted by the computational efficiency of the surrogate model. This work presents a framework for integrating the machine learning-based surrogate model into a Monte Carlo simulation to perform the reliability analysis with a satisfying accuracy and efficiency. An ML model is established and screened from four candidate ML models: as ANN, DT, RF, and XGBoost; the prediction performances of these are examined through three performance measures such as RMSE, MAE, and $R^2$. Furthermore, the advantages of ML models are embodied by comparison with five empirical models. The final prediction model is used as the surrogate model of MCS, and an RC slab-column joint in an actual structure is introduced as the object of reliability analysis. The following conclusions can be drawn from this paper:

The punching shear resistance of RC slab-column joints is influenced mainly by seven influential factors: $s$, $A$, $d$, $f'_c$, $f_y$, $\rho$, and $\lambda$ [38]. The capture of the mapping relationship between them can guarantee the construction of the ML model. With the help of the grid search method and 10-fold cross validation, four ML models with optimal hyperparameters are established. After comparison, XGBoost has the best prediction performance reflected in RMSE, MAE, and $R^2$, and is selected as the final prediction model and used for reliability analysis.

To facilitate the understanding of the prediction process of ML, SHAP is utilized to quantify the contribution of input variables to punching shear resistance, and to visualize the prediction process. According to the importance sorting of input variables, $d$ and $s$ have the greatest and least impacts, respectively, on punching shear resistance. Furthermore, feature dependency plots display the specific impact of each input variable by marginalizing the impacts of other variables. The analysis of the influential factors provides not only the understanding of prediction process, but also the suitable optimization sorting in structural design.

The actual structure adopted for the case study is an RC slab-column shear wall office building. The punching shear resistance of 1,000,000 samples produced by random sampling is calculated through XGBoost. The reliability analysis of the interior joint selected from the prototype building is conducted through MCS, and the final reliability index $\beta$ meets the requirement of the design provisions of GB 50068-2018 [55]. Moreover, the sensitivity analysis reveals the impact of the stochastic context and the values of stochastic variables on structural reliability. Based on these, the computational efficiency of the reliability analysis of the slab-column joints can be enhanced on the premise of high computational accuracy. In future reliability analysis, some advanced sampling methods, such as Latin hypercube sampling and importance sampling, can be used to reduce the number of simulations appropriately. Furthermore, a program with some input windows of influential factors can be designed as a practical tool for reliability analysis.

**Author Contributions:** Conceptualization, L.S. and S.L.; software, L.S. and Y.S.; validation, L.S., Y.S. and S.L.; formal analysis, L.S.; writing—original draft preparation, Y.S.; writing—review and editing, S.L.; visualization, Y.S.; supervision, L.S. and S.L.; project administration, S.L. All authors have read and agreed to the published version of the manuscript.

**Funding:** This research was funded by Science Foundation of Zhejiang Province of China, grant number LY22E080016; National Science Foundation of China, grant number 51808499; Science Foundation of Zhejiang Sci-Tech University (ZSTU), grant number 19052460-Y; and the Education of Zhejiang Province, grant number 20050061-F.

**Institutional Review Board Statement:** Not applicable.

**Informed Consent Statement:** Not applicable.

**Data Availability Statement:** Not applicable.

**Acknowledgments:** This study is supported by the Engineering Research Centre of Precast Concrete of Zhejiang Province. The help of all members of the Engineering Research Centre is sincerely appreciated. We would also like to express our sincere appreciation to the anonymous referee for valuable suggestions and corrections.

**Conflicts of Interest:** The authors declare no conflict of interest.

## Appendix A

To facilitate the acquirement of data, the entire database has been uploaded to GitHub: https://github.com/shenyx0126/Database-used-for-reliability-analysis.git (accessed on 3 October 2022).

## References

1. Kotsovos, M.D. Design for Punching of Flat Slabs. In *Compressive Force-Path Method*, 1st ed.; Springer: Berlin, Germany, 2014; Volume 5, pp. 83–107.
2. Yang, Y.Z.; Diao, M.Z.; Li, Y.; Guan, H.; Lu, X.Z. Post-punching failure mechanism and resistance of flat plate-column joints with in-plane constraints. *Eng. Fail. Anal.* **2022**, *138*, 106360. [CrossRef]
3. King, S.; Delatte, N.J. Collapse of 2000 Commonwealth Avenue: Punching Shear Case Study. *J. Perform. Constr. Facil.* **2004**, *18*, 54–61. [CrossRef]
4. Schellhammer, J.; Delatte, N.J.; Bosela, P.A. Another Look at the Collapse of Skyline Plaza at Bailey's Crossroads, Virginia. *J. Perform. Constr. Facil.* **2013**, *27*, 354–361. [CrossRef]
5. Kinnunen, S.; Nylander, H. *Punching of Concrete Slabs without Shear Reinforcement*; KTH Royal Institute of Technology: Stockholm, Sweden, 1960.
6. Broms, C.E. Elimination of flat plate punching failure mode. *ACI Struct. J.* **2000**, *97*, 94–101.
7. Broms, C.E. Concrete Flat Slabs and Footings-Design Method for Punching and Detailing for Ductility. Ph.D. Thesis, KTH Royal Institute of Technology, Stockholm, Sweden, 2005.

8. Tian, Y.; Jirsa, J.O.; Bayrak, O. Strength Evaluation of Interior Slab-Column Connections. *ACI Struct. J.* **2008**, *105*, 692–700.
9. Stasio, D.; Buren, M.R.V. Transfer of Bending Moment between Flat Plate Floor and Column. *ACI J. Proc.* **1960**, *57*, 299–314.
10. Moe, J. *Shearing Strength of Reinforced Concrete Slabs and Footings under Concentrated Loads*; Portland Cement Association: Portland, Oregon, 1961.
11. MOHURD (Ministry of Housing and Urban-Rural Development of the People's Republic of China). *GB 50010—2010 Code for Design of Concrete Structures*; China Architecture & Building Press: Beijing, China, 2015; pp. 232–236.
12. ACI (American Concrete Institute). *Building Code Requirements for Structure Concrete (ACI 318-19) and Commentary*; American Concrete Institute: Farmington Hills, MI, USA, 2019.
13. Muttoni, A. Punching shear strength of reinforced concrete slabs without transverse reinforcement. *ACI Struct. J.* **2008**, *105*, 440–450.
14. Wu, L.F.; Huang, T.C.; Tong, Y.L.; Liang, S.X. A Modified Compression Field Theory Based Analytical Model of RC Slab-Column Joint without Punching Shear Reinforcement. *Buildings* **2022**, *12*, 226. [CrossRef]
15. Chetchotisak, P.; Ruengpim, P.; Chetchotsak, D.; Yindeesuk, S. Punching Shear Strengths of RC Slab-Column Connections: Prediction and Reliability. *KSCE J. Civ. Eng.* **2018**, *22*, 3066–3076. [CrossRef]
16. Nguyen, H.D.; Truong, G.T.; Shin, M. Development of extreme gradient boosting model for prediction of punching shear resistance of r/c interior slabs. *Eng. Struct.* **2021**, *235*, 112067. [CrossRef]
17. Mangalathu, S.; Shin, H.; Choi, E.; Jeon, J.S. Explainable machine learning models for punching shear strength estimation of flat slabs without transverse reinforcement. *J. Build. Eng.* **2021**, *39*, 102300. [CrossRef]
18. Chen, H.G.; Li, X.; Wu, Y.Q.; Zuo, L.; Lu, M.J.; Zhou, Y.S. Compressive Strength Prediction of High-Strength Concrete Using Long Short-Term Memory and Machine Learning Algorithms. *Buildings* **2022**, *12*, 302. [CrossRef]
19. Barkhordari, M.S.; Armaghani, D.J.; Mohammed, A.S.; Ulrikh, D.V. Data-Driven Compressive Strength Prediction of Fly Ash Concrete Using Ensemble Learner Algorithms. *Buildings* **2022**, *12*, 132. [CrossRef]
20. Shatnawi, A.; Alkassar, H.M.; Al-Abdaly, N.M.; Al-Hamdany, E.A.; Bernardo, L.F.A.; Imran, H. Shear Strength Prediction of Slender Steel Fiber Reinforced Concrete Beams Using a Gradient Boosting Regression Tree Method. *Buildings* **2022**, *12*, 550. [CrossRef]
21. Jiang, Y.M.; Li, H.Y.; Zhou, Y.S. Compressive Strength Prediction of Fly Ash Concrete Using Machine Learning Techniques. *Buildings* **2022**, *12*, 690.
22. Shen, Y.X.; Wu, L.F.; Liang, S.X. Explainable machine learning-based model for failure mode identification of RC flat slabs without transverse reinforcement. *Eng. Fail. Anal.* **2022**, *141*, 106647. [CrossRef]
23. Shen, Y.X.; Sun, J.H.; Liang, S.X. Interpretable Machine Learning Models for Punching Shear Strength Estimation of FRP Reinforced Concrete Slabs. *Crystals* **2022**, *12*, 259. [CrossRef]
24. Truong, G.T.; Hwang, H.J.; Kim, C.S. Assessment of punching shear strength of FRP-RC slab-column connections using machine learning algorithms. *Eng. Struct.* **2022**, *255*, 113898. [CrossRef]
25. Fu, B.; Feng, D.C. A machine learning-based time-dependent shear strength model for corroded reinforced concrete beams. *J. Build. Eng.* **2021**, *36*, 102118. [CrossRef]
26. Feng, D.C.; Ren, X.D.; Li, J. Stochastic damage hysteretic model for concrete based on micromechanical approach. *Int. J. Non-Liner Mech.* **2016**, *83*, 15–25. [CrossRef]
27. Chojaczyk, A.A.; Teixeira, A.P.; Neves, L.C.; Cardoso, J.B.; Soares, C.G. Review and application of Artificial Neural Networks models in reliability analysis of steel structures. *Struct. Saf.* **2015**, *52*, 78–89. [CrossRef]
28. Huang, B.Q.; Du, X.P. Probabilistic uncertainty analysis by mean-value first order Saddlepoint Approximation. *Reliab. Eng. Syst. Saf.* **2008**, *93*, 325–336. [CrossRef]
29. Feng, D.C.; Xie, S.C.; Li, Y.; Jin, L. Time-dependent reliability-based redundancy assessment of deteriorated RC structures against progressive collapse considering corrosion effect. *Struct. Saf.* **2021**, *89*, 102061. [CrossRef]
30. Xu, J.; Feng, D.C. Stochastic dynamic response analysis and reliability assessment of non-linear structures under fully non-stationary ground motions. *Struct. Saf.* **2019**, *79*, 94–106. [CrossRef]
31. Nassim, K.; Bouafia, Y.; Khalil, B. Reliability and punching shear resistance of slabs in non linear domain. *Gradevinar* **2016**, *67*, 1051–1062.
32. Olmati, P.; Sagaseta, J.; Cormie, D.; Jones, A.E.K. Simplified reliability analysis of punching in reinforced concrete flat slab buildings under accidental actions. *Eng. Struct.* **2017**, *130*, 83–98. [CrossRef]
33. Ricker, M.; Feiri, T.; Nille-Hauf, K.; Adam, V.; Hegger, J. Enhanced reliability assessment of punching shear resistance models for flat slabs without shear reinforcement. *Eng. Struct.* **2021**, *226*, 111319. [CrossRef]
34. Qiu, Z.P.; Huang, R.; Wang, X.J.; Qi, W.C. Structural reliability analysis and reliability-based design optimization: Recent advances. *Sci. China-Phys. Mech. Astron.* **2013**, *56*, 1611–1618. [CrossRef]
35. Rashki, M.; Ghavidel, A.; Arab, H.G.; Mousavi, S.R. Low-cost finite element method-based reliability analysis using adjusted control variate technique. *Struct. Saf.* **2018**, *75*, 133–142. [CrossRef]
36. Afshari, S.S.; Enayatollahi, F.; Xu, X.Y.; Liang, X.H. Machine learning-based methods in structural reliability analysis: A review. *Reliab. Eng. Syst. Saf.* **2022**, *219*, 108223. [CrossRef]
37. Wakjira, T.G.; Ebead, U.; Alam, M.S. Machine learning-based shear capacity prediction and reliability analysis of shear-critical RC beams strengthened with inorganic composites. *Case Stud. Constr. Mater.* **2022**, *16*, e01008. [CrossRef]

38. Deifalla, A. A comparative study and a simplified formula for punching shear design of concrete slabs with or without membrane tensile forces. *Structures* **2021**, *33*, 1936–1953. [CrossRef]
39. Xu, H.H.; Deng, Y. Dependent Evidence Combination Based on Shearman Coefficient and Pearson Coefficient. *IEEE Access* **2018**, *6*, 11634–11640. [CrossRef]
40. Feng, D.C.; Wang, W.J.; Mangalathu, S.; Taciroglu, E. Interpretable XGBoost-SHAP Machine-Learning Model for Shear Strength Prediction of Squat RC Walls. *J. Struct. Eng.* **2021**, *147*, 04021173. [CrossRef]
41. Salehi, H.; Burgueno, R. Emerging artificial intelligence methods in structural engineering. *Eng. Struct.* **2018**, *171*, 170–189. [CrossRef]
42. Chen, T.Q.; Guestrin, C. XGBoost: A Scalable Tree Boosting System. In Proceedings of the 22nd ACM SIGKDD International Conference on Knowledge Discovery and Data Mining (KDD), San Francisco, CA, USA, 13–17 August 2016.
43. Feng, D.C.; Wang, W.J.; Mangalathu, S.; Hu, G.; Wu, T. Implementing ensemble learning methods to predict the shear strength of RC deep beams with/without web reinforcements. *Eng. Struct.* **2021**, *235*, 111979. [CrossRef]
44. Liang, S.X.; Shen, Y.X.; Ren, X.D. Comparative study of influential factors for punching shear resistance/failure of RC slab-column joints using machine-learning models. *Structures* **2022**, *45*, 1333–1349. [CrossRef]
45. Rahman, J.; Ahmed, K.S.; Khan, N.I.; Islam, K.; Mangalathu, S. Data-driven shear strength prediction of steel fiber reinforced concrete beams using machine learning approach. *Eng. Struct.* **2021**, *233*, 111743. [CrossRef]
46. Mangalathu, S.; Karthikeyan, K.; Feng, D.C.; Jeon, J.S. Machine-learning interpretability techniques for seismic performance assessment of infrastructure systems. *Eng. Struct.* **2022**, *250*, 112883. [CrossRef]
47. Lundberg, S.M.; Lee, S.I. A Unified Approach to Interpreting Model Predictions. In Proceedings of the 31st Annual Conference on Neural Information Processing Systems (NIPS), Long Beach, CA, USA, 4–9 December 2017.
48. Lundberg, S.M.; Erion, G.; Chen, H.; DeGrave, A.; Prutkin, J.M.; Nair, B.; Katz, R.; Himmelfarb, J.; Bansal, N.; Lee, S.I. From local explanations to global understanding with explainable AI for trees. *Nat. Mach. Intell.* **2020**, *2*, 56–67. [CrossRef]
49. Yitzhaki, D. Punching strength of reinforced concrete slabs. *ACI J. Proc.* **1966**, *63*, 527–542.
50. Hawkins, N.M.; Fallsen, H.B.; Hinojosa, R.C. Influence of Column Rectangularity on the Behavior of Flat Plate Structures. *ACI Spec. Publ.* **1971**, *30*, 127–146.
51. Long, A.E. A Two-Phase Approach to the Prediction of the Punching Strength of Slabs. *ACI J. Proc.* **1975**, *72*, 37–45.
52. Regan, P.E. Symmetric punching of reinforced concrete slabs. *Mag. Concr. Res.* **1986**, *38*, 115–128. [CrossRef]
53. Tang, M.; Yi, W.J.; Liu, L.W. Investigation on seismic performance of interior slab-column connections subjected to reversed cyclic loading. *Earthq. Eng. Eng. Dynam.* **2019**, *39*, 109–121.
54. MOHURD (Ministry of Housing and Urban-Rural Development of the People's Republic of China). *GB 50011—2010 Code for Seismic Design of Buildings*; China Architecture & Building Press: Beijing, China, 2016.
55. MOHURD (Ministry of Housing and Urban-Rural Development of the People's Republic of China). *GB 50068—2018 Unified Standard for Reliability Design of Building Structures*; China Architecture & Building Press: Beijing, China, 2018.
56. Wakjira, T.G.; Ibrahim, M.; Ebead, U.; Alam, M.S. Explainable machine learning model and reliability analysis for flexural capacity prediction of RC beams strengthened in flexure with FRCM. *Eng. Struct.* **2022**, *255*, 113903. [CrossRef]
57. Shooman, M.L. *Probabilistic Reliability: An Engineering Approach*; McGraw-Hill: New York, NY, USA, 1986.
58. Hadianfard, M.A.; Malekpour, S.; Momeni, M. Reliability analysis of H-section steel columns under blast loading. *Struct. Saf.* **2018**, *75*, 45–56. [CrossRef]
59. Xia, Z.Y.; Quek, S.T.; Li, A.Q.; Li, J.H.; Duan, M.J. Hybrid approach to seismic reliability assessment of engineering structures. *Eng. Struct.* **2017**, *153*, 665–673. [CrossRef]
60. Strauss, A.; Hoffmann, S.; Wendner, R.; Bergmeister, K. Structural assessment and reliability analysis for existing engineering structures, applications for real structures. *Struct. Infrastruct. Eng.* **2009**, *5*, 277–286. [CrossRef]
61. Gesund, H.; Kaushik, Y.P. *Analysis of Punching Shear Failures in Slabs*; International Association for Bridge and Structural Engineering: Zurich, Switzerland, 1970; pp. 41–60.
62. Ramdane, K.E. Punching shear of high performance concrete slabs. In Proceedings of the 4th International Symposium on Utilization of High Strength/High Performance Concrete, Paris, France, 29 May 1996.
63. Xiao, R.Y.; Chin, C.S. Flat Slabs at Slab-Column Connection: Nonlinear Finite Element Modelling and Punching Shear Capacity Design Criterion. *Adv. Struct. Eng.* **2007**, *10*, 567–579. [CrossRef]

buildings

MDPI

*Article*

# Real-Time Monitoring for Monolithic Movement of a Heritage Curtilage Using Wireless Sensor Networks

**Lulu Shen [1], Bo Yang [1], Yingwu Yang [2,*], Xuelin Yang [3], Wenwei Zhu [3] and Qingzhong Wang [4]**

1 School of Civil Engineering and Architecture, Zhejiang Sci-Tech University, Hangzhou 310018, China
2 College of Landscape and Architecture, Zhejiang A&F University, Hangzhou 310023, China
3 Zhejiang Province Institute of Architectural Design and Research, Hangzhou 310006, China
4 Hangzhou Senkee Construction Special Engineering Co., Ltd., Hangzhou 310030, China
* Correspondence: yangyw06@163.com

**Abstract:** Since monolithic movement is considered a promising technology to relocate historical buildings, corresponding real-time monitoring is of great interest due to the buildings' age and poor structural integrity. However, the related paperwork and practical applications are still limited. This paper describes a wireless sensor network (WSN)-based strategy as a non-invasive approach to monitor heritage curtilage during monolithic movement. The collected data show that the inclination of the curtilage is almost negligible. With the aid of finite element simulation, it was found that the crack displacement curves changed from −0.02 to 0.07 mm, which is affected by moving direction while the value is not enough to cause structural cracks. The deformation of the steel underpinning beam, which is used to reinforce masonry walls and wooden pillars, is obviously related to the stiffness in different directions. Additionally, the strain variations of the steel chassis, which bear the vertical loads from wooden pillars and masonry walls, are less than 0.04%. This indicates that they are kept within the elastic range during monolithic movement. This work has proved that the WSN-based approach has the potential to be applied as an effective route in real-time monitoring of the monolithic movement of an historic building.

**Keywords:** wireless sensor network; historic building; real-time monitoring

**Citation:** Shen, L.; Yang, B.; Yang, Y.; Yang, X.; Zhu, W.; Wang, Q. Real-Time Monitoring for Monolithic Movement of a Heritage Curtilage Using Wireless Sensor Networks. *Buildings* **2022**, *12*, 1785. https://doi.org/10.3390/buildings12111785

Academic Editor: Jurgita Antucheviciene

Received: 27 September 2022
Accepted: 20 October 2022
Published: 25 October 2022



## 1. Introduction

Building monolithic movement, a kind of procedure to relocate the position or direction of an existing building whilst maintaining its integrity and availability, has attracted much interest due to its resource-saving, environmentally friendly and low-cost advantages [1]. Since 1998, a series of related protection statutes and regulations with an emphasis on the protection of preserved historical buildings have been published in mainland China, aiming at materializing historical influences and differences [2]. The corresponding monolithic movement technology is also considered as a new resolution to conserve historical buildings whilst adapting them to new conditions and uses [3].

Compared with traditional demolition and reconstruction, the monolithic movement of historical buildings has the particular advantage of maintaining the original humanistic value and the overall structural integrity. The embryonic form of modern movement-engineering technology can be traced back to 1983, when a school building with a masonry structure (weighing 8000 kN) was moved a distance of 15 m in the city of Warrington, England [4]. With the advancement in technology, the monolithic movement techniques of historical buildings are being developed and implemented worldwide [5]. For example, Kossakowski et al. [6] described the case of the relocation of the Rogatka Grochowska building, which was carried out in Warsaw. The related work and projects in China came late to developed countries. Shan et al. [7] implemented a complex monolithic movement of the Ci-yuan temple with a brick-wooden structure, which was built in the Tang Dynasty and was located in Anyang City in China. Moreover, the documented application projects, including Jinlun Guild Hall (Qing Dynasty), Shanghai Concert Hall (1930), and Centennial Minli high school of Shanghai (1920), were also realized through historic building moving [8,9]. The existing cases signify that three general routes, including moving with rolling bars, moving with a slide layer, and moving by trailer transportation, have been performed successfully as monolithic movement applications of historical buildings in China. To date, due to the buildings' intrinsic rarity and structural aging, only limited paperwork has been disclosed, and few practical applications of the monolithic movement technology of an overall heritage curtilage courtyard have been carried out; however, this technology is slowly becoming more widespread and pervasive in modern civil infrastructure.

During the monolithic movement procedure, it is well known that prioritizing the protection of the historical building, with its intrinsic cracks and cavities, against destructive loads is essential. A program of non-invasive monitoring needs to be undertaken to inform about the ongoing structural status [10–12]. As the moving of historical building means there are some risk factors, the traditional inspection methods for structural cracks, deterioration, and damage tend to be inconvenient and dangerous. With the progress of wireless communication and sensor miniaturization, the wireless sensor network (WSN) system—with a series of smart sensors in a self-organizing and multi-hop manner for monitoring structural deformation—has been developed in recent years [13,14]. Compared with a customary wired network, the WSN boasts easy deployment, a dramatically lower cost of installation and maintenance, and it provides a flexible and manageable approach to monitoring remotely in real time [15,16]. Dong et al. [17] compared the WSN system with the wired sensing system for the performance of a 2-story, 2-bay concrete frame building, and they found that the WSN achieved the same quality of data as that of the wired system. Furthermore, the existing research has proved that the WSN-based approaches could identify the existence and location of damage for long- and short-term monitoring to achieve structural health and safety assessments of historic buildings whilst maintaining their structural integrity and functionalities [18,19]. For example, Wu et al. [20] introduced a dedicated WSN into Torre Aquila, built in the 13th century and located in the city of Trento, Italy, to evaluate its static and dynamic state by utilizing accelerometers and deformation sensors. They also found that the collected data are in agreement with the prediction from the three-dimensional finite element (FE) results. Samuels et al. [21] developed a WSN for monitoring the tilt in the walls of the St. Paul Lutheran church, an historic masonry church with a timber-framed roof, under rehabilitation. Potenza et al. [22] undertook the deployment, test, and management of a WSN for the structural monitoring of the Basilica S. Maria di Collemaggio with masonry structure, which is designed for seismic and dynamic response analyses via acceleration, crack opening, and wall inclination measurements. Mesquita et al. [23] adopted temperature, relative humidity, and displacement sensors to perform a one-year monitoring of the Foz Côa Church (in Portugal), a damaged historic structure from the 16th century based on the WSN system. Barsocchi et al. [24] presented an application of the WSN technology on the Matilde Tower in Livorno (Italy), an historic masonry tower built in the Livorno harbor, to monitor the structural health over the long term and detect potential damages in real time. However, given the uniqueness and the preservation of each historical building, real-time monitoring applications based on the WSN system for the case of its monolithic movement are still very challenging, and few related works have yet been documented.

Therefore, this paper aims to explore the real-time monitoring ability of the WSN in a monolithic movement project for a heritage curtilage with a masonry–timber structure. Considering the age and the poor structural integrity of the heritage curtilage, four kinds of smart sensors were adopted with the aid of practical engineering experiences and FE simulation to real-time monitor the structural deformation and deterioration. Then, the acquisition system receives the processed data and transmits them to a cloud platform via wireless remote communication (3G/4G/GPRS). Finally, all the data can be accessed directly at the preferred time with wireless communication in locations where internet access is available. This paper not only develops a comprehensive scheme for monolithic movement monitoring of a heritage curtilage, but also provides an in-depth understanding of the structural deformation and deterioration during monolithic movement.

## 2. Methods

### 2.1. Case Study

The south-facing curtilage (Figure 1a) was a traditional courtyard with two-story masonry–timber structure, which was located in Shengzhou City, Zhejiang Province, China, with cover an area of 685.23 m$^2$, ridge height of 7.45 m, and gross weight of $3.65 \times 10^6$ kN. According to the architectural style, it was identified as being built in the late Qing Dynasty or the Republic of China. The combination of beam-lifted and Chuan-Dou frame was adopted as the load-bearing structure, and there were stone foundations and capstones under the wooden columns. The primary and secondary entrances and exits of the inner courtyard were both located in the south, the surrounding exterior walls were built with a rowlock cavity wall, and the wall foundation was stacked with rock blocks. In coordination with the emergency project of flood prevention improvements of Chengtan River within Cangyan village, the curtilage movement was planned to be conducted over a one-week period and included as follows: monolithic moving 15 m in the western direction over four days, and then 40 m in the southern direction to the new location in three days, shown in Figure 1b.

(a)

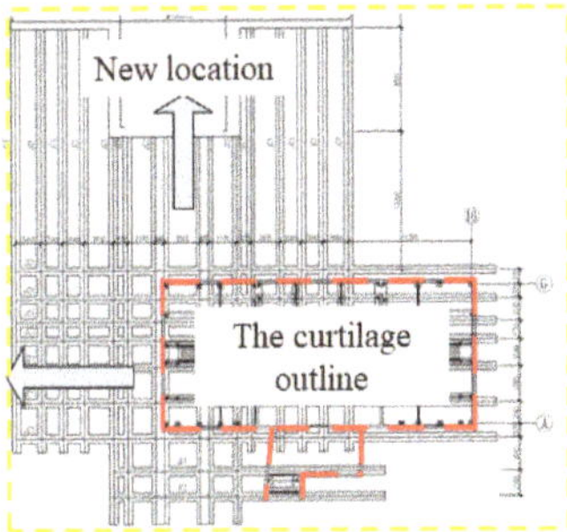


(b)

**Figure 1.** (**a**) The aerial view of the curtilage and its new location; (**b**) the schematic diagram of the monolithic moving procedure.

The monolithic movement of the heritage curtilage was achieved based on lifting jack technology, and the speed was set as 0.8–1.6 mm/s [3] under a horizontal pushing load provided by the push-in jack (SCLRG-100-500-T) with a rated pushing load of 1000 kN and a pushing distance of 500 mm, as shown in Figure 2a. Due to the complicated structural form and the weak structural integrity, the bottom of the masonry wall and wooden pillar were reinforced by a two-clip steel beam with block stone fillers between them to form underpinning beams. In addition, 9 longitudinal and 16 transverse H-shaped steel beams were assembled in Figure S1 (Supplementary Materials) as the integral, and the underpinning chassis was built in a grid pattern with appropriate diagonal crossing beams to enhance the horizontal stiffness The protective supporting platform was also adopted for the superstructure of the curtilage. Thereby, the vertical loads were transferred from wooden pillars and masonry walls to the steel structural chassis, and the top major structure was separated from the original foundation to form a mobile body, which was displayed in Figure 2b. The monolithic movement was achieved by the sliding movement between upper and bottom rail beams using the floating jack as the special sliding support, which could reduce the horizontal resistance and regulate vertical deformation. The upper and bottom rail beams were connected to the steel chassis and were placed on the foundation, respectively, shown in Figure 2b.

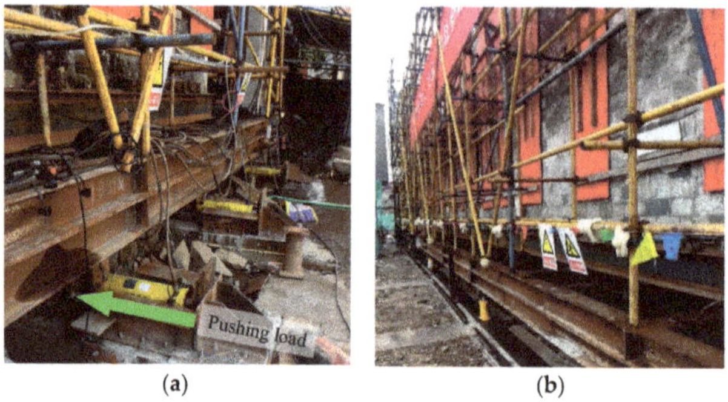


**Figure 2.** The photos of (**a**) the push-in moving process and (**b**) the mobile body.

### 2.2. Monolithic Movement Monitoring Scheme Based on the WSN System

The proposed WSN formed a data collection network of sensors, which were driven by the power supply to serve a specific target-oriented application. The real-time monitoring for the monolithic movement of the heritage curtilage consisted of the parameters of measurements such as cracks, strain, and deformation, thereby obtaining the structural response. Figure 3 presents the four kinds of sensor nodes used in this paper: (1) inclinometer (XKKJQJ-77, ±15°, 0.01°), showing the inclinations of the timber column and masonry wall; (2) displacement transducer (XKKJZX-2109, 0–20 mm, 0.1%F.S/0.5%F.S), monitoring the crack behaviors of the masonry walls and the masonry–timber connections; (3) series inclinometer (XKKJWY-5675, ±30°, 0.015°), to reveal the deformation of the steel beam at the bottom of the masonry wall; (4) strain gauge (XKKJZX-212, 3000 με, ≤0.5%F.S.), attached to the steel chassis to obtain their strain variations. It is noted that the first item in the above brackets is the sensor type, the second is the measuring range, and the third is the precision.

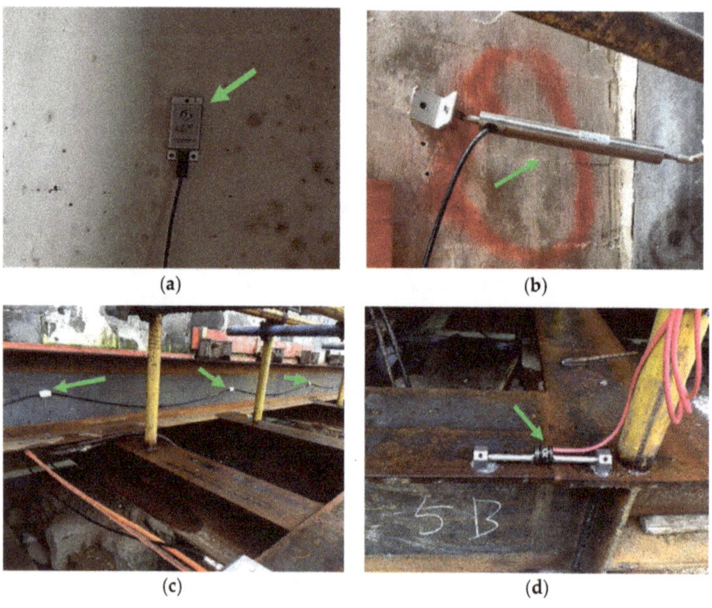


**Figure 3.** Four kinds of sensor nodes used: (**a**) inclinometer, (**b**) displacement transducer, (**c**) series inclinometer, and (**d**) strain gauge.

Figure 4 demonstrates the WSN system schematic to collect and observe the monitoring data in real-time and remotely. Taking displacement transducers as an example for simplicity in Figure 4a, the sensor nodes were deployed on the primary locations. The corresponding acquisition devices received processed data from the sensor nodes and transmitted these data to a cloud platform within the monitoring server (Figure 4b) via wireless remote communication (3G/4G/GPRS). Then, all the data could be accessed directly at preferred times with a personal computer (Figure 4c), mobile phone, or other wireless communication in locations where internet access is available.

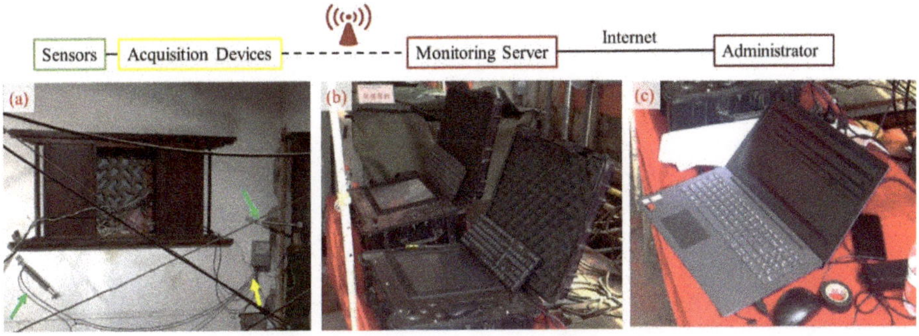


**Figure 4.** The WSN system schematic: (**a**) sensors (pointed with yellow arrows) and the corresponding acquisition devices (pointed with green arrows); (**b**) monitoring server; (**c**) wireless communication.

## 2.3. Numerical Method

In order to understand the structural behaviors and obtain the rational distributions of the displacement transducers, three-dimensional FE models of the exterior walls were built in Abaqus software, consisting of masonry assembly, window frame made of rock, steel reinforcing beams, and block stone fillers (shown in Figure 5). Due to the same geometric structure of the east and west walls, only the west wall was simulated in this paper; half of the north wall was built considering its symmetric geometric structure. Based on computational efficiency, the masonry assembly was regarded as an isotropic composite material, and it followed a nonlinear elastic–plastic constitutive relation [25], which is described by the embedded Concrete Smeared Cracking (CSC) model in Abaqus. The used CSC model, detailed by Lotfi and Shing [26,27], adopted a *J*2-plasticity model with nonlinear isotropic strain hardening and softening to demonstrate the mechanical behaviors of uncracked materials, and a nonlinear orthotropic model to describe the behaviors of cracked materials. The top structure of the studied FEA model was fully fixed, and an incremental displacement loading was applied at the bottom to simulate the deformation during monolithic moving. The basic mechanical parameters of the exterior walls are listed in Table S1.

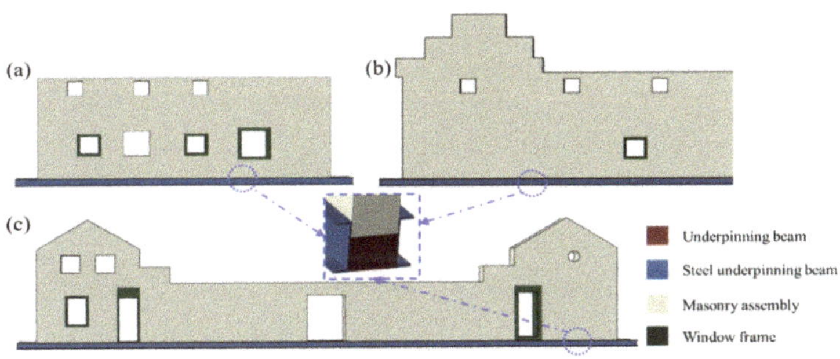


**Figure 5.** FE models of exterior walls: (**a**) west side, (**b**) north side, and (**c**) south side.

## 3. Results and Discussion

### 3.1. Inclination

The near-zero data of the angular variations, collected from all inclinometer sensors, can be found in this case, which indicate that the inclination of the timber column and masonry wall was negligibly small during the complete monolithic movement process. Figure 6 takes the W01 and W02 inclinometers as the typical examples, which are installed on the west wall at 1.5 m above the first-floor ground. Each anchorage point could provide two datasets along east–west (E–W) and south–north (S–N) directions, respectively. Whether moving to the west or north, the inclination angle changed from approximately −0.0025 to 0.0025 degree in both the E–W and S–N directions. It can be deduced that the monolithic movement of the heritage curtilage is expected to realize the aspect of almost negligible inclination.

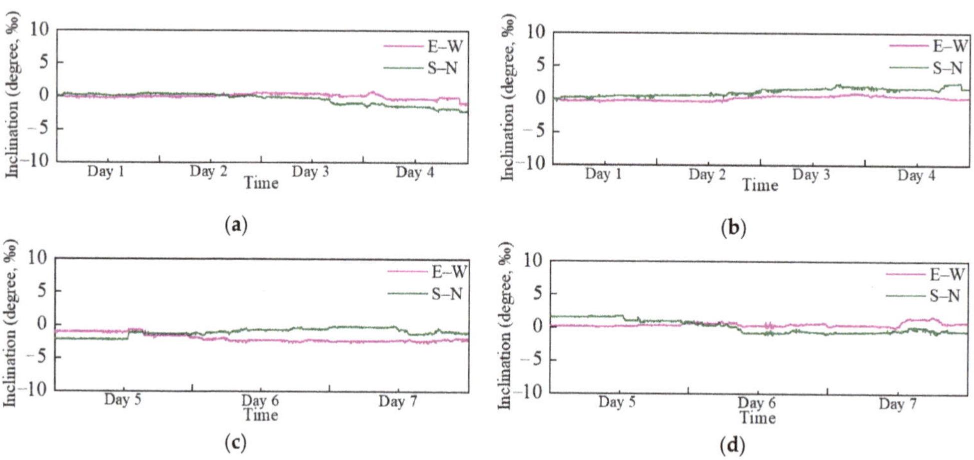


**Figure 6.** The obtained inclination angle: (**a**) W01 and (**b**) W02 during westward movement; (**c**) W01 and (**d**) W02 during northward movement.

### 3.2. Crack Behaviors

As already highlighted, a numerical simulation based on the CSC model in Figure 7 is performed to predict the crack locations on the exterior walls, in which the smeared crack is studied instead of an individual crack. Figure 7a–c displays the equivalent plastic strain (PEEQ) distributions of the exterior walls to reveal the crack locations most easily; thereby obtaining the corners of the window and door that tend to crack under a relative deformation between the upper and lower parts of the walls. Accordingly, just four displacement transducers, denoted as E01, S01, W01, and N01 in Figure 7d, were attached to the exterior walls on the east, south, west, and north sides, respectively. Moreover, masonry-timber connections were also taken into consideration due to the masonry-timber structure of the curtilage. As illustrated in Figure 8d, the displacement transducers at the primary entrance (A01) and on the interior wall (A02) were taken as typical examples in this case.

Figure 8 shows the crack displacement variations along with the moving procedure. For the exterior walls, the crack displacements of the S01 and N01 sensors are slightly higher than that of the E01 and W01 sensors when moving west in Figure 8a, which means that the south and north walls are easier to crack under the horizontal west-pushing loads. When moving north, the east and west walls become easier to crack, as deduced from Figure 8b. As for the masonry-timber connections, the crack displacement variations of the A01 and A02 sensors are also plotted, which shows that the changing range of the A01 curve is larger than A02 when moving west; when moving north, the A02 curve changes more obviously. Moreover, the FE results based on the CSC model also show that when the displacement variations within the areas of the displacement transducers are below 0.17 mm, there is no plastic deformation that can be observed. It is worth noting that all of the crack displacement curves are changed in the range of −0.02 to 0.07 mm in Figure 8, which indicated that the crack displacement variations are very small and insufficient to give rise to structural cracks during the monolithic movement.

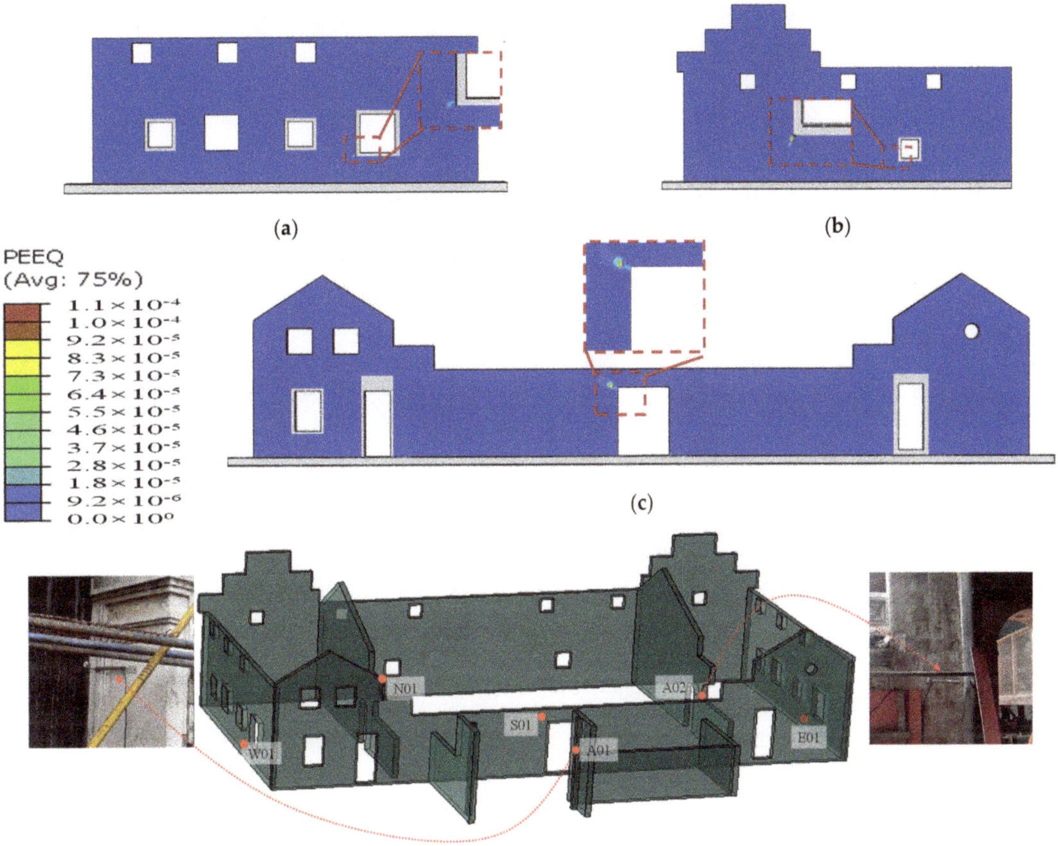


**Figure 7.** PEEQ distributions of the exterior walls: (**a**) west (**b**) north and (**c**) south sides; (**d**) layout of the displacement transducers.

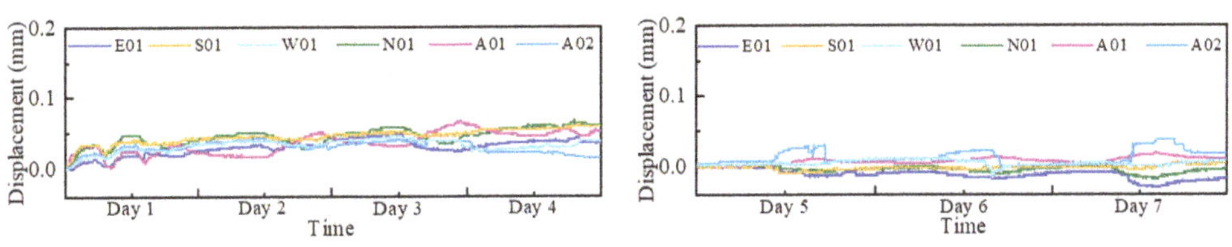


**Figure 8.** Crack displacements obtained from the displacement transducers: (**a**) westward; (**b**) northward.

### 3.3. Deformation of the Steel Underpinning Beams

In order to transfer the vertical loads to the steel chassis, the reinforced underpinnings are constructed at the bottom of the masonry walls with two-clip steel beams. As shown in Figure 8a, three serious inclinometers are stuck on the outside surfaces of the steel beams at the bottom of the east, north, and west walls in this case, assigned as D01, D02, and D03, respectively. Here, D01 and D02 are selected as representative sensors for simplicity, and the ranges of their measured lengths are 12 m along the east–west direction and 30 m along the south–north direction, respectively. The used series inclinometer with ten measuring points could output displacement data along the $Y$ and $Z$ directions, and their local coordinates were plotted in Figure 9a. Figure 9b presents the displacements of D01 along the $Y_1$ directions. Whether moving westward or northward during these 7 days, the displacement curves almost followed the same trend with the range from −0.5 to 1.2 mm, which indicates that there is almost no deformation along the $Y_1$ direction under the external horizontal loads provided by the push-in jacks (Figure S1). This is because the moving speed is very low and the stiffness of the underpinning beams in the $Y_1$ direction is high enough to resist the deformation. When it comes to the deformation along $Z_1$ direction in Figure 9c, the variation trends of the displacement curves are also similar; meanwhile, the displacement in the $Z_1$ direction decreased from 6.0 to 1.5 mm with the increased distance from the north end to the south end of the series inclinometer when moving west in the first to the fourth days. The displacement variations mainly originate from the difference of the external pushing loads. Additionally, it is obvious that the amplitude of deformation is larger than that during the movement north in the fifth to seventh days and the D01 data, which is caused by the direction of the pushing loads and the lower stiffness of the underpinning beam in the $Z_1$ direction than that in the $Y_1$ direction. Similarly, the displacement curves obtained via the D02 sensor are quite stable (small gaps of 1.2 to 2.0 mm and 0.3 to 1.2 mm for the westward and northward movements) in Figure 9d, which proves the homogeneous deformation in $Y_2$ direction of the underpinning beam during the whole monolithic movement. Additionally, the stable displacement curves in $Z_2$ direction distracted from the D02 sensor with a small variation of −0.1 to 0.9 mm can be found when moving west in Figure 9e. The larger deformation with the displacement varying from −5 to −2 mm is revealed when moving north due to a combination of the significantly lower stiffness in the $Z_2$ direction than the $Y_2$ direction and the external pushing-load direction.

### 3.4. Strain of the Steel Chassis

The strain variations in the steel chassis, which bears the vertical loads of the whole mobile body, are collected by the strain gauges and automatically accessed by the computer in this case. Taking one strain gauge on each side of the chassis for the typical examples, their locations are presented in Figure S1, and the corresponding strain variations are shown in Figure 9, where SE, SW, SN, and SS denote the strain gauge attached on the east, west, north, and south steel chassis beams. During both the westward and northward movements, a good correlation in the strain variations can be found for the SE and SW curves, and also for the SN and SS curves, which indicates the synchronous moving process of the opposite steel beams and further provides a guarantee for maintaining the integrity and availability of the curtilage. The strains of the north and south beams are higher than those of the east and west beams in Figure 10a, which are induced by the horizontal pushing force provided by the push-in jack during the first westward monolithic movement stage. Moreover, the strains of the east and west beams are nearly zero, which proves that there is little or no deformation of them. Similarly, the higher strain of the north and south beams and the near-zero strain of the east and west beams are shown in Figure 10b during the northward stage. Furthermore, it is interesting to notice that, despite the reasonable amplitude of these curves, the strains for all the studied chassis beams are less than 0.04%, which is far less than the yield strain of the steel material, helping to confirm that steel chassis are keeping in the elastic range during moving to the new location.

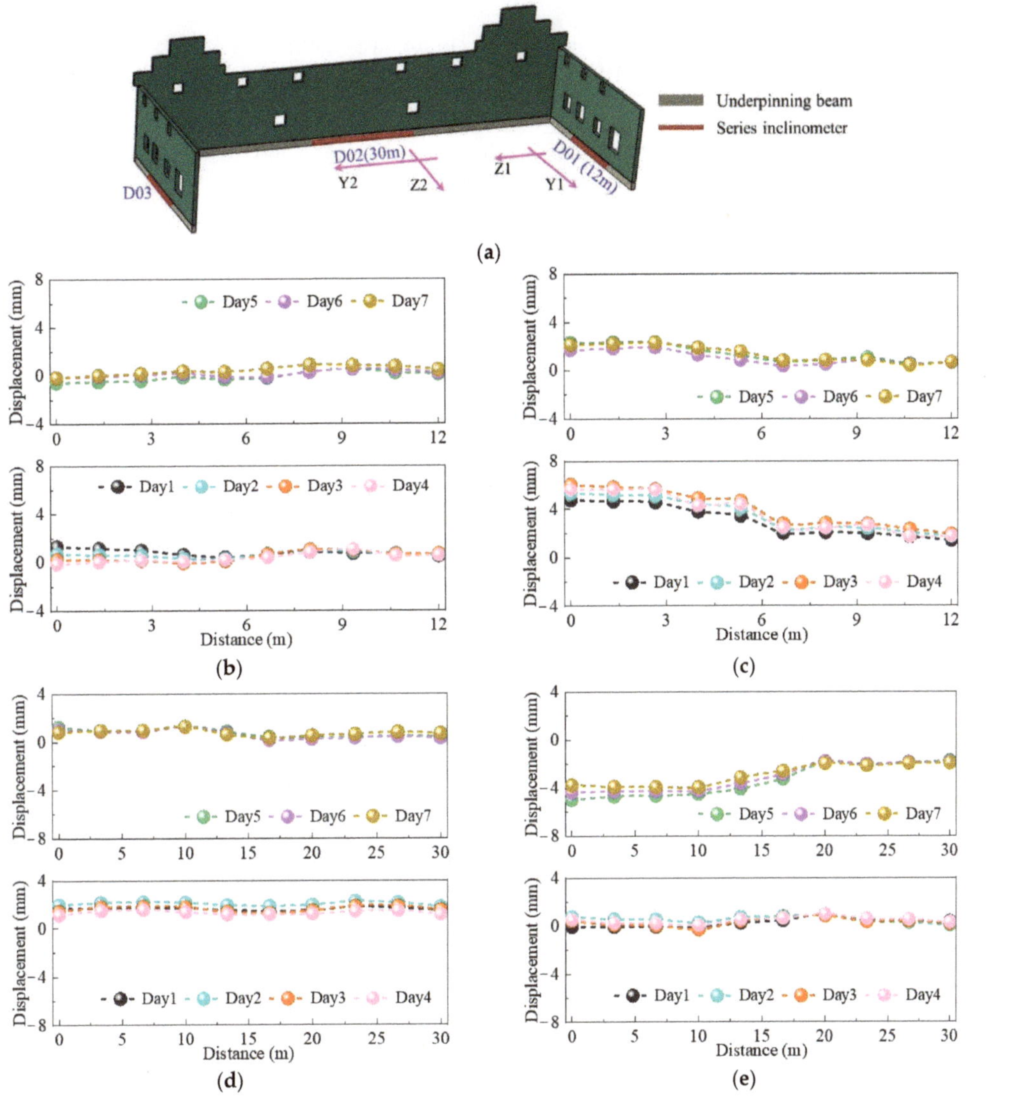


**Figure 9.** (**a**) The distribution of the serious inclinometers; the displacements measured by: D01 along the (**b**)$Y_1$ and (**c**) $Z_1$ directions and D02 along the (**d**) $Y_2$ and (**e**) $Z_2$ directions.

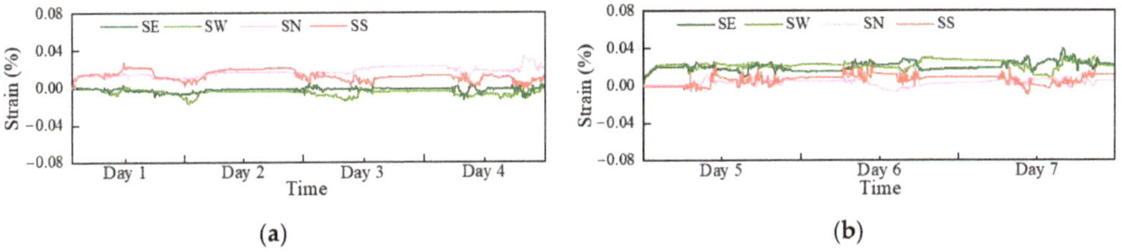


**Figure 10.** Strain variations during the (**a**) westward and (**b**) northward movement.

## 4. Conclusions

The study undertaken herein involves the case study of a WSN-based approach enabling real-time monitoring for the monolithic movement of a heritage curtilage. In consideration of the lifting jack technology to implement the monolithic movement, the deformation of the steel underpinning beams and the steel chassis are monitored simultaneously in addition to the inclination and the crack behaviors during translocation.

The following conclusions are drawn:

1. The inclinometer results show that the inclination of the timber pillars and the exterior masonry walls is almost negligible, confirming that the used lifting-jack technology has great potential in the monolithic movement of the historic building;
2. FE simulation and engineering experience are combined to reveal that the corners of the window and the door as well as the masonry-timber connections tend to crack more easily, thereby determining the distributions of the displacement transducers. The corresponding displacement curves reveal that the crack behaviors are affected by the moving direction. The value range is also changed from −0.02 to 0.07 mm, which is not enough to initiate structural cracks;
3. The strain gauges are attached to the steel underpinning beams to monitor their deformations. It is obvious that the deformation of the steel underpinning beam is related to the moving direction and its stiffness in different directions;
4. The strain variations of the steel chassis, obtained from series inclinometers, are less than 0.04%, which provides evidence of keeping the elastic range during monolithic movement.

In brief, this work provides useful insights into developing non-invasive, real-time, and remote monitoring strategies for historic buildings.

**Supplementary Materials:** The following supporting information can be downloaded at: https://www.mdpi.com/article/10.3390/buildings12111785/s1, Figure S1: The plane view of the steel chassis with 9 longitudinal and 16 transverse H-shaped steel beams, floating jacks, pushing loads, and typical strain gauge locations; Table S1: The input parameters of the exterior walls.

**Author Contributions:** Conceptualization, Y.Y., X.Y. and Q.W.; methodology, B.Y. and L.S.; software, L.S. and B.Y.; investigation, L.S. and Y.Y.; writing—original draft preparation, L.S.; writing—review and editing, Y.Y. and B.Y.; visualization, L.S. and W.Z.; supervision, B.Y.; project administration, Q.W. and Y.Y. All authors have read and agreed to the published version of the manuscript.

**Funding:** This research was funded by the Natural Science Foundation of Zhejiang Province (Grant No. LGG20E080002, LQ21A020009) and the Science Foundation of Zhejiang Sci-Tech University (ZSTU) under Grant No. 19052460-Y.

**Institutional Review Board Statement:** Not applicable.

**Informed Consent Statement:** Not applicable.

**Data Availability Statement:** Not applicable.

**Acknowledgments:** Not applicable.

**Conflicts of Interest:** The authors declare no conflict of interest.

## References

1. Guo, T.; Li, A.; Wei, L.; Gu, Y. Horizontal Translocation of a High-Rise Building: Case Study. *J. Perform. Constr. Fac.* **2013**, *27*, 235–243. [CrossRef]
2. Huan, J.; Guo, X.; Guan, Z.; Yan, T.; Chu, T.; Sun, Z. An Experimental Study of the Hysteresis Model of the Kanchuang Frame Used in Chinese Traditional Timber Buildings of the Qing Dynasty. *Buildings* **2022**, *12*, 887. [CrossRef]
3. Zhao, D.M.; Yao, Y.H. Study on Key Technology of Frame Structure Building Monolithic Movement. *Adv. Mater. Res.* **2012**, *450–451*, 856–860. [CrossRef]
4. Pryke, J.F.S. Relevelling, Raising and Re-Siting Historic buildings. In *Strengthening of Building Structures—Diagnosis and Therapy: Final Report*; Proceedings Symposium IABSE; IABSE: London, UK, 1983.

5. Kozlu, H.H.; Batukan Dördüncü, B.F. Moving Techniques for Traditional Buildings as an Architectural Preservation Method. *ICONARP J. Archit. Plan.* **2019**, *7*, 286–313. [CrossRef]
6. Kossakowski, P.G. Relocation of a Historic Building at the Old Norblin Factory in Warsaw. *Arch. Civ. Eng.* **2021**, *67*, 351733.
7. Shan, R.; Zhang, X.; Lu, M.; Zhao, H. The Complex Monolithic Movement for the Brick-wooden Building in Deformation Analysis and Reinforcement. *Open Civ. Eng. J.* **2016**, *10*, 884–890.
8. Zhang, H.Y.; Pan, R.S. Techniques of Building Monolithic Moving that Avoiding Building Rubbish. *Appl. Mech. Mater.* **2015**, *737*, 603–607. [CrossRef]
9. Lu, X.L.; Lu, W.S.; Zhang, X. New Resolution for Historic Building Conservation by Building Moving Technology. *Adv. Mater. Res.* **2010**, *133–134*, 19–25. [CrossRef]
10. Zonta, D.; Wu, H.; Pozzi, M.; Zanon, P.; Ceriotti, M.; Mottola, L.; Picco, G.P.; Murphy, A.L.; Guna, S.; Corra, M. Wireless sensor networks for permanent health monitoring of historic buildings. *Smart Struct. Syst.* **2010**, *6*, 595–618. [CrossRef]
11. Kilic, G. Using advanced NDT for historic buildings: Towards an integrated multidisciplinary health assessment strategy. *J. Cult. Herit.* **2015**, *16*, 526–535. [CrossRef]
12. Lorenzoni, F.; Casarin, F.; Caldon, M.; Islami, K.; Modena, C. Uncertainty quantification in structural health monitoring: Applications on cultural heritage buildings. *Mech. Syst. Signal Processing* **2016**, *66–67*, 268–281. [CrossRef]
13. Shi, C.; Wei, R.; Zhang, Y. Application Research of Wireless Sensor Network in Building Structure Safety Monitoring. *IOP Conf. Ser. Mater. Sci. Eng.* **2018**, *366*, 012084. [CrossRef]
14. Sivasuriyan, A.; Vijayan, D.S.; Górski, W.; Wodzyński, Ł.; Vaverková, M.D.; Koda, E. Practical Implementation of Structural Health Monitoring in Multi-Story Buildings. *Buildings* **2021**, *11*, 263. [CrossRef]
15. Chintalapudi, K.; Fu, T.; Paek, J.; Kothari, N.; Rangwala, S.; Caffrey, J.; Govindan, R.; Johnson, E.; Masri, S. Monitoring civil structures with a wireless sensor network. *IEEE Internet Comput.* **2006**, *10*, 26–34. [CrossRef]
16. Sofi, A.; Jane Regita, J.; Rane, B.; Lau, H.H. Structural health monitoring using wireless smart sensor network—An overview. *Mech. Syst. Signal Processing* **2022**, *163*, 108113. [CrossRef]
17. Dong, X.; Liu, X.; Wright, T.; Wright, T.; Wang, Y.; DesRoches, R. Validation of wireless sensing technology densely instrumented on a full-scale concrete frame structure. In Proceedings of the International Conference on Smart Infrastructure and Construction, London, UK, 27–29 June 2016.
18. Gopinath, V.K.; Ramadoss, R. Review on structural health monitoring for restoration of heritage buildings. *Mater. Today Proc.* **2021**, *43*, 1534–1538. [CrossRef]
19. Jiang, W.; Zhang, F.; Lin, Q.; Li, Q. Application of Sensing Technology in the Protection of Architectural Heritage: A Review. In Proceedings of the 2021 IEEE International Conference on Artificial Intelligence and Computer Applications (ICAICA), Dalian, China, 28–30 June 2021.
20. Wu, H.Y.; Zonta, D.; Pozzi, M.; Zanon, P.; Corrà, M. Historic Buildings: Long Term Stability Evaluation Using Wireless Sensor Networks. *Adv. Mater. Res.* **2010**, *133–134*, 235–240. [CrossRef]
21. Samuels, J.M.; Reyer, M.; Hurlebaus, S.; Lucy, S.H.; Woodcock, D.G.; Ozbulut, O.E. Monitoring of the St. Paul Lutheran Church using a Wireless Sensor Network (WSN) During Foundation Stabilization. *Int. J. Archit. Herit.* **2013**, *7*, 68–88. [CrossRef]
22. Potenza, F.; Federici, F.; Lepidi, M.; Gattulli, V.; Graziosi, F.; Colarieti, A. Long-term structural monitoring of the damaged Basilica S. Maria di Collemaggio through a low-cost wireless sensor network. *J. Civ. Struct. Health* **2015**, *5*, 655–676. [CrossRef]
23. Mesquita, E.; Arêde, A.; Pinto, N.; Antunes, P.; Varum, H. Long-term monitoring of a damaged historic structure using a wireless sensor network. *Eng. Struct.* **2018**, *161*, 108–117. [CrossRef]
24. Barsocchi, P.; Bartoli, G.; Betti, M.; Girardi, M.; Mammolito, S.; Pellegrini, D.; Zini, G. Wireless Sensor Networks for Continuous Structural Health Monitoring of Historic Masonry Towers. *Int. J. Archit. Herit.* **2021**, *15*, 22–44. [CrossRef]
25. Stavridis, A.; Shing, P.B. Finite-Element Modeling of Nonlinear Behavior of Masonry-Infilled RC Frames. *J. Struct. Eng.* **2010**, *136*, 285–296. [CrossRef]
26. Lotfi, H.R.; Shing, P.B. An appraisal of smeared crack models for masonry shear wall analysis. *Comput. Struct.* **1991**, *41*, 413–425. [CrossRef]
27. Lotfi, H.R.; Shing, P.B. Interface Model Applied to Fracture of Masonry Structures. *J. Struct. Eng.* **1994**, *120*, 63–80. [CrossRef]

buildings

MDPI

*Article*

# Polyethylene (PE) Waste Minimization Study of Cement Mortar with Adding PE Content under Different W/B Ratios

**Keng-Ta Lin [1], Chang-Chi Hung [2], Her-Yung Wang [3,*] and Fu-Lin Wen [3]**

1 Department of Civil and Environmental Engineering, National Kaohsiung University, Kaohsiung 811726, Taiwan
2 School of Architecture and Civil Engineering, Huizhou University, Huizhou 516007, China
3 Department of Civil Engineering, National Kaohsiung University of Science and Technology, Kaohsiung 807618, Taiwan
* Correspondence: wangho@nkust.edu.tw

**Abstract:** Wastes can be effectively used in concrete and the characteristics of concrete can be maintained or enhanced, the economy of waste management can be greatly increased, and the pollution of the earth can be reduced. This study aimed to research the durability of cement mortar prepared using different W/B ratios and different percentages of waste PE content. The cement mortar was mixed with 0%, 1%, 2%, 3%, and 4% waste PE and 20% ground-granulated blast-furnace slag (GGBFS) in W/B ratios of 0.4, 0.5, and 0.6. The results sw that the slump and flow decrease as the waste PE content is increased and increase with increasing W/B ratio, and the setting time is shortened as the waste PE content is increased. In terms of hardened properties, the specimen strength is slightly decreased as the waste PE content is increased, but the hardened properties are better at a later age due to the pozzolanic reaction of slag, which can be verified by microscopic SEM.

**Keywords:** recycling of waste materials; water-to-binder ratio (W/B ratio); waste PE; cement mortar; durability

**Citation:** Lin, K.-T.; Hung, C.-C.; Wang, H.-Y.; Wen, F.-L. Polyethylene (PE) Waste Minimization Study of Cement Mortar with Adding PE Content under Different W/B Ratios. *Buildings* **2022**, *12*, 2117. https://doi.org/10.3390/buildings12122117

Academic Editors: Huazhe Jiao and Abdelhafid Khelidj

Received: 16 October 2022
Accepted: 30 November 2022
Published: 2 December 2022



## 1. Introduction

Cement manufacturing can be identified as a major source of $CO_2$ emissions, from production to emissions, making it the largest source of industrial emissions [1]. The cement industry is the second-largest source of $CO_2$ emissions, accounting for 27% of $CO_2$ emissions from the industrial sector and 8% of global $CO_2$ emissions [2,3]. It is estimated that by 2050, cement production will increase by approximately 12–22% compared to 2014. Therefore, one way to reduce $CO_2$ emissions is to replace cement with materials that use waste materials or industrial by-products to reduce $CO_2$ emissions [4].

Blast-furnace slag is an industrial by-product of iron extraction. Depending on the cooling method, blast-furnace slag can be divided into air-cooled blast-furnace slag and ground-granulated blast-furnace slag (GGBFS), which is then dried and ground into a powder of comparable fineness to replace cement as a cementitious material [5]. While the chemical composition of GGBFS is very similar to that of Portland cement, its composition consists of varying proportions of lime and alumina. The cementitious properties of GGBFS are controlled by the type of ore, the type of flux used, and the contaminants in the charged coke. Magnesium, silicon, calcium, aluminum, and oxygen account for 95% of the total GGBFS content [6]. Therefore, proper use of ground-granulated blast-furnace slag (GGBFS) to replace cement can not only reduce cement use but also reduce slag emissions, and the properties of ground-granulated blast-furnace slag can also be used to improve its engineering properties.

In the 1950s, plastic or synthetic organic polymers were mass-produced and used. Although the rapid growth in the production of plastic man-made materials still does not surpass that of steel and cement materials widely used in civil construction, the impact of plastic waste on the environment and how to eliminate it is still an important issue that cannot be ignored [7]. The largest market for plastics is packaging, where the use of single-use containers has grown so rapidly that the proportion of solid waste generated from the use of these containers has increased from less than 1% in 1960 for middle- and high-income countries to more than 10% in 2005 [8].

The vast majority of monomers used to make plastics, such as ethylene and propylene, are derived from fossil hydrocarbons. None of the common plastics are biodegradable [9]. The only way to permanently eliminate plastic waste is through destructive heat treatment, such as combustion or pyrolysis, which often causes secondary pollution of the environment due to the subsequent emission of $CO_2$. If such waste is buried in a landfill, it will only accumulate in the natural environment and will not decompose [9,10]. Therefore, the near-permanent contamination of the natural environment by plastic waste is a growing problem [11].

Disposable or single-use paper containers are often coated with a plastic film, such as polyethylene (PE), on the inner surface to meet waterproof or oil-proof requirements. Therefore, most single-use paper containers are regarded as plastic waste and not waste paper. Recycling is an effective way to reuse or regenerate waste into useful products, materials, or components, especially with regard to recycling waste made from composite materials such as lunch boxes and beverage cups [12–14]. With the continuous increase in plastic waste, the inclusion of recycled plastic waste in building materials, such as bricks and concrete, has been studied by researchers [15,16]. In addition, after recycled plastic waste is combined with wood and other plant fibers, it can also be used in plastic–wood building materials [17].

The cement in concrete consumes 2% of global energy in the production process, and every ton of cement produced will emit 0.85 tons of carbon dioxide [18,19]. Therefore, many researchers continue to investigate sustainable, environmentally friendly, and cost-effective alternatives. Recycled plastic waste can be transformed into a suitable shape or size and added to the concrete as an aggregate [20], which can reduce structural weight, increase design flexibility, reduce total construction costs, reduce the structural gravity load, and improve seismic performance. Due to its advantages, such as structural reaction under action and enhanced structural thermal insulation, it is one of the alternatives to decontaminate waste [16,21].

Plastic waste is incorporated into mortars and cements as aggregates, and most studies have focused on the microplastic and neoplastic range [11,22,23]. Taiwan has the second-highest density of convenience stores in the world [24], and most disposable food packaging contains plastic waste composed of polyethylene (PE) components [14], a high-quality material with good chemical stability, resistance to impact, and low temperature resistance. According to the Environmental Protection Agency, in Taipei City alone, food delivery packaging increased by 85% in May 2021 [24], perhaps due to the impact of the COVID-19 outbreak. Usually, recycled food packaging is composed of different polymers and complex materials, which makes the separation of each material difficult [24]. It is necessary to separate the PE components in the packaging container through processes such as buoyancy sorting before recycling for reuse. The focus of this study is to investigate the durability of recycled polyethylene (PE) plastic waste added to cement mortar and the benefits of PE waste reuse under different W/B ratios.

PE polymers are one of many fibers, whether they are polymers or metals, and they are widely used in concrete engineering because of their advantages [25–28]. The compressive strength and toughness of concrete can be improved by adding steel fibers due to the high modulus of elasticity and stiffness [25]. Although adding steel fibers to concrete can improve the properties of concrete, the fiber content must be high. This increases the structural gravity load of concrete and has a balling effect during mixing, thus reducing workability [29,30]. There have been many studies on the use of PE as a substitute for some natural aggregates in cementitious materials. Basha et al. found that the amount of recycled plastic aggregate leads to a decrease in compressive strength, flexural strength, elastic modulus, and adhesive strength but is useful for concrete thermal insulation [23]. By incorporating high-density polyethylene (HDPE) and low-density polyethylene (LDPE) into concrete, Rumsys et al. found that the compressive strength of concrete is decreased [31]. Yoo and Kim found that replacing steel fibers with PE fibers results in a decrease in compressive strength due to uneven dispersion of high aspect ratio PE fibers [32].

Single-use paper containers or tableware coated with polyethylene (PE) plastic films for waterproofing and oil stain prevention have become indispensable in people's lives because of their low cost and convenience. After use, it is sent to a professional waste paper container treatment plant for recycling, and its plastic film can be reused. However, due to inaccurate recycling classification or high cost, it is mixed with waste paper and sent to general nonprofessional waste paper factories for processing. The separated plastic film will be identified as garbage and sent to incinerators for incineration, which increases the processing cost of waste paper factories, secondary air pollution, and energy consumption. In addition, in other studies related to the acquisition of plastic wastes to concrete, such as polyethylene terephthalate (PET), some researchers pointed out that the accumulation of this waste did not contribute to the improvement of concrete compressive strength [33–39]. However, the issue of the behavior of cement mortars containing PE wastes is new, and research in this context is limited. Research in this context should continue to highlight the important hardening and durability properties of concrete containing PE plastic wastes coating disposable containers. Therefore, this study used different W/B ratios and different contents of waste PE to make cement mortar, with the GGBFS content being fixed at 20%, to discuss the durability and the benefits of energy savings and carbon reduction. The preliminary study of cement mortar specimens was used to investigate the feasibility of waste reuse and to suggest the appropriate PE addition ratio for the reference of concrete mixture proportioning design, so as to achieve waste minimization by recycling PE film on the surface of disposable containers.

## 2. Experiment Plan

### 2.1. Experimental Materials

Cement: Portland Type I cement produced by Taiwan Cement Corporation was used; the properties of this cement conformed to ASTM C150, the specific gravity was 3.15, and the fineness was 3450 $cm^2/g$. Waste PE: The waste PE was provided by the manufacturer, with a specific gravity of 0.92 and a water content of 8.2%. It had the appearance of spherical large granules. After it was decomposed by pulverization, it appeared as flocculent plastic fibers, as shown in Figure 1a, with a specific gravity of 0.92. Figure 1 shows the Fourier-transform infrared spectroscopy (FTIR) spectrum for waste PE. GGBFS: GGBFS was provided by CHC Resources, conforming to CNS12549, with a specific gravity of 2.9 and a fineness of 4000 $cm^2/g$. The fine aggregate was derived from the river sand of the Ligang River, and the specific gravity was tested according to ASTM C127. The specific gravity was 2.65, and the water absorption was 1.48%. The chemical compositions of the test materials are shown in Table 1.

**Table 1.** Physical properties and chemical composition of test materials.

| Materials | Chemical Contents (%) | | | | | | | | | | | | | | | |
|---|---|---|---|---|---|---|---|---|---|---|---|---|---|---|---|---|
| | $SiO_2$ | $Al_2O_3$ | $Fe_2O_3$ | CaO | MgO | $SO_3$ | $K_2O$ | $Na_2O$ | $TiO_2$ | $P_2O_5$ | f-CaO | $C_3S$ | $C_2S$ | $C_3A$ | $C_4AF$ | L.O.I |
| Cement | 19.6 | 4.4 | 3.1 | 62.5 | 4.9 | 2.2 | - | - | 0.5 | 0.11 | 0.7 | 56 | 14 | 7 | 9 | 2.5 |
| PE | 42.5 | 36 | 1.1 | 11.2 | 2.9 | - | 0.4 | 0.7 | 5.2 | | - | - | - | - | - | 42 |
| GGBFS | 33.5 | 14.7 | 0.4 | 41.2 | 6.4 | 0.6 | 0.3 | 0.2 | 0.5 | 0.01 | - | - | - | - | - | 0.6 |

### 2.2. Test Variables and Mix Proportions

The tested fresh properties included the slump and flow. Cement mortar specimens were prepared and cured in saturated limewater. Their hardened properties, durability, and microscopic properties were discussed at the ages of 3, 7, 28, 56, and 91 days. The ratio of unit weight and the test methods and specifications are shown in Tables 2 and 3.

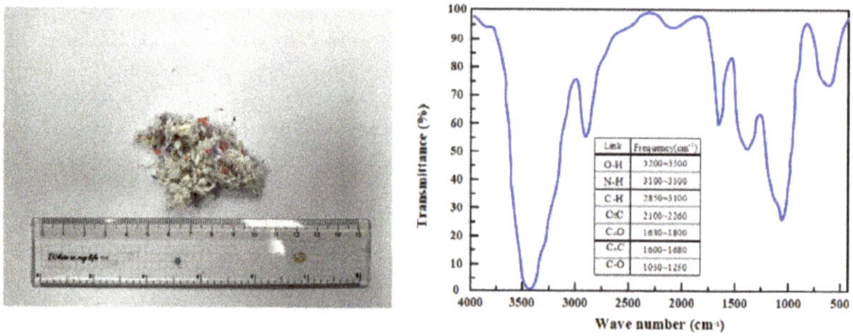


(**a**) material appearance (**b**) FTIR analyze

**Figure 1.** Material appearance and Fourier-transform infrared spectroscopy (FTIR) analysis of polyethylene.

**Table 2.** Mixture proportions of cement mortar (unit: kg/m$^3$).

| W/B | GGBFS (%) | GGBFS | Cement | PE (%) | PE | Sand | Water |
|---|---|---|---|---|---|---|---|
| 0.4 | 20 | 172 | 749 | 0 | 0 | 2302 | 375 |
| | | | | 1 | 13.8 | | |
| | | | | 2 | 27.7 | | |
| | | | | 3 | 41.5 | | |
| | | | | 4 | 55.4 | | |
| 0.5 | | 151 | 658 | 0 | 0 | | 411 |
| | | | | 1 | 13.8 | | |
| | | | | 2 | 27.7 | | |
| | | | | 3 | 41.5 | | |
| | | | | 4 | 55.4 | | |
| 0.6 | | 135 | 586 | 0 | 0 | | 439 |
| | | | | 1 | 13.8 | | |
| | | | | 2 | 27.7 | | |
| | | | | 3 | 41.5 | | |
| | | | | 4 | 55.4 | | |

**Table 3.** Test method and regulations.

| No. | Test Items | Test Regulations |
|---|---|---|
| 1 | Slump | ASTM C143 |
| 2 | Flow | ASTM C230 |
| 3 | Setting time | ASTM C403 |
| 4 | Compressive strength | ASTM C109 |
| 5 | Flexural strength | ASTM C348 |
| 6 | Tensile strength | ASTM C190 |
| 7 | Water absorption rate | ASTM C1585 |
| 8 | Ultrasonic velocity | ASTM C597 |
| 9 | Resistivity | ASTM C876 |
| 10 | Resistance to sulfate attack | ASTM C1012 |

For micro and component analysis, a scanning electron microscope (SEM) is used to generate secondary electrons by striking the specimen with an electron beam, and the cathode ray tube is used to observe the surface microstructure of the specimen. An energy-dispersive spectrometer (EDS) can be used for qualitative and quantitative analysis of the composition of the experimental specimen by scanning a high-energy focused electron beam on the surface of the specimen.

## 3. Results and Discussion

### 3.1. Slump

As shown in Table 4 and Figure 2, with a fixed GGBFS of 20%, 1% of waste PE, and W/B ratio of 0.4, 0.5, and 0.6, corresponding slumps of cement mortar were 1.5, 2.6, and 3.8 cm, respectively. The slump of cement mortar reveals an upward trend. Furthermore, the slump was 1.9~1.2 cm with a W/B ratio of 0.4 and 1~4% addition of waste PE fiber material, and it can be observed that the slump tends to decrease with the increase in the addition of waste PE fiber material. The slump is observed to increase with the W/B ratio. As the waste PE absorbs water, it absorbs the free water in the mortar; when the content of waste PE in the mortar increases, the slump decreases accordingly. Therefore, the slump increases as the W/B ratio increases and the waste PE content decreases, as shown in Table 4.

**Table 4.** Slump and slump ratio of cement mortar with different W/B ratios and waste PE (unit: cm).

| RM (%) | AM (%) | W/B | | | | | |
|---|---|---|---|---|---|---|---|
| GGBFS | PE | 0.4 | | 0.5 | | 0.6 | |
| | | Slump | SR * | Slump | SR * | Slump | SR * |
| 20 | 0 | 1.9 | 1.00 | 2.8 | 1.00 | 4.1 | 1.00 |
| | 1 | 1.5 | 0.79 | 2.6 | 0.93 | 3.8 | 0.92 |
| | 2 | 1.3 | 0.68 | 2.5 | 0.89 | 3.7 | 0.90 |
| | 3 | 1.3 | 0.68 | 2.3 | 0.82 | 3.5 | 0.85 |
| | 4 | 1.2 | 0.63 | 2.1 | 0.75 | 3.4 | 0.83 |

* SR: Slump ratio of PE material added specimen to the control specimen without PE material added.

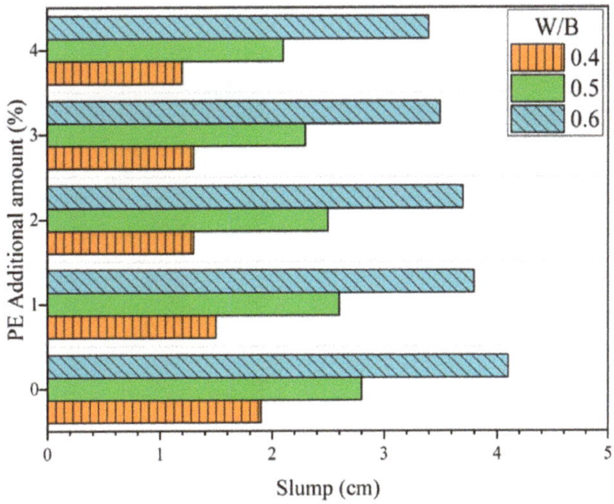


**Figure 2.** Slump of cement mortar with different W/B ratios and waste PE.

Regarding W/B = 0.4, the slump ratio of the specimen with 1% PE fiber material added to the control specimen without PE fiber material added was 0.79, as shown in Table 4. Still, the decreasing trend of the slump was slowed down with the increase in PE material added, and the maximum slump ratio was 0.63, which occurred in the specimen with 4% PE material added. Similarly, at W/B = 0.5 and 0.6 series, the slump also showed a decreasing trend with the increase in PE addition. Therefore, the effect of PE addition on the deterioration of fresh mix properties could be improved significantly by W/B. With 2% PE addition, the slump of W/B = 0.5 increased significantly from 1.3 cm to 2.5 cm compared with that of W/B = 0.4, which was 1.92 times. The results showed that the PE fiber material addition has little effect on the fresh mix properties. Therefore, the fresh mix property test results show that the cement mortar can be added with PE fiber material to achieve the impact of waste minimization.

### 3.2. Flow

As shown in Table 5 and Figure 3, with a W/B ratio of 0.4 and a fixed GGBFS content of 20%, the flow is 15.6~14.2 cm; with a W/B ratio of 0.5, the flow is 19.9~18.7 cm; with a W/B ratio of 0.6 and a GGBFS content of 20%, the flow value is 21.7~20.9 cm. The flow value is greatly increased with the W/B ratio, the mortar flow increases as the water consumption increases, and the overall workability of the mortar is enhanced.

**Table 5.** Flow and flow ratio of cement mortar with different W/B ratios and waste PE (unit: cm).

| RM (%) | AM (%) | W/B | | | | | |
|---|---|---|---|---|---|---|---|
| GGBFS | PE | 0.4 | | 0.5 | | 0.6 | |
| | | Flow | FR * | Flow | FR * | Flow | FR * |
| 20 | 0 | 15.6 | 1.00 | 19.9 | 1.00 | 21.7 | 1.00 |
| | 1 | 15.2 | 0.97 | 19.7 | 0.99 | 21.3 | 0.98 |
| | 2 | 14.7 | 0.94 | 19.4 | 0.97 | 21.2 | 0.98 |
| | 3 | 14.4 | 0.92 | 18.9 | 0.95 | 21.0 | 0.97 |
| | 4 | 14.2 | 0.91 | 18.7 | 0.94 | 20.9 | 0.96 |

* FR: Flow ratio of PE fiber material added specimen to the control specimen without PE fiber material added.

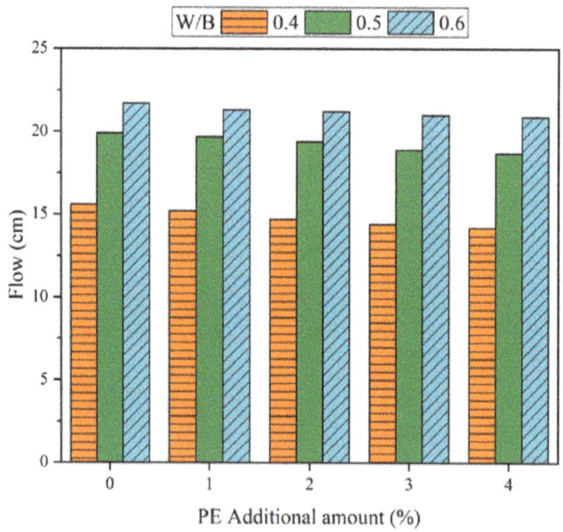


**Figure 3.** Flow of cement mortar with different W/B ratios and waste PE.

With a W/B ratio of 0.5 and 0%, 1%, 2%, 3%, and 4% for the waste PE, the cement mortar shows flow values of 19.9, 19.7, 19.4, 18.9, and 18.7 cm, respectively. When the content is increased from 1% to 4%, the flow is reduced by 1~6%, meaning the flow is decreased as the content of waste PE is increased.

### 3.3. Setting Time

As shown in Table 6 and Figure 4, with a fixed furnace slag content of 20%, a W/B ratio of 0.4, and 1% waste PE, the initial setting time and final setting time of the cement mortar are 315 min and 382 min, respectively; when the W/B ratio is 0.5, the initial setting time is 387 min and the final setting time is 505 min; when the W/B ratio is 0.6, the initial setting time is 468 min and the final setting time is 625 min. This indicates that the setting time increases with the W/B ratio. As the total water consumption is increased with the W/B ratio, the hydration heat reaction slows down and the setting time increases.

**Table 6.** Setting time of cement mortar with different W/B ratios and waste PE.

| W/B | RM (%) | AM (%) | Initial Setting | Final Setting |
|---|---|---|---|---|
| | GGBFS | PE | min | min |
| 0.4 | 20 | 0 | 324 | 394 |
| | | 1 | 315 | 382<br>334 |
| | | 2 | 302 | 365 |
| | | 3 | 284<br>268 | 347 |
| | | 4 | 268 | 334 |
| 0.5 | 20 | 0 | 401 | 518 |
| | | 1 | 387<br>356 | 505<br>474 |
| | | 2 | 375 | 496 |
| | | 3 | 367 | 488 |
| | | 4 | 356 | 474 |
| 0.6 | 20 | 0 | 479 | 633 |
| | | 1 | 468 | 625 |
| | | 2 | 460 | 617 |
| | | 3 | 447<br>439 | 609<br>598 |
| | | 4 | 439 | 589 |

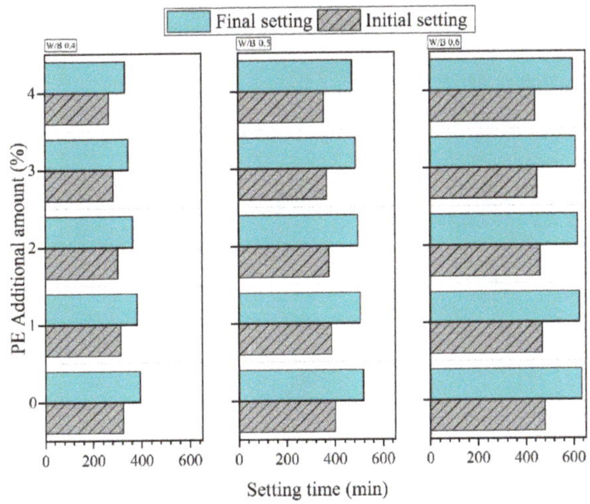


**Figure 4.** Setting time of cement mortar with different W/B ratios and waste PE.

When the W/B ratio is 0.5, the content of waste PE is 0%, 1%, 2%, 3%, and 4%, the slag content is 20%, the initial setting time is 401~356 min, and the final setting time is 518~474 min, meaning that the setting time is shortened with increasing content of waste PE.

### 3.4. Compressive Strength

As shown in Table 7 and Figure 5, at the age of 28 days, when the W/B ratio is 0.4, the content of GGBFS is 20%, the content of waste PE is 0%~4%, and the compressive strength is 54.1~48.9 MPa; when the W/B ratio is increased to 0.5, the compressive strength is 48.5~44.9 MPa (reduced by 8.2~10.4%); when the W/B ratio is increased from 0.5 to 0.6, the compressive strength is 37.2~34.7 MPa; the compressive strength is reduced by 11.9~16.6% when the W/B ratio is increased from 0.4 to 0.5 and reduced by 30.1~31.2% when the W/B ratio is increased from 0.5 to 0.6, indicating that the compressive strength is decreased with increasing W/B.

**Table 7.** Compressive strength of cement mortar with different W/B ratios and waste PE (unit: MPa).

| W/B | GGBFS (%) | PE (%) | 3 Days | 7 Days | 28 Days | 56 Days | 91 Days |
|---|---|---|---|---|---|---|---|
| 0.4 | 20 | 0 | 26.5 | 36.3 | 54.1 | 59.8 | 65.3 |
| | | 1 | 25.1 | 35.6 | 52.4 | 59.0 | 64.4 |
| | | 2 | 24.3 | 34.2 | 51.0 | 58.1 | 61.9 |
| | | 3 | 23.6 | 33.7 | 50.4 | 57.3 | 62.5 |
| | | 4 | 22.5 | 32.4 | 48.9 | 55.6 | 60.0 |
| 0.5 | 20 | 0 | 24.7 | 33.3 | 48.5 | 53.5 | 59.1 |
| | | 1 | 23.0 | 33.1 | 47.8 | 52.5 | 57.9 |
| | | 2 | 21.6 | 30.0 | 46.8 | 50.0 | 55.5 |
| | | 3 | 21.0 | 29.4 | 45.9 | 51.1 | 56.9 |
| | | 4 | 20.3 | 28.6 | 44.9 | 49.3 | 53.8 |
| 0.6 | 20 | 0 | 20.0 | 27.1 | 37.2 | 45.7 | 50.4 |
| | | 1 | 19.0 | 25.4 | 35.3 | 44.0 | 50.2 |
| | | 2 | 18.4 | 24.7 | 35.1 | 42.6 | 48.2 |
| | | 3 | 17.6 | 24.2 | 34.6 | 42.0 | 47.6 |
| | | 4 | 17.0 | 23.5 | 34.7 | 40.3 | 46.6 |

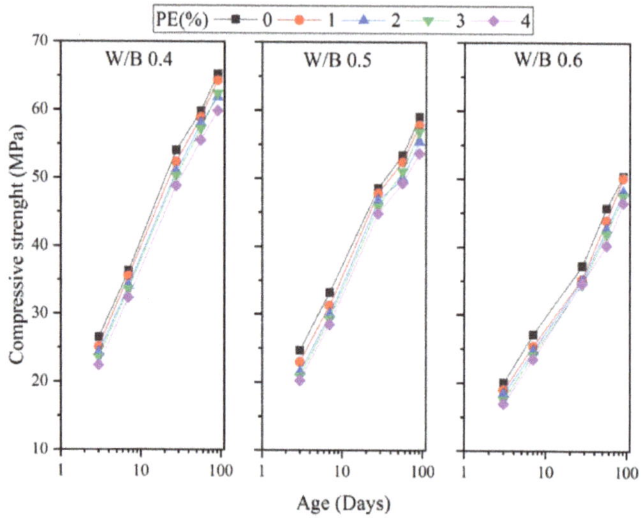


**Figure 5.** Compressive strength of cement mortar with different W/B ratios and waste PE.

When the W/B ratio is 0.5 and the content of slag is 20% at the age of 3 days, the strength of the control group is 24.7 MPa, and the compressive strength is 23.0~20.3 MPa (reduced by 6.9~17.8%) for a waste PE content of 1~4%. At the age of 28 days, the strength of the control group is 48.5 MPa and the compressive strength is 47.8~44.9 MPa (reduced by 1.4~7.5%) for a waste PE content of 1~4%. At the age of 91 days, the strength of the control group is 59.1 MPa and the compressive strength is 57.4~53.8 MPa (reduced by 2.0~9.1%) for a waste PE content of 1~4%, meaning that the compressive strength is decreased as the waste PE fiber material content is increased. As the waste PE is fibrous, the balling phenomenon is likely to occur in the specimen when the specimen is compacted. As this leads to the formation of fiber agglomerates inside the specimen, the pores inside the specimen are enlarged, and the waste PE is relatively soft, which cannot provide effective compressive strength for the specimen. When the content increases, the compressive strength decreases.

When the W/B ratio is 0.5 and the substitution amount of GGBFS is 20%, at the age of 28 days, the strength is reduced by 1.4~3.5% for a waste PE content of 1~2%, and the strength is reduced by 5.4~7.3% for a waste PE content of 3~4%. At the age of 56 days, when the waste PE content is 1~2% and 3~4%, the compressive strength is reduced by 1.9~4.5% and 6.5~7.9%, respectively. When the waste PE content is 1~2% and the strength reduction is controlled within 5%, the waste can be effectively eliminated, and the economy of the waste is enhanced, with a reduction in waste generation and a reduction in environmental pollution.

The compressive strength of the PE fiber material added at 2% showed a decreasing trend at 28 days of age with W/B ratios of 0.4, 0.5, and 0.6 compared to the control specimen without PE fiber material. The ratios were 0.94, 0.96, and 0.94, respectively. The decrease in compressive strength of the PE material added at 2% could be controlled within 5%. However, at 56 days of age, the compressive strength development in the three different W/B series exceeded the compressive strength of the control specimens at 28 days of age. Therefore, although the hardening properties of the compressive strength tended to decrease with the addition of 2% PE material, the compressive strength at late ages was 58.1, 50.0, and 42.6 MPa with W/B ratios of 0.4, 0.5, and 0.6, respectively, by adding GGBFS, and all of them exceeded the compressive strength of the control specimens. In addition, with the addition of PE fiber material to 4% of the specimen, the compressive strength at 56 days of age was also the same. The test results showed that by adding PE material to eliminate the waste, the compressive strength could be no lower than the test value of the control specimen without PE material by adding GGBFS to extend the concrete curing age.

### 3.5. Flexural Strength

As shown in Figure 6, at the age of 28 days, for a GGBFS content of 20%, W/B ratios of 0.4, 0.5, and 0.6, and a waste PE content of 0~4%, the flexural strengths are 20.2~17.4 MPa, 18.1~15.5 MPa, and 12.3~9.9 MPa, respectively, meaning that the flexural strength is decreased with increasing W/B ratio.

At the age of 3 days, when the W/B ratio is 0.5 and the GGBFS content is 20%, the flexural strength is 7.7~5.3 MPa, indicating that the flexural strength decreases as the content increases (6.5~29.8%). At the age of 28 days, the flexural strength of the mortar without waste PE is 18.1 MPa, and as the waste PE content is increased to 4%, the flexural strength is 15.5 MPa (reduced by 14.1%). At the age of 91 days, the flexural strength of the mortar without waste PE is 26.3 MPa; the flexural strength of the mortar with 4% waste PE is 24.3 MPa, and when the waste PE content is increased from 1% to 4%, the flexural strength is reduced by 3.1~7.6%, indicating that the flexural strength is decreased with increasing waste PE content. At the age of 56 days, when the amount of substituted GGBFS is 20% and the waste PE content is 1~2%, the strength is 2.2~5.2%, but the strength is reduced by 9.5~10.3% when the waste PE content is 3~4%. At the age of 91 days, the strength is reduced by 3~5.3% for a waste PE content of 1~2%, and the strength is reduced by 6.5~7.6% for a waste PE content of 3~4%. This indicates that the strength reduction can be controlled to within 6% when the waste PE content is lower than 2%. This value is the closest to that of the control group; hence, the waste can be effectively eliminated to reduce environmental pollution.

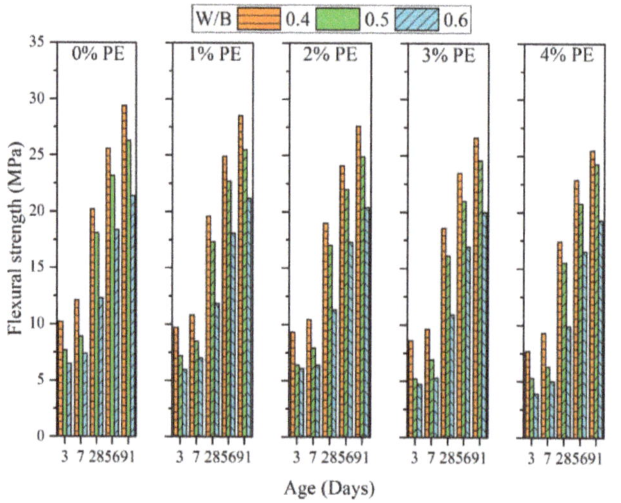


**Figure 6.** Flexural strength of cement mortar with different W/B ratios and waste PE.

### 3.6. Tensile Strength

As shown in Figure 7, at the age of 28 days, with a W/B ratio of 0.4, a GGBFS content of 20%, and a waste PE content of 0~4%, the tensile strength is 12.0~9.9 MPa; when the W/B ratio is increased to 0.5, the tensile strength is 11.2~8.6 MPa (reduced by 6.7~13.1%); when the W/B ratio is increased to 0.6, the tensile strength is 7.8~5.8 MPa, and the tensile strength is reduced by 35~41.4% in comparison to a W/B ratio of 0.4, meaning that the tensile strength is decreased with increasing W/B ratio.

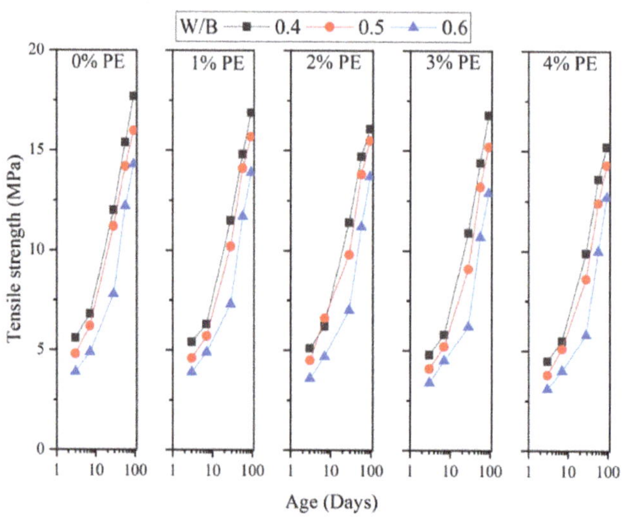


**Figure 7.** Tensile strength of cement mortar with different W/B ratios and waste PE.

At the age of 56 days, when the W/B ratio is 0.5, the GGBFS content is 20% and the tensile strength is 14.2~12.4 MPa for a waste PE content of 0~4%, indicating that the tensile strength is slightly decreased with the increased waste PE content; at the age of 91 days, the tensile strength is 16~14.3 MPa. As the waste PE is soft and the tensile strength cannot be effectively increased, the tensile strength is decreased with increasing waste PE content. Meanwhile, the tensile strength, flexural strength, and compressive strength show the same trend.

### 3.7. Ultrasonic Pulse Velocity

As shown in Figure 8, at the age of 28 days, with a W/B ratio of 0.4, GGBFS of 20%, and different percentages of waste PE, the ultrasonic pulse velocity is 4720~4501 m/s; the pulse velocity is 4386~4173 m/s for a W/B ratio of 0.5, which is reduced by 7.1~7.4% in comparison to that at a W/B ratio of 0.4; when the W/B ratio is 0.6, the pulse velocity is 3825~3611 m/s, which is reduced by 12.8~13.5% in comparison to that at a W/B ratio of 0.5, meaning that the ultrasonic pulse velocity is decreased with increasing W/B ratio.

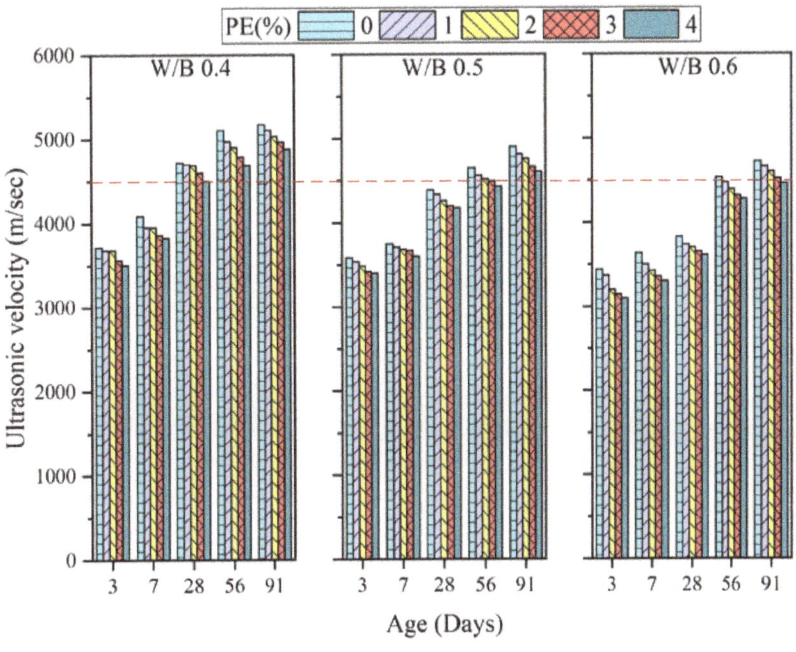


**Figure 8.** Ultrasonic velocity of cement mortar with different W/B ratios and waste PE.

At the age of 91 days, the control group shows a better ultrasonic pulse velocity of 4903 m/s; the ultrasonic pulse velocity is 4816 m/s for a waste PE content of 1%; and the ultrasonic pulse velocity is decreased to 4607 m/s for a waste PE content of 4%, which represents a 6.1% reduction in comparison to that for the control group. At the age of 56 days, when the substitution amount of GGBFS is 20% and the waste PE fiber material content is 1~2%, the ultrasonic pulse velocity is higher than 4500 m/s, meaning that the concrete quality is good.

When the waste PE content is increased, pores are formed as the specimen absorbs free water so that the ultrasonic pulse velocity is decreased. When the substitution amount of GGBFS is increased, the pulse velocity slowly increases at an earlier age; however, under the effect of pozzolanic reaction, the specimen pores are filled up at a later age, leading to a better density for the specimen, and the pulse velocity grows slowly at the later stage.

The internal porosity of the specimen is increased with increasing W/B ratio, resulting in a decrease in the pulse velocity. Furthermore, the ultrasonic pulse velocity is decreased as the content of waste PE fiber material is increased.

### 3.8. Water Absorption Rate

As shown in Figure 9, at the age of 28 days, with W/B ratios of 0.4, 0.5, and 0.6, a GGBFS content of 20%, and a waste PE content of 1~4%, the water absorption rates are 5.2~6.1%, 8.7~10.5%, and 13.1~14.2%, respectively. The water absorption rate is increased with an increasing W/B ratio.

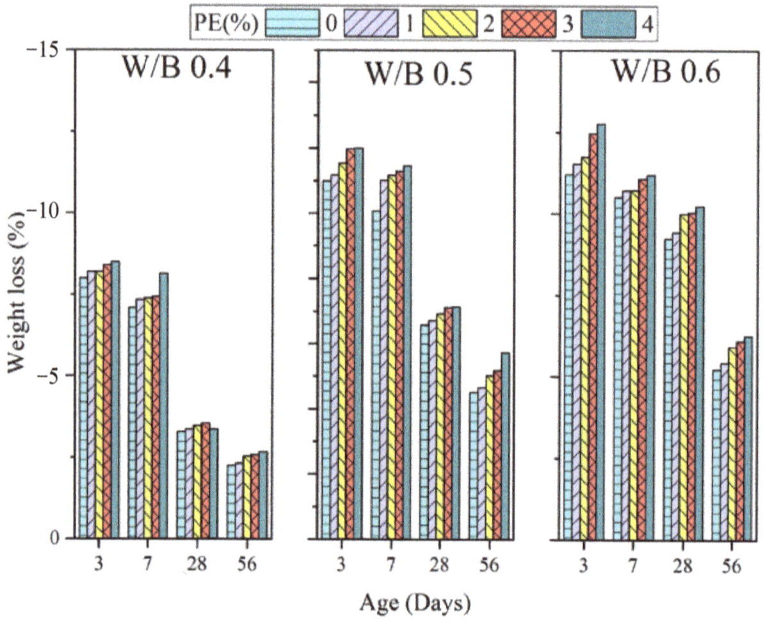


**Figure 9.** Water absorption rate of cement mortar with different W/B ratios and waste PE.

At the age of 28 days, with a W/B ratio of 0.5, a GGBFS content of 20%, and different percentages of waste PE added to the cement mortar, the water absorption rates are 8.7%, 9.0%, 9.4%, 9.9%, and 10.5%, respectively. The water absorption rate is increased with the waste PE content because the addition of waste PE induces the balling phenomenon, which increases the internal porosity of the specimen, and the water absorption rate is increased with PE fiber material content.

Pores are formed inside the specimen as the W/B ratio and waste PE content are increased so that the internal porosity of the specimen is increased. The hydration of GGBFS occurs slowly at an earlier age, so the specimen has a higher water absorption rate than the control group; however, under the effect of pozzolanic reaction, the specimen pores are filled up, and the specimen has a better density.

### 3.9. Resistivity

As shown in Table 8 and Figure 10, at the age of 28 days, for a W/B ratio of 0.4 and waste PE content of 1%, the resistivity is 26.6 kΩ-cm; for a W/B ratio of 0.5, the resistivity is 25.4 kΩ-cm (reduced by 4.5%); for a W/B ratio of 0.6, the resistivity is 20.9 kΩ-cm, which is reduced by 21.4% in comparison to that for a W/B ratio of 0.4, meaning that the resistivity is obviously decreased as the W/B ratio is increased for a waste PE content of 1%.

For a W/B ratio of 0.5 and a waste PE content of varying percentages, at the age of 3 days, the resistivity is 10~8.1 kΩ-cm, and the resistivity is reduced by 5~20% as the waste PE content is increased. At the age of 28 days, the resistivity of the specimen without waste PE is 26.3 kΩ-cm, and as the waste PE content is increased to 4%, the resistivity is 23.3 kΩ-cm (reduced by 11.4%), meaning that the resistivity is decreased with increasing waste PE content. At the age of 91 days, the resistivity of the control group is 38.9 kΩ-cm; the resistivity of the specimen with a waste PE content of 4% is 33.8 kΩ-cm, and when the PE material content is increased from 0 to 4%, the resistivity is reduced by 13.1%, meaning the specimen structure has a better density at the later age.

**Table 8.** Resistivity of cement mortar with different W/B ratios and waste PE (unit: kΩ-cm).

| W/B | GGBFS (%) | PE (%) | 3 Days | 7 Days | 28 Days | 56 Days | 91 Days |
|---|---|---|---|---|---|---|---|
| 0.4 | 20 | 0 | 11.7 | 15.8 | 26.6 | 35.6 | 39.0 |
| | | 1 | 11.0 | 15.4 | 26.8 | 34.8 | 38.8 |
| | | 2 | 10.4 | 14.4 | 25.2 | 34.0 | 37.3 |
| | | 3 | 9.6 | 14.0 | 24.6 | 33.8 | 36.4 |
| | | 4 | 8.9 | 13.2 | 24.0 | 33.3 | 35.7 |
| 0.5 | 20 | 0 | 10.0 | 14.7 | 26.3 | 35.5 | 38.9 |
| | | 1 | 9.5 | 14.2 | 25.4 | 34.5 | 37.3 |
| | | 2 | 9.1 | 13.7 | 24.9 | 34.0 | 36.2 |
| | | 3 | 8.6 | 13.4 | 24.0 | 33.6 | 36.3 |
| | | 4 | 8.0 | 12.2 | 23.3 | 32.6 | 33.8 |
| 0.6 | 20 | 0 | 8.4 | 13.2 | 21.6 | 32.8 | 35.4 |
| | | 1 | 8.0 | 12.1 | 20.9 | 33.2 | 35.2 |
| | | 2 | 7.7 | 12.1 | 20.9 | 32.6 | 34.9 |
| | | 3 | 7.2 | 11.5 | 20.0 | 31.8 | 33.8 |
| | | 4 | 6.9 | 11.6 | 19.4 | 30.9 | 32.2 |

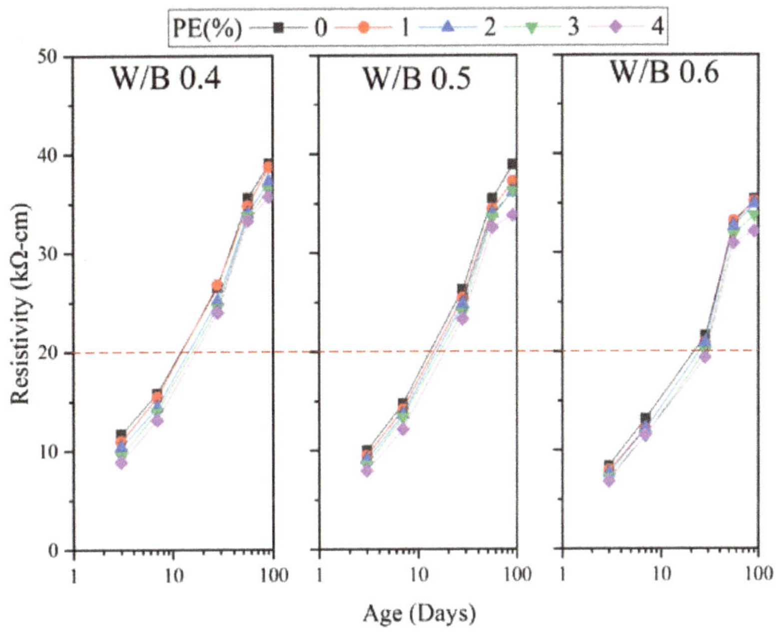


**Figure 10.** Resistivity of cement mortar with different W/B ratios and waste PE.

When the waste PE content is increased, fiber agglomerates are formed; hence, the resistivity is decreased; at the age of 28 days, the resistivity of various mix proportions is higher than 20 kΩ-cm, showing durability.

### 3.10. Resistance to Sulfate Attack

As shown in Figure 11, at the age of 28 days, with W/B ratios of 0.4, 0.5, and 0.6, GGBFS content of 20%, and waste PE content of 1~4%, the weight loss rates are −3.3~3.4%, −6.6~7.1%, and −9.2~10.2%, respectively. When the W/B ratio is increased, the resistance to sulfate attack is degraded, and the weight loss is increased. When the W/B ratio is low, the internal porosity of the specimen is reduced, the specimen is denser, and the resistance to sulfate attack is better.

At the age of 28 days, when the W/B ratio is 0.5, the weight loss rates for cement mortar specimens with a waste PE content of 0%, 1%, 2%, 3%, and 4% are −6.6%, −6.7%, −6.9%, −7.1%, and −7.1%, respectively, after five cycles, meaning that when the waste PE content is increased, the weight loss rate is increased, and the resistance to sulfate attack is degraded. This phenomenon was observed because the internal porosity of the material increases with the waste PE content, and the sulfate solution more easily erodes the specimen. When the W/B ratio and the waste PE content are increased, the internal porosity of the specimen and the weight loss rate are increased.

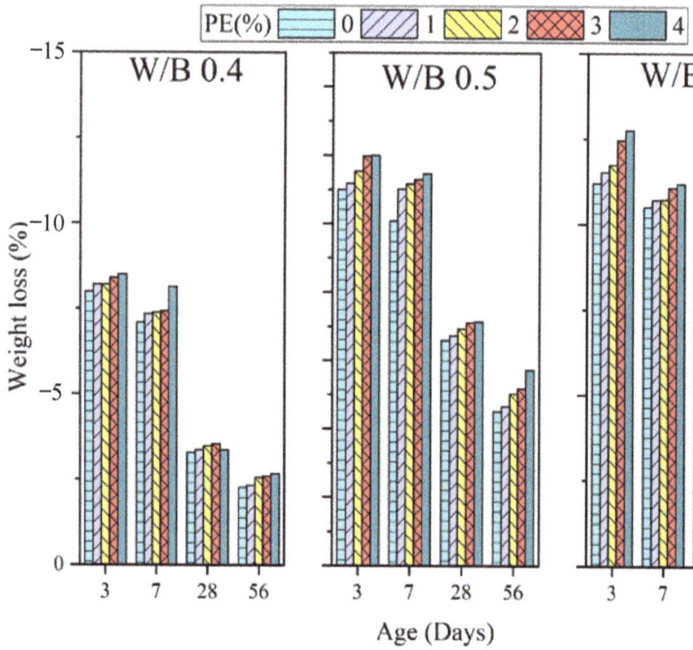


**Figure 11.** Resistance to sulfate attack of cement mortar with different W/B ratios and waste PE.

### 3.11. Microscopic Analysis

As shown in Figure 12, at the age of 7 days, for a W/B ratio of 0.5 and waste PE contents of 1%, 2%, 3%, and 4%, the number of fibers inside the cement mortar specimen is significantly increased. According to the EDS analysis, the main elements were O, Ca, C, and Si. The waste PE is found to increase the specimen porosity, verifying that the compressive strength is decreased with increasing waste PE content.

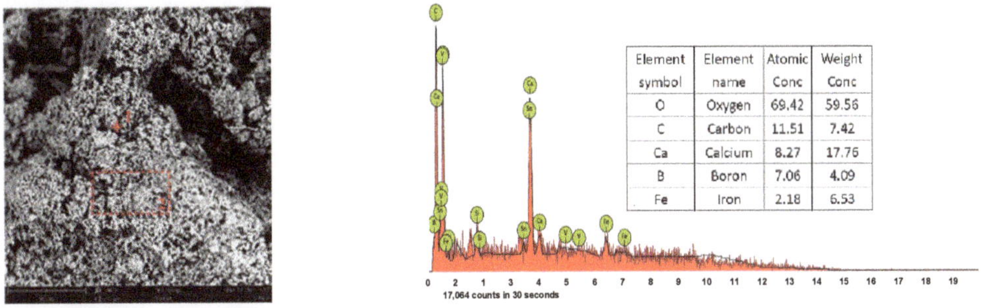


| Element symbol | Element name | Atomic Conc | Weight Conc |
|---|---|---|---|
| O | Oxygen | 69.42 | 59.56 |
| C | Carbon | 11.51 | 7.42 |
| Ca | Calcium | 8.27 | 17.76 |
| B | Boron | 7.06 | 4.09 |
| Fe | Iron | 2.18 | 6.53 |

(**a**) PE 1%

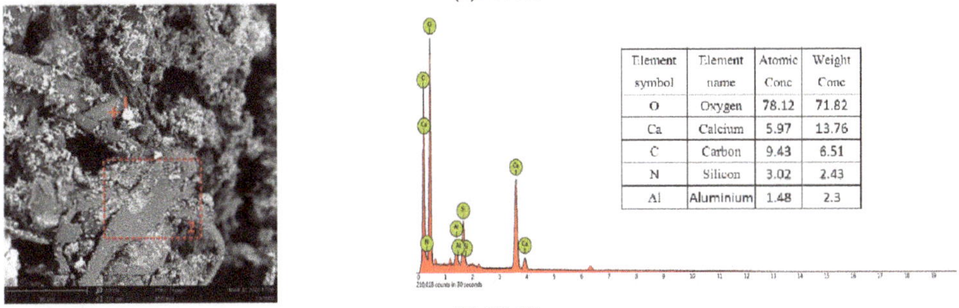

| Element symbol | Element name | Atomic Conc | Weight Conc |
|---|---|---|---|
| O | Oxygen | 78.12 | 71.82 |
| Ca | Calcium | 5.97 | 13.76 |
| C | Carbon | 9.43 | 6.51 |
| N | Silicon | 3.02 | 2.43 |
| Al | Aluminium | 1.48 | 2.3 |

(**b**) PE 2%

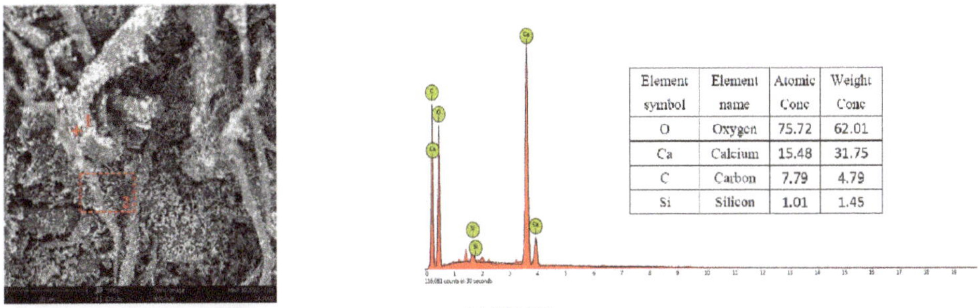

| Element symbol | Element name | Atomic Conc | Weight Conc |
|---|---|---|---|
| O | Oxygen | 75.72 | 62.01 |
| Ca | Calcium | 15.48 | 31.75 |
| C | Carbon | 7.79 | 4.79 |
| Si | Silicon | 1.01 | 1.45 |

(**c**) PE 3%

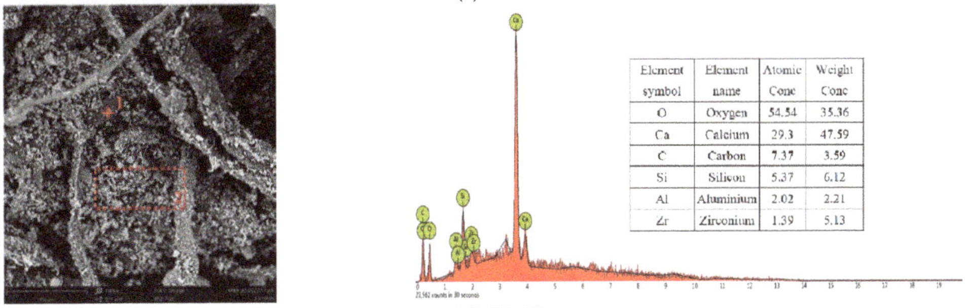

| Element symbol | Element name | Atomic Conc | Weight Conc |
|---|---|---|---|
| O | Oxygen | 54.54 | 35.36 |
| Ca | Calcium | 29.3 | 47.59 |
| C | Carbon | 7.37 | 3.59 |
| Si | Silicon | 5.37 | 6.12 |
| Al | Aluminium | 2.02 | 2.21 |
| Zr | Zirconium | 1.39 | 5.13 |

(**d**) PE 4%

**Figure 12.** SEM of cement mortar with different waste PE (×1000) (day 7, W/B ratio of 0.5).

## 4. Conclusions

1. The slump and flow properties of a specimen decrease as the waste PE content is increased so that the overall workability is degraded; the slump is increased with increasing W/B ratio.
2. The setting time is shortened as the waste PE content is increased. When the W/B ratio is increased, the water consumption is increased so that the overall setting time is increased.
3. The compressive strength, flexural strength, and tensile strength decrease as the waste PE content is increased. For a waste PE content of 2%, the waste can be eliminated most effectively; meanwhile, the strength is decreased with increasing W/B ratio and increased with increasing age.
4. The ultrasonic pulse velocity is reduced by 6.9%~8.7% as the age and the waste PE content are increased. Since the waste PE fiber material absorbs the free water in cement mortar, it causes voids after cement mortar hardening and reduces the density of the specimen. Therefore, in the ultrasonic test, the ultrasonic wave speed showed a decreasing trend with the addition of PE material from 1% to 4%, and the decrease increased from 6.9% to 8.7%. With the addition of 20% GGBFS, which has the property of delaying the reaction of the Portland, the ultrasonic wave speed can exceed 4500 m/s at the age of 28 days only if the W/B ratio is 0.4. With the increase in the W/B ratio, the test results show that at the age of 56 days and 91 days, the amount of waste PE material added within 2%, although the density of the cement mortar specimen will be reduced, it can still meet the ultrasonic wave speed of 4500 m/s. The density requirement of wave speed is over 4500 m/s. The addition of 2% of waste PE material can be used as a reference for concrete mixture proportioning design, which is helpful to remove the waste PE material.
5. As the W/B ratio and waste PE content are increased, pores are formed inside the specimen, and the water absorption rate is increased. The specimen has a better density due to hydration at a later age, so the water absorption rate is decreased. Regardless of whether the cement mortar specimen has added waste PE fiber material or not, its water absorption rate will show a decreasing trend with the development of age, and because of the addition of 20% GGBFS, at a late age, it delays the hydration reaction of the Brahmin, so that the specimen shows a better dense condition internally at a late age. At the same age, for the series with a W/B ratio of 0.4, the effect of the addition of waste PE fiber material on the water absorption was minimal, and the same was true for W/B ratios of 0.5 and 0.6. The correlation between water absorption and the reaction time of hydration, i.e., the development of age, indicates that the addition of waste PE material is feasible as long as the hydration reaction is developed until the cement mortar meets the requirement of density.
6. The resistivity is decreased as the waste PE content and W/B ratio are increased. The resistivity shows the same trend as that for the ultrasonic pulse velocity. At the age of 28 days, the durability resistance test values of all specimens exceeded 20 kΩ-cm, indicating that although the addition of 20% GGBFS delayed the hydration effect, the resistance values met the requirement of compactness. Compared with the age of 28 days, the difference in resistance values between the PE materials added to the specimens at the same W/B was more minor. Still, at the late age of 91 days, the difference in resistance values between the PE materials added to the specimens at the same W/B was more significant. With the addition of PE materials, the resistance values decreased, indicating that the waste PE materials could not produce hydration. After a complete hydration reaction, the voids were formed. In addition, microscopic analysis by an electron microscope revealed that the PE fibers in the specimen made tangled masses, which was also the reason for the increase in voids.
7. The resistance to sulfate attack is degraded for increasing W/B ratio and waste PE content; however, with the pozzolanic reaction, as the specimen pores are filled with

hydrates, the specimen has a better density, and the decrease in weight loss rate is not obvious at a later stage.

8. Under the condition of a W/B ratio of 0.4, the addition of 2% waste PE and 20% GGBFS in the mix proportion leads to better hardening properties and durability to achieve the goal of waste reuse. In addition, it suggests the appropriate PE addition ratio for the reference of concrete mixture proportioning design, so as to achieve waste minimization by recycling PE film on the surface of disposable containers.

**Author Contributions:** Conceptualization, K.-T.L., C.-C.H., and H.-Y.W.; methodology, H.-Y.W.; formal analysis, K.-T.L.; resources, C.-C.H.; data curation, F.-L.W.; writing—original draft preparation, K.-T.L.; writing—review and editing, C.-C.H.; supervision, H.-Y.W. All authors have read and agreed to the published version of the manuscript.

**Funding:** This research received no external funding.

**Data Availability Statement:** Not applicable.

**Conflicts of Interest:** The authors declare no conflict of interest.

## References

1. Zhaurova, M.; Soukka, R.; Horttanainen, M. Multi-criteria evaluation of $CO_2$ utilization options for cement plants using the example of Finland. *Int. J. Greenh. Gas Control* **2021**, *112*, 103481. [CrossRef]
2. Andrew, R.M. Global $CO_2$ emissions from cement production, 1928–2018. *Earth Syst. Sci. Data* **2019**, *11*, 1675–1710. [CrossRef]
3. Jang, J.G.; Kim, G.M.; Kim, H.J.; Lee, H.K. Review on recent advances in $CO_2$ utilization and sequestration technologies in cement-based materials. *Constr. Build. Mater.* **2016**, *127*, 762–773. [CrossRef]
4. Abdulkareem, M.; Havukainen, J.; Horttanainen, M. How environmentally sustainable are fibre reinforced alkali-activated concretes. *J. Clean. Prod.* **2019**, *236*, 117601. [CrossRef]
5. Islam, M.J.; Islam, K.; Shahjalal, M.; Khatun, E.; Islam, S.; Razzaque, A.B. Influence of different types of fibers on the mechanical properties of recycled waste aggregate concrete. *Constr. Build. Mater.* **2022**, *337*, 127577. [CrossRef]
6. Ramakrishnan, K.; Pugazhmani, G.; Sripragadeesh, R.; Muthu, D.; Venkatasubramanian, C. Experimental study on the mechanical and durability properties of concrete with waste glass powder and ground granulated blast furnace slag as supplementary cementitious materials. *Constr. Build. Mater.* **2017**, *156*, 739–749. [CrossRef]
7. Geyer, R.; Jambeck, J.R.; Law, K.L. Production and use, and fate of all plastics ever made. *Sci. Adv.* **2017**, *3*, e1700782. [CrossRef]
8. Jambeck, J.R.; Geyer, R.; Wilcox, C.; Siegler, T.R.; Perryman, M.; Andrady, A.; Narayan, R.; Law, K.L. Plastic waste inputs from land into the ocean. *Science* **2015**, *347*, 768–771. [CrossRef]
9. Barnes, D.K.A.; Galgani, F.; Thompson, R.C.; Barlaz, M. Accumulation and fragmentation of plastic debris in global environments. *Philos. Trans. R. Soc. B Biol. Sci.* **2009**, *364*, 1526. [CrossRef]
10. Khalid, M.Y.; Arif, Z.U.; Ahmed, W.; Arshad, H. Recent trends in recycling and reusing techniques of different plastic polymers and their composite materials. *Sustain. Mater. Technol.* **2022**, *31*, e00382. [CrossRef]
11. Ghuzlan, K.A.; Al-Khateeb, G.G.; Qasem, Y. Rheological properties of polyethylene/modified asphalt binder. *Athens J. Technol. Eng.* **2015**, *2*, 75–88. [CrossRef]
12. Ferdous, W.; Manalo, A.; Siddique, R.; Mendis, P.; Zhuge, Y.; Wong, H.S.; Lokuge, W.; Aravinthan, T.; Schubel, P. Recycling of landfill wastes (tyres, plastics and glass) in construction—A review on global waste generation, performance, application and future opportunities. *Resour. Conserv. Recycl.* **2021**, *173*, 105745. [CrossRef]
13. Triantafillopoulos, N.; Koukoulas, A.A. The future of single-use paper coffee cups: Current progress and outlook. *BioResources* **2020**, *15*, 7260–7287. [CrossRef]
14. Dele, R. Application of plastics and paper as food packaging materials—An overview. *Emir. J. Food Agric.* **2013**, *25*, 177–188.
15. Islam, J.; Shahjalal, M. Effect of polypropylene plastic on concrete properties as a partial replacement of stone and brick aggregate. *Case Stud. Constr. Mater.* **2021**, *15*, e00627. [CrossRef]
16. Kumar, R.; Kumar, M.; Kumar, I.; Srivastava, D. A review on utilization of plastic waste materials in bricks manufacturing process. *Mater. Today Proc.* **2021**, *46*, 6775–6780. [CrossRef]
17. Chanhoun, M.; Padonou, S.; Adjovi, E.C.; Olodo, E.; Doko, V. Study of the implementation of waste wood, plastics and polystyrenes for various applications in the building industry. *Constr. Build. Mater.* **2018**, *167*, 936–941. [CrossRef]
18. Zhang, C.Y.; Han, R.; Yu, B.; Wei, Y.M. Accounting process-related $CO_2$ emissions from global cement production under Shared Socioeconomic Pathways. *J. Clean. Prod.* **2018**, *184*, 451–465. [CrossRef]
19. Adina, B.; Masek, O.; Oakey, J.E. $CO_2$ capture technologies for cement industry. *Energy Procedia* **2009**, *1*, 133–140.
20. Hopewell, J.; Dvorak, R.; Kosior, E. Plastics recycling: Challenges and opportunities. *Philos. Trans. R. Soc. B Biol. Sci.* **2009**, *364*, 2115–2126. [CrossRef]

21. Yesilata, B.; Yusuf, I.; Paki, T. Thermal insulation enhancement in concretes by adding waste PET and rubber pieces. *Constr. Build. Mater.* **2009**, *23*, 1878–1882. [CrossRef]
22. Abu-Saleem, M.; Zhuge, Y.; Hassanli, R.; Ellis, M.; Rahman, M.; Levett, P. Evaluation of concrete performance with different types of recycled plastic waste for kerb application. *J. Build. Eng.* **2021**, *293*, 123477. [CrossRef]
23. Basha, S.I.; Ali, M.R.; Al-Dulaijan, S.U.; Maslehuddin, M. Mechanical and thermal properties of lightweight recycled plastic aggregate concrete. *J. Build. Eng.* **2020**, *32*, 101710. [CrossRef]
24. Wang, C.H.; Lin, H.T. Quantification of food packaging generation and material loss from major retailers in Taipei, Taiwan. *Waste Manag.* **2022**, *137*, 139–149. [CrossRef]
25. Taerwe, L.; Van Gysel, A. Influence of steel fibers on design stress-strain curve for high-strength concrete. *J. Mech. Eng.* **1996**, *122*, 695–704. [CrossRef]
26. Kakemi, M.; Hannant, D.J. Effect of autoclaving on cement composites containing polypropylene, glass and carbon fibres. *Cem. Concr. Compos.* **1996**, *18*, 61–66. [CrossRef]
27. Song, P.S.; Hwang, S.; Sheu, B.C. Strength properties of nylon- and polypropylene-fiber-reinforced concretes. *Cem. Concr. Res.* **2005**, *35*, 1546–1550. [CrossRef]
28. Wang, W.; Wu, S.; Dai, H. Fatigue behaviour and life prediction of carbon fiber reinforced concrete under cyclic flexural loading. *Mater. Sci. Eng. A* **2006**, *434*, 347–351. [CrossRef]
29. Hsie, M.; Tu, C.; Song, P.S. Mechanical properties of polypropylene hybrid fiber-reinforced concrete. *Mater. Sci. Eng. A* **2008**, *494*, 153–157. [CrossRef]
30. Song, P.S.; Hwang, S. Mechanical properties of high-strength steel fiber-reinforced concrete. *Constr. Build. Mater.* **2004**, *18*, 669–673. [CrossRef]
31. Rumšys, D.; Bačinskas, D.; Spudulis, E.; Meškėnas, A. Comparison of material properties of lightweight concrete with recycled polyethylene and expanded clay aggregates. *Procedia Eng.* **2017**, *172*, 937–944. [CrossRef]
32. Yoo, D.Y.; Kim, M.J. High energy absorbent ultra-high-performance concrete with hybrid steel and polyethylene fibers. *Constr. Build. Mater.* **2019**, *209*, 354–363. [CrossRef]
33. Saikia, N.; De Brito, J. Waste polyethylene terephthalate as an aggregate in concrete. *Mat. Res.* **2013**, *16*, 341–350. [CrossRef]
34. Albano, C.; Camacho, N.; Hernadez, M.; Matheus, A.; Gutierrez, A. Influence of content and particle size of waste pet bottles on concrete behavior at different w/c ratios. *Waste Manag.* **2009**, *29*, 2707–2716. [CrossRef]
35. Choi, Y.W.; Moon, D.J.; Chung, J.S.; Cho, S.K. Effect of waste PET bottles aggregate on properties of concrete. *Cem. Concr. Res.* **2005**, *35*, 776–781. [CrossRef]
36. Frigione, M. Recycling of PET bottles as fine aggregate in concrete. *Waste Manag.* **2010**, *30*, 1101–1106. [CrossRef]
37. Hanna, T.H.; Mosa, I.H. Effect of using waste polyethelene production on mechanical properties of concrete. *J. Duhok Univ.* **2012**, *15*, 37–42.
38. Irwan, J.M.; Asyraf, R.M.; Othman, N.; Koh, H.B.; Annas, M.M.K.; Faisal, S.K. The mechanical properties of PET fiber reinforced concrete from recycled bottle wastes. *Adv. Mat. Res.* **2013**, *795*, 347–351. [CrossRef]
39. Juki, M.I.; Muhamad, K.; Mahamad, M.K.A.; Boon, K.H.; Othman, N.; Kadir, A. Abdul; Roslan, M.A.; Khalid, F.S. Development of concrete mix design nomograph containing polyethylene terephthalate (PET) as fine aggregate. *Adv. Mat. Res.* **2013**, *701*, 12–16.

buildings

MDPI

Article

# Real-Time Monitoring for Effects of Vibration and Temperature of Construction Site on Steel Assembly Bracing of Foundation Pit

Yingwu Yang [1], Hangbin Zeng [2], Xingxi Liu [2], Bo Yang [2,*] and Ying Li [3,*]

1 College of Landscape and Architecture, Zhejiang A&F University, Hangzhou 311300, China
2 School of Civil Engineering and Architecture, Zhejiang Sci-Tech University, Hangzhou 310018, China
3 Zhejiang Province Institute of Architectural Design and Research, Hangzhou 310006, China
* Correspondence: youngbo@zstu.edu.cn (B.Y.); liying3104@163.com (Y.L.)

**Abstract:** In this paper, a real-time monitoring system—including vibration acceleration sensors, temperature sensors, and static and dynamic strain sensors—is used to monitor the safety status of a steel assembly bracing in a practical project. It uses 5G wireless networking technology to transmit monitoring data to a cloud server for early warning of abnormal changes and development trends. Real-time monitoring data obtained in a construction site are used as the inputs of the finite element model and the corresponding results of numerical simulation are compared with the results from the real-time monitoring. It can be concluded that: (1) the stress caused by environmental temperature is very significant which can be higher than the initial prestress of the steel assembly bracing; (2) the stress caused by the vertical vibration mainly from construction vehicles is not remarkable, however, the vertical frequency-weighted acceleration of support vibration is relatively large which can affect the sense of safety of engineering technicians on site; (3) the combination of the environmental temperature and vertical vibration does not affect the safety of the steel assembly bracing.

**Keywords:** steel assembly bracing; foundation pit; environmental temperature; vibration analysis; real-time monitoring

**Citation:** Yang, Y.; Zeng, H.; Liu, X.; Yang, B.; Li, Y. Real-Time Monitoring for Effects of Vibration and Temperature of Construction Site on Steel Assembly Bracing of Foundation Pit. *Buildings* **2023**, *13*, 450. https://doi.org/10.3390/buildings13020450

Academic Editor: Humberto Varum

Received: 31 December 2022
Revised: 31 January 2023
Accepted: 2 February 2023
Published: 6 February 2023



## 1. Introduction

Reasonable setting of internal bracing is an important guarantee for the construction safety of deep and large foundation pits. Due to the temporary nature of foundation pit engineering, it is necessary to consider three factors including safety, economy and construction period in the design of a bracing system. A reinforced concrete structure is a traditional and mature support form for a bracing system. However, it has a long construction period of structure-forming and can easily cause excessive environmental noise and produce more construction waste after the bracing system has been removed. In recent years, engineering research companies in Southeast China have developed steel assembly bracing as a new type of bracing system. Compared with traditional bracing technology, its on-site construction period is short, and there are no noise and construction waste problems. Hence, steel assembly bracing is more environmentally friendly and has a wide application market in engineering.

As far as general engineering experience is concerned, even if the reinforced concrete structure is used as the internal support of a foundation pit, the influence of temperature on the internal force of the support must be considered when the length exceeds 40 m. Furthermore, the steel assembly bracing is more sensitive to temperature than the concrete bracing. Affected by higher thermal conductivity and lower specific heat capacity, the temperature of steel coated with dark antirust coating is usually higher than that of concrete in the same environment. The surface temperature of a steel structure bridge which is similar to steel assembly bracing in structural form is 15~20 °C higher than the environmental temperature, and the stress of each component is also different under solar radiation [1,2].

The steel inner brace of a supporting system will produce large additional axial force and axial deformation under the influence of environmental temperature [3,4]. In engineering cases, the method of covering the supporting system with the surface layer is sometimes used for thermal insulation treatment. Other studies related to steel truss structures show that the dynamic characteristics of steel structures will also change under the influence of environmental temperature, and even cause structural damage in severe cases [5–7]. Therefore, the effect of temperature on the steel assembly bracing is significant and cannot be ignored.

Steel assembly bracing has been designed as one or multiple supports which consist of H-shaped steel parts [8,9], among which high-strength bolts are usually adopted to connect the steel. The compression bar analysis method of steel structure is used to design the bracing. Steel assembly bracing consisting of steel pipes or section steels has been used to support the vertical retaining wall structure around the foundation pit with a certain prestress applied in the axial direction of members to reduce the inward deformation. The study reported that the steel assembly bracing with appropriate prestress can significantly reduce the horizontal displacement of the retaining structure of the foundation pit [10]. When the environmental temperature changes greatly, sometimes engineers even use steel diagonal assembly bracing for internal support [11]. In fact, the axial force of each horizontal support transfers with the deformation of the retaining wall structure and becomes more complicated if the impact of environmental temperature is also considered [12,13]. With the increase of support length, the influence of temperature will become more and more significant. Sometimes it is even necessary to reinforce the soil near the support to ensure that there is no separation between the soil and the retaining structure when the support shrinks [14].

Calculation of steel bracing of the foundation pit is usually based on the elastic foundation beam model to consider the effect of temperature, or based on the Winkler model to establish the empirical model of the effect of temperature on the bracing system of the foundation pit. Due to its complexity, inappropriate simplified models sometimes underestimate the temperature effect on steel bracing, which eventually leads to bracing failure [15]. For steel assembly bracing, owing to the influence of lateral deformation along the depth direction of the foundation pit, the influence of temperature on the second layer bracing is sometimes greater than the first layer [16]. In the low-temperature area of North China, the reduction of axial force caused by the temperature drop of steel bracing will also cause the additional deformation of the foundation pit retaining to the pit [17]. Meanwhile, earthwork trucks can cause obvious vibration of steel bracing when they pass through concrete trestles near the end of steel bracing in which vehicle load, speed and distance are very important influencing factors [18].

To the authors' knowledge, at present, the research on steel assembly bracing is not enough and a safety real-time monitoring system has not been reported to be used to study steel assembly bracing. In the practical project studied in the paper, another problem also deserves attention: on-site technicians reflected that the vibration of the steel assembly bracing was sometimes obvious and thus complained about the insecurity. Therefore, the safety of steel assembly bracing under multiple working conditions is studied by the safety real-time monitoring system combined with numerical simulation in the paper.

## 2. Monitoring and Numerical Method

### 2.1. Technical Information of Foundation Pit Support

The size of the foundation pit of Huidong No.14 plot in Hangzhou City is 144 m × 215 m × 15 m, which is shown in Figure 1.

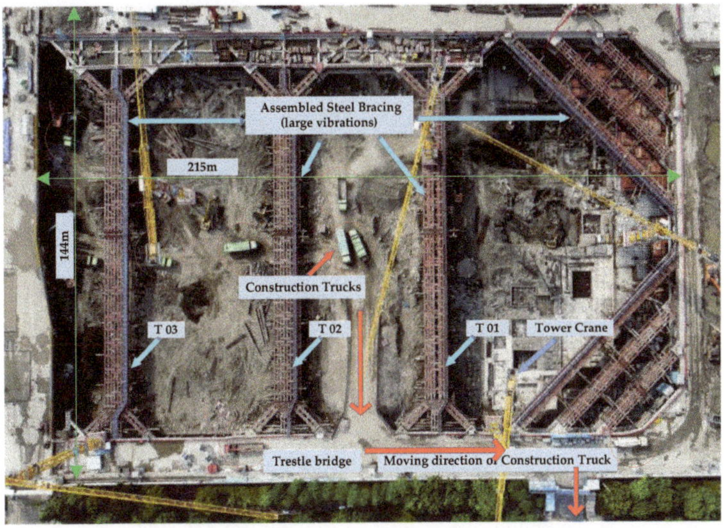


**Figure 1.** Steel assembly bracing of foundation pit and truck driving route.

Around the foundation pit, the reinforced concrete diaphragm walls and the triaxial cement-soil mixing piles form the lateral enclosure, steel assembly bracing are used in the horizontal direction, and some reinforced concrete purlins are used as trestle bridges on the earthwork transportation path. Figure 1 shows the driving route of earthwork trucks at the foundation pit construction site and the plane layout of the steel assembly bracing. The transportation channel for the earthwork truck is closest to the steel assembly bracing T01. As can be seen from Figure 2, there are upper and lower steel assembly internal bracings along the height direction of the foundation pit, and both ends are supported on the reinforced concrete trestle bridge for earthwork truck transportation. In addition, vertical steel lattice columns are set at a certain distance in the foundation pit to provide vertical support along the length direction of the steel assembly bracing.

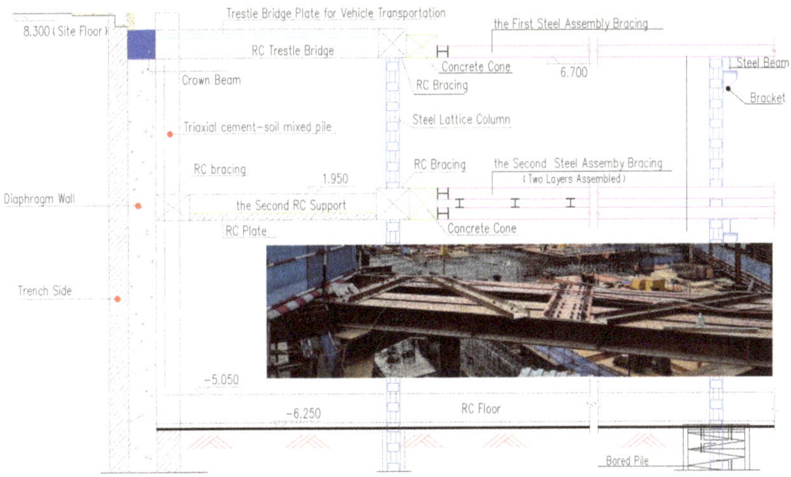


**Figure 2.** Side view of steel assembly bracing of foundation pit.

In the process of excavating and transporting soil, the steel assembly internal bracing, which is also used as pedestrian passageway, has obvious vibration when earthwork vehicles drive through the pit. This phenomenon makes engineering technicians pay attention to the safety of the foundation pit. Because the project has been under construction, and the site project progress and working conditions are changeable, it is difficult for the traditional once-a-day monitoring technology to obtain axial force, temperature and support vibration data at the same time and realize safety early warning in case of an emergency. In this study, a real-time monitoring network is established by virtue of the wireless sensor monitoring technology in the construction site to support safety monitoring through real-time monitoring data.

### 2.2. Monitoring Points Locations

Steel tube internal bracing is a common form of the foundation pit, which has been widely used in a narrow and regular foundation pit. In order to monitor the safety of soil, buildings and structures, an underground pipe network and facilities around the foundation pit during excavation and construction of a deep and large foundation pit or special foundation pit, the designers must determine the safety monitoring points and monitoring contents of the steel pipe support in the design documents according to national standards, design documents and engineering experience [19]. In this study, the monitoring experience of reinforced concrete support and steel tube support is used for reference, to set up the monitoring contents and layout of various measuring points for steel assembly bracing.

According to the site investigation, the vertical vibration of T01 reflected by engineers and construction workers can be obviously felt, especially when several earthwork trucks pass through the trestle bridge one after another. In order to master the relationship between the bracing vibration and running vehicle, and study its influence on the safety of the bracing system, the resistive dynamic strain gauges (precision $\pm 0.1$ με, range $\pm 3000$ με, type XKKJ-DTA100) and vibration acceleration sensors (sensitivity 0.3 V*s/m, range of scale 20 $m/s^2$, range of frequency 0.25~80 Hz, type BY-S07) have been arranged on the first track (initial pre-axial force $P_{01}$ = 3200 kN, compressive stress 18.22 MPa) and the second track (initial pre-axial force $P_{01}$ = 8000 kN, compressive stress 22.78 MPa) of T01 which is the closest to the trestle passage and has obvious vibration sense to monitor the dynamic response. Vibrating wire static strain gauges (strain: precision $\pm 7$ με, range $\pm 1500$ με, type T-900, temperature: precision $\pm 0.2$ °C (0~50 °C), range −40~60 °C, type CG-01) have been arranged on the surface of an H-beam because they are widely used in monitoring for foundation pit support and can be utilized to measure the steel surface temperature and surface strain of an H-beam at the same time [20]. In addition, in order to analyze the correlation between vehicle and the response of structural dynamics, video surveillance is used to monitor earthwork truck transportation. A 5G network is used for wireless networking and remotely transmits data to the cloud database. Figure 3 shows the locations and numbers of various types of sensors on the T01 steel assembly internal bracing.

As can be seen in Figure 3, the vibration acceleration sensors are fixed on the beam for the reinforced concrete trestle, while all kinds of measuring points are installed on the upper flange of the H-beam for the steel assembly bracing. Ten dynamic strain measuring points (DSS-01~DSS-10) are arranged on the first steel assembly bracing, and six dynamic strain measuring points (DSS-11~DSS16) are arranged on the second steel assembly bracing. Two static strain measuring points (SSTS-01, SSTS-02) are arranged to master the static strain and steel temperature of steel assembly bracing. At the same time, acceleration sensors, numbered with 3AS-01 and 3AS-02, are arranged on the trestle bridge and steel assembly bracing, respectively. It is noted that the sampling interval of static strain and temperature data is 5 min; the sampling frequency of dynamic strain is 50 Hz; and the sampling frequency of vibration acceleration is 200 Hz.

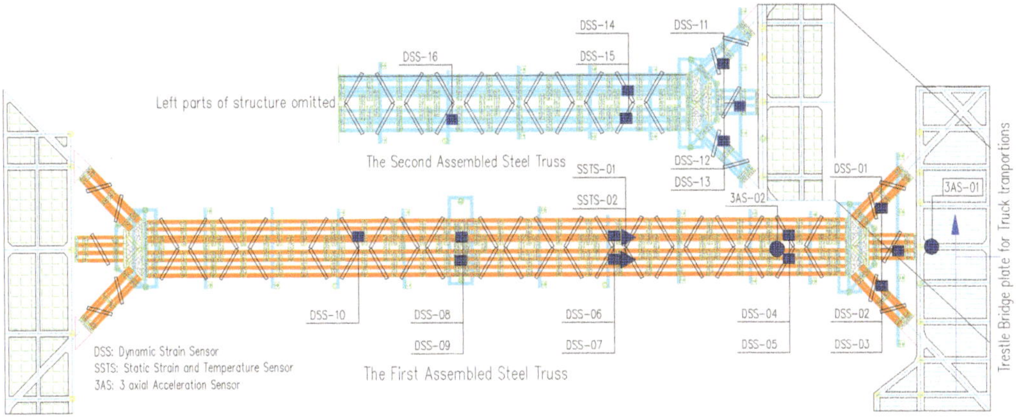


**Figure 3.** Layout of monitoring points in steel assembly bracing.

According to the monitoring and analysis of real-time data for more than one month, the responses of vibration acceleration of the reinforced concrete trestle bridge and the steel assembly bracing are significantly larger when the earthwork truck passes through the trestle bridge, and the amplitude of vertical vibration acceleration on the first steel assembly bracing is larger than that of the concrete trestle bridge. In addition, the steel assembly bracing is obviously affected by environmental temperature. In order to analyze the influence of vehicle vibration and temperature on the safety of steel assembly bracing, the finite element method is adopted to analyze the structural responses under the combined action of vibration and temperature.

### 2.3. Numerical Method

In this section, a three-dimensional finite element method is adopted to analyze the steel assembly bracing via ABAQUS software. The calculation results can be used as a reference to compare with the results from the real-time monitoring [21]. The finite element analysis involves modal analysis, steady-state thermal analysis and dynamic analysis. The vertical natural frequency and mode of the steel assembly bracing are obtained by modal analysis; the steady-state thermal analysis is used to analyze the temperature effect on the assembly bracing; the dynamic analysis is used to analyze the vibration effect of transport vehicles. Finally, the most unfavorable combination of temperature and vibration is considered to analyze the steel assembly bracing.

It should be emphasized that the highest temperature and the maximum vibration response of the vehicle driving during the monitoring period are simultaneously considered in the numerical simulation which cannot occur at the same time in practice. Therefore, the numerical simulation results in the paper will be conservative.

#### 2.3.1. Numerical Simulation of Soil Layer

In order to consider the influence of the surrounding soil layer on foundation pit structure, the length and width of the soil layer are taken as four times that of the foundation pit size, and the thickness of the soil layer is taken as twice that of the foundation pit depth. The total size of the ABAQUS finite element model is set as 576 m × 860 m × 30 m [22]. An eight-node C3D8R volumetric element is adopted for the soil layer and the number of elements is about 334,000. Considering that there is a concrete trestle bridge, concrete crown beam and diaphragm wall around the steel assembly bracing to be considered, a linear elastic model is adopted to simplify the calculation of soil layer parameters, which are shown in Table 1. In the process of calculation and analysis, the soil layer adopts C3D8R element type with hourglass control. At the same time, in order to improve the calculation accuracy, the soil elements inside and around the foundation pit are encrypted, and the size of the elements far away from the foundation pit is gradually increased to improve the calculation speed. The bottom boundary of soil is constrained by three-dimensional displacement, the side of soil is constrained by normal displacement, and the upper surface is free. Because the mechanical properties of soil are usually little affected by environmental temperature, the temperature of the soil is set at a normal temperature of 25 °C.

**Table 1.** Physical and mechanical parameters of soil layer.

| Name of Soil Layer | Thickness of Soil Layer $h$ (m) | Unit Weight $\gamma$ (N/m$^3$) | Elastic Modulus $E$ (MPa) | Poisson's Ratio $v$ |
|---|---|---|---|---|
| Miscellaneous fill | 6.0 | 18500 | 7.4 | 0.40 |
| Sandy silt | 10.6 | 19400 | 9.9 | 0.35 |
| Silt with sand | 3.5 | 19700 | 11.3 | 0.34 |
| Sandy silt | 2.4 | 19400 | 9.9 | 0.35 |
| Muddy clay | 5.0 | 19200 | 3.6 | 0.41 |
| Silty clay | 2.5 | 19800 | 7.7 | 0.35 |

2.3.2. Numerical Simulation of Supporting Structure

The supporting structure around the foundation pit is a concrete diaphragm wall with concrete C35 and a thickness of 1000 mm, which extends into the soil layer at the bottom of the foundation pit to twice the depth of the foundation pit by underwater pouring technology. A concrete crown beam is set at the top of the concrete diaphragm wall and a concrete structure is adopted for the purlin of the foundation pit with concrete C30. A steel crown beam and a steel waist beam are arranged along the vertical height of the steel assembly bracing. Both the crown beam and waist beam adopt the section steel with the same specification, i.e., H400 × 400 × 13 × 21 (Q355b), and the upright column also adopts section steel H400 × 400 × 13 × 21 (Q355b).

Based on the above situation, the S4R element, that is, a four-node curved thin shell element with hourglass control, is used for the diaphragm wall and the surrounding concrete purlin plate where the global element size is 1.0 m and the number of the elements is about 28,000. The B31 element, i.e., a two-node spatial linear beam element, is adopted in the model for the steel assembly bracing, steel purlin beam, section steel column and concrete crown beam in which the global element size is 1.0 m and the number of the elements is about 52,000. The thermal expansion coefficient of steel is $\alpha = 1.2 \times 10^{-5}$, the damping coefficient of steel is $\zeta_S = 0.02$, and that of reinforced concrete is $\zeta_C = 0.02$. The material parameters of each component of the foundation pit supporting structure are shown in Table 2, and the finite element model is shown in Figure 4. The outer soil and diaphragm wall have face-to-face contact, and the normal contact is set as hard contact which means that the normal pressure can be transmitted between the two contact surfaces only when they are not separated [22,23]. The penalty function is adopted in the tangential direction, and the friction coefficient is taken as 0.2.

**Table 2.** Physical and mechanical parameters of steel assembly bracing.

| Material of Supporting Structure | Elastic Modulus $E$ (MPa) | Poisson's Ratio $\mu$ | Density $\rho$ (kg/m$^3$) |
|---|---|---|---|
| Diaphragm wall | 31,000 | 0.2 | 2400 |
| Surrounding purlin | 31,000 | 0.2 | 2400 |
| Steel assembly bracing | 210,000 | 0.3 | 7850 |
| Section steel column | 210,000 | 0.3 | 7850 |

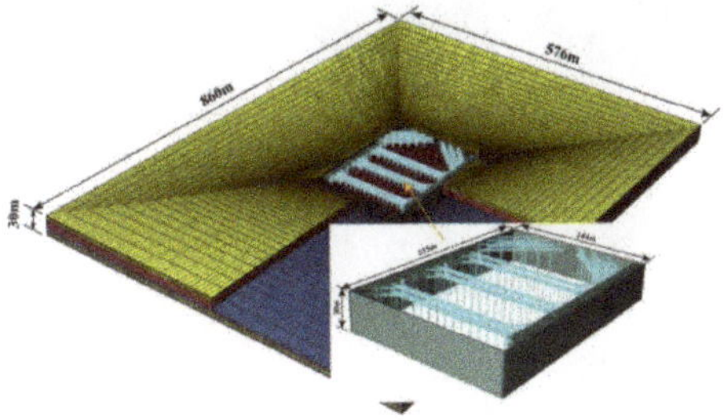

**Figure 4.** Numerical model of foundation pit braced by section steel assembly.

With the help of engineering experience, it can be concluded that the stiffness of reinforced concrete diaphragm walls, cement mixing pile and reinforced concrete brace around the wall is much larger than the vertical stiffness of steel assembly bracing, and the low-order natural frequency of the foundation pit support system should occur in the steel assembly bracing, which has been also proved by numerical results. The low-order vertical natural frequencies of the steel assembly bracing system are mainly concentrated in (6.16 Hz~7.02 Hz), and the first four natural frequencies and the corresponding vertical vibration modes are shown in Figures 5–8 as follows.

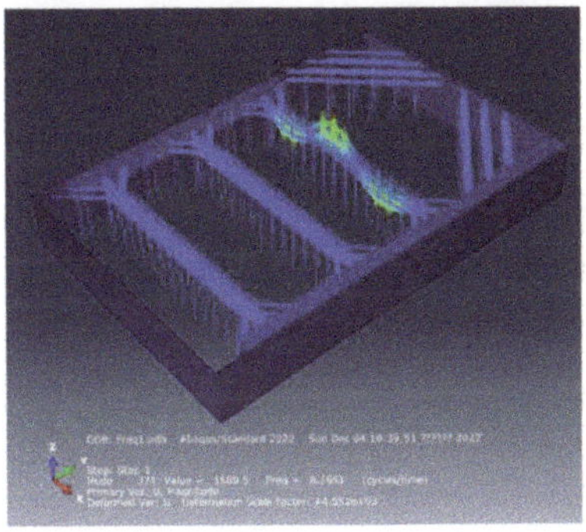

**Figure 5.** Vertical first-order vibration mode (6.16 Hz).

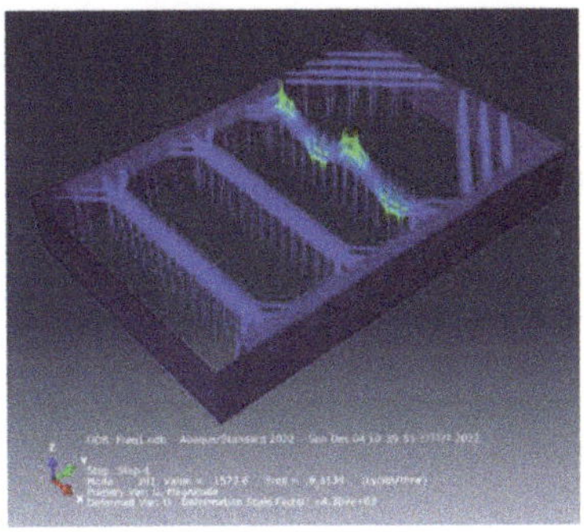

**Figure 6.** Vertical second-order vibration mode (6.31 Hz).

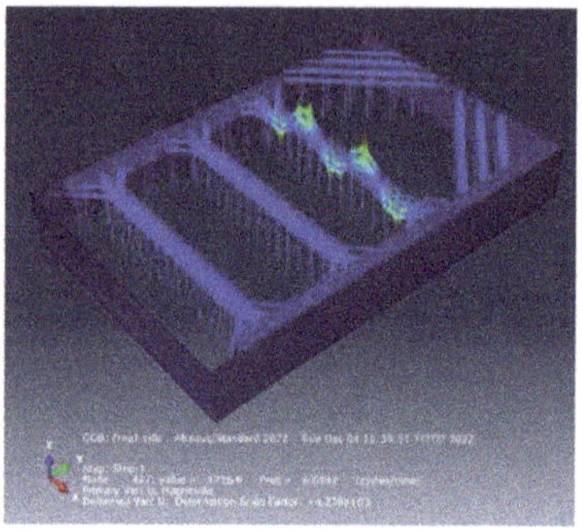

**Figure 7.** Vertical third-order vibration mode (6.59 Hz).

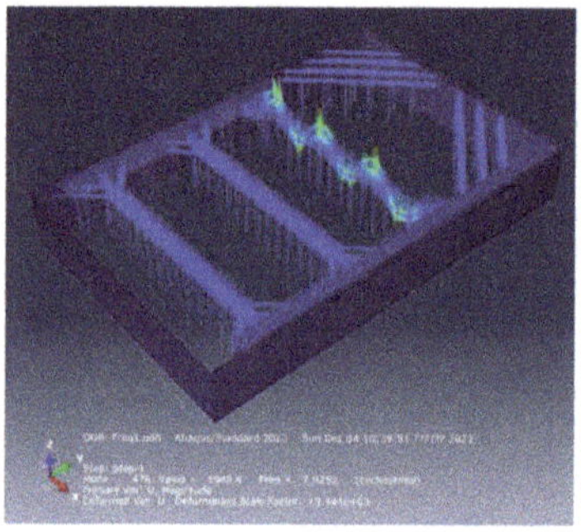

**Figure 8.** Vertical fourth-order vibration mode (7.02 Hz).

It can be found that the low-order natural frequencies of the steel assembly bracing are mainly concentrated on the T01 steel support shown in Figure 1, and it is easier to cause T01 vibration when the earthwork truck drives over the concrete trestle bridge. The numerical analysis results preliminarily prove the problems reflected by construction workers.

### 2.3.3. Acting Load

It has been proved that the internal force changes of steel bracing under the action of temperature cannot be ignored. Considering the abnormal vibration of T01 complained of by workers in this project, it is necessary to consider the combined action problem of temperature and vibration of the steel assembly bracing in both field monitoring and calculation analysis. When the site-monitoring system was established, the construction stage was in the binding steel bars and pouring concrete on the right foundation pit bottom plate near T01, while the left side was in the stage of outward soil transportation. Therefore, the working conditions of the construction site were complicated. As far as temperature changes are concerned, from August to November 2022, the 100 consecutive days of monitoring showed that the strain and temperature monitoring data of the steel assembly bracing changed little at night, while the steel surface temperature of the first steel assembly bracing of T01 ranged from 10.9 °C to 51.6 °C during the day, and the compressive strain of the brace decreased or increased with the temperature obviously. During the 100 days of real-time online monitoring, even in October when the temperature drops, for example, the lowest temperature of steel was 26.0 °C at 4 am on October 3, the highest temperature of steel was 51.6 °C at 14 pm, and the temperature difference between high and low was 25.6 °C. On that day, the axial static compressive strain SSTS-01 increased by 156.3 $\mu\varepsilon$ with the temperature, and the similar high temperature environment frequently occurred within three months. The temperature changes of the steels from 12 August to 22 November is shown in Figure 9.

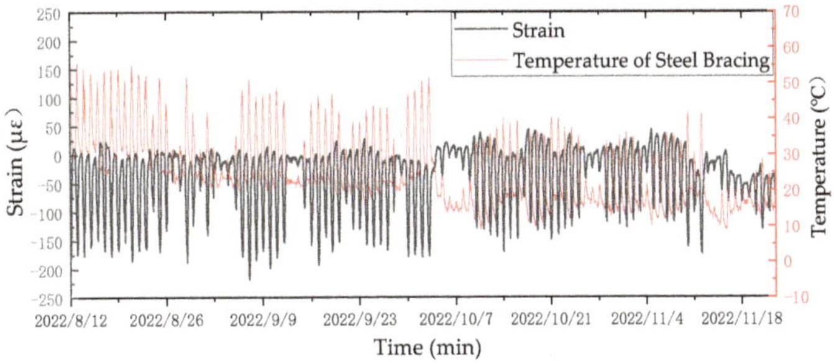


**Figure 9.** Temperature–strain diagram of steel assembly bracing during monitoring period (12 August to 22 November 2022).

As for the vertical vibration problem of steel bracing, since the site is in the construction stage, the vibration sources of structure include: (1) the vibration caused by earthwork truck running; (2) the vibration caused by pouring concrete in the foundation pit bottom plate; (3) the vibration caused by tower crane running and hoisting construction materials, and vibration caused by various construction machines and tools. Frankly speaking, it is extremely difficult to accurately analyze the steel assembly bracing under the action of each vibration source. The multi-day vibration monitoring data show that the vertical vibration of the reinforced concrete trestle bridge and steel assembly bracing increases obviously when the earthwork truck passes through the trestle bridge, which further shows that the earthwork truck passing through the trestle bridge is the main reason for the significant vibration of T01. Video monitoring of the speed of earthwork trucks on the trestle is about 10 km/h. The truck is fully loaded and the roof soil layer is neatly covered, and the average total mass weighed is 20,000 kg. In order to obtain the moving load of the truck running on the trestle, a vertical vibration sensor is installed on one of the trucks to collect the vertical vibration monitoring data of the vehicle. The vibration action during the driving process of the vehicle is shown in Figures 10 and 11 below. The vertical vibration responses of trestle ground vibration acceleration monitoring point 3AS-01 and steel support vibration acceleration monitoring point 3AS-02 at the same time are shown in Figures 12 and 13. It can be seen from the vibration time history and frequency spectrum curves in Figures 10 and 12 that the first-order frequency (Figure 10) of the vehicle action when the truck passes through the concrete trestle is basically consistent with the first-order dominant frequency of the vertical trestle response. The trestle vibration contains forced vibration components, and the frequency components are complex. The first frequency of the vibration response in Figure 13 is 6.661 Hz, which is close to the vertical natural frequency of the bracing in Figure 7 of numerical analysis. The vertical vibration of the bracing is mainly caused by resonance.

During numerical analysis, the vibration effect of earthwork truck as the unique vibration source is considered and the static and dynamic strain of steel assembly bracing at the highest temperature difference are only considered. With regard to the vibration calculation input, the amplitude modulation input model of Figure 10 is calculated based on the maximum value Atbt.max of the vibration response of the concrete trestle bridge (purlin) in Figure 14, such that Atbt.max of Figure 12 is equal to Atbt.max of Figure 14. The time history curve of vehicle vibration acceleration after amplitude modulation in Figure 10 was taken as the calculation input.

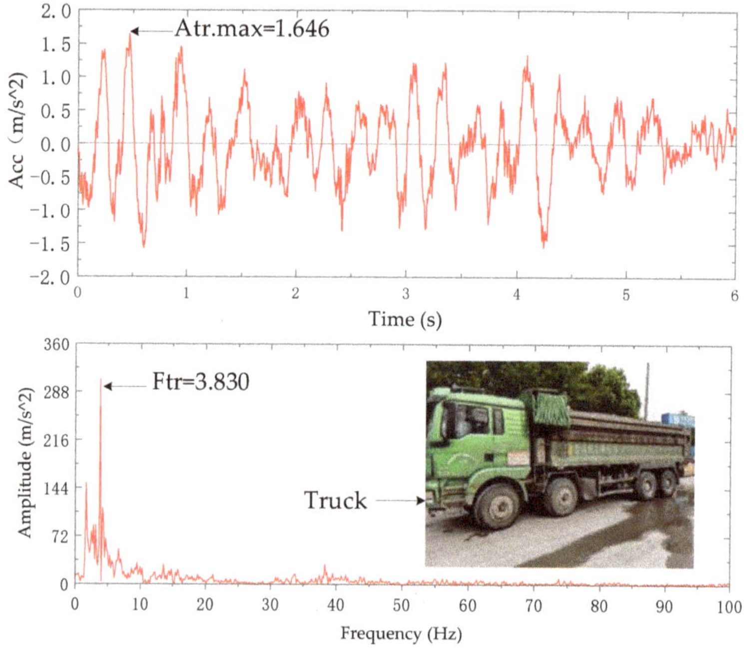


**Figure 10.** Vibration action of earthwork truck.

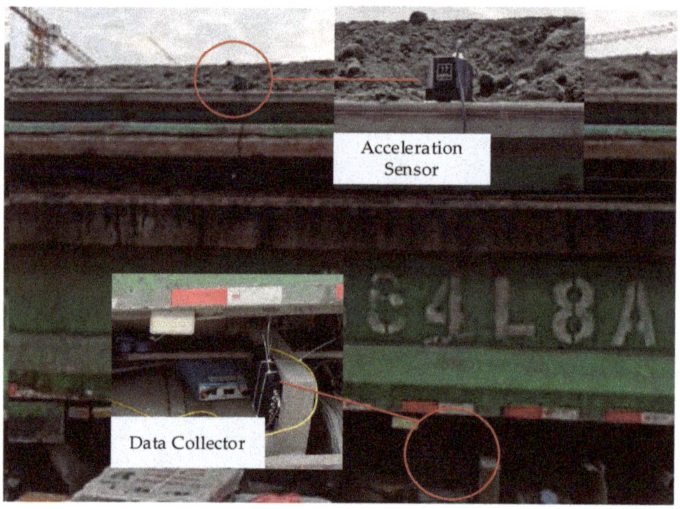


**Figure 11.** Test system of vehicle vibration.

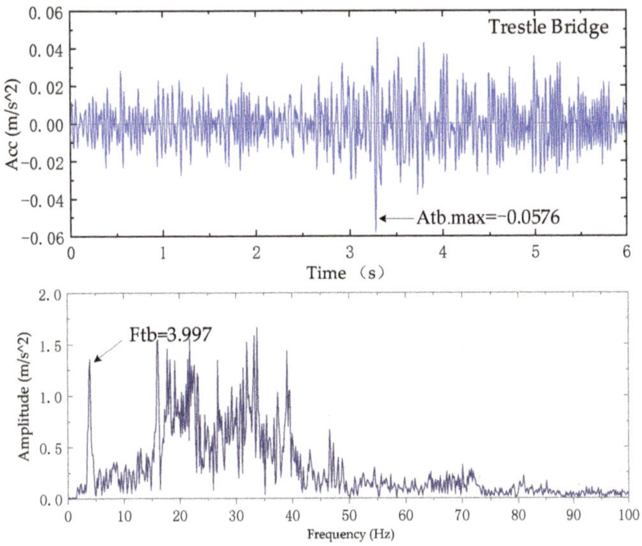


**Figure 12.** Vibration response of concrete trestle bridge.

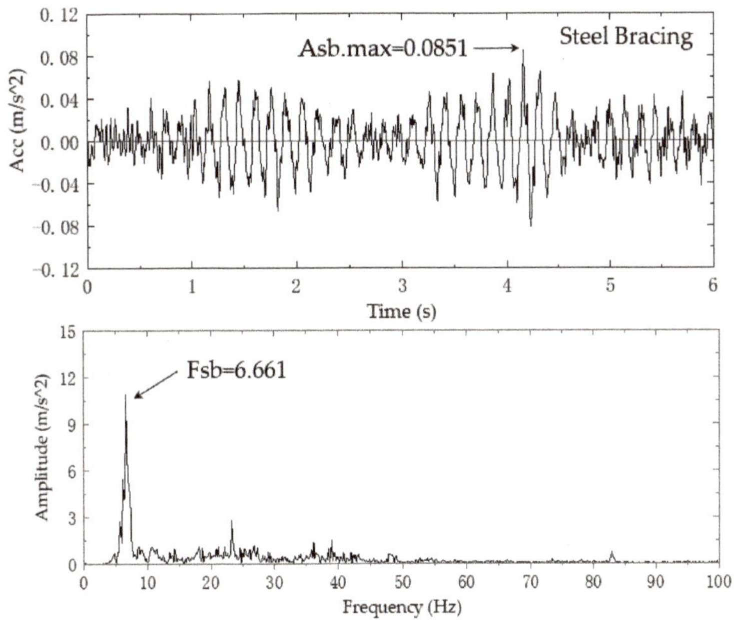


**Figure 13.** Vibration response of steel assembly bracing.

In order to simulate the influence of the vertical vibration of vehicles, eight concentrated forces are set in this model to represent the mass distribution of vehicles on the trestle bridge. The numerical values are calculated according to the vehicle weight and the vehicle mass distribution percentage. The first half of the earthwork vehicle accounts for 36% of the total mass of the vehicle, and the second half accounts for 64% of the total mass of the vehicle, and then the mass is evenly distributed to the concentrated force in the area located. During numerical analysis, the total mass of the vehicle is taken as 20,000 kg, and the front and rear wheels of the vehicle are distributed in proportion, in which the concentrated forces exerted by the front wheels RP-1 to RP-4 are all 1800 kg, and the concentrated forces exerted by the rear wheels RP-5 to RP-8 are all 3200 kg, as shown in Figure 15. The time history curve of vehicle vibration acceleration after amplitude modulation is assigned as the amplitude curve which is multiplied by the above mass of the vehicle to obtain the dynamic load-time history curve. Finally, the dynamic load action can be analyzed. It is assumed that the most unfavorable working condition is considered, that is a linear combination of the maximum temperature difference of steel assembly bracing and the maximum vibration response amplitude of the trestle which occur at the same time.

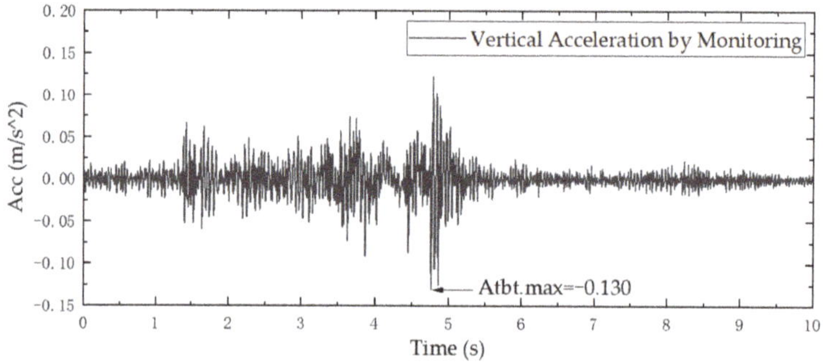


**Figure 14.** Maximum vibration response of trestle bridge during monitoring period.

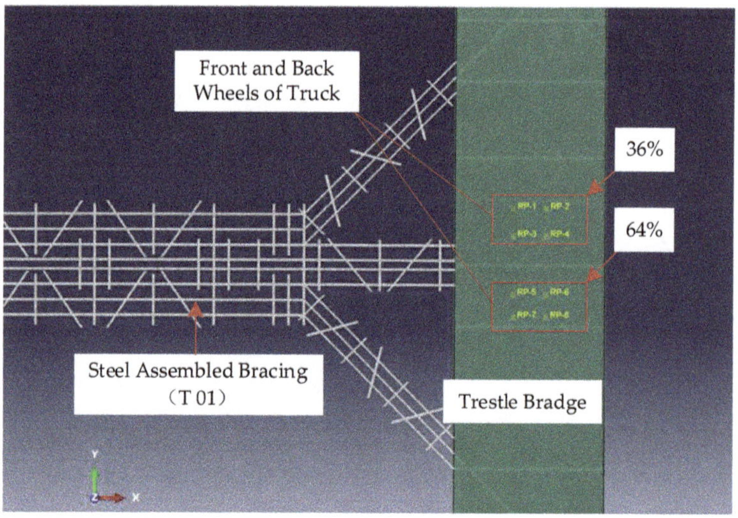


**Figure 15.** Mass distribution of front and rear wheels of earthwork trucks.

## 3. Results and Discussion

### 3.1. Strain of the Steel Assembly Bracing

According to the change of steel assembly bracing with temperature for more than 100 days in Figure 9, the surface temperature changes of section steel collected in typical weather environments of low temperature, medium temperature and high temperature are selected and input into the finite element model, and the temperature influence analysis of static strain at the SSTS-01 measuring point of T01 steel assembly bracing is carried out. SSTS-01 steel surface temperature, monitoring strain, and model in-situ calculated strain are shown in Figure 16, and monitoring strain and calculated strain information are listed in Table 3.

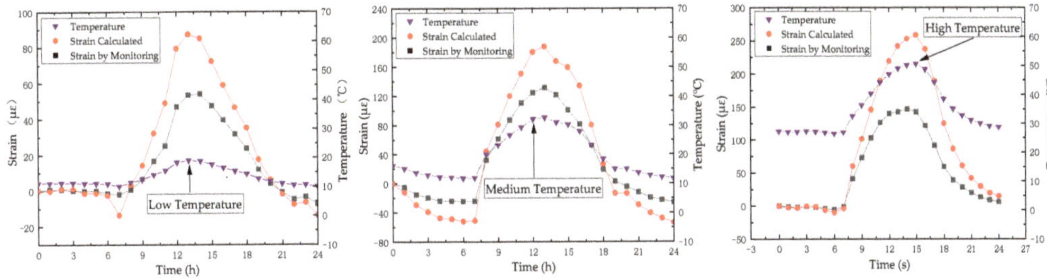


**Figure 16.** Influence of typical ambient temperature change on steel brace strain.

**Table 3.** Strain of steel brace at different ambient temperatures.

| Temperature Condition | Steel Temperature (°C) | Temperature Difference (°C) | Calculated Strain (με) | Measured Strain (με) | Calculated Strain –Measured Strain (με) |
|---|---|---|---|---|---|
| Low temperature environment | 10~18.9 | 8.9 | 87.40 | 53.55 | 33.85 |
| Moderate temperature environment | 11.7~32.7 | 21.0 | 187.94 | 131.69 | 56.25 |
| High temperature environment | 26.5~50.6 | 24.1 | 258.44 | 146.44 | 115.88 |

According to the data listed in Figure 16 and Table 3, it can be concluded that the calculated strain is larger than the actual monitored strain, and such a difference at high temperature is larger. In addition, the temperature of steel in the steel assembly bracing is affected by environmental temperature and sunlight radiation, and the surface temperature of each member varies with the degree of sunlight exposure. The temperature of steel used in calculation and simulation comes from SSTS-01, which is located on the upper surface of the first support of the steel assembly bracing. During the monitoring period, it receives more sunlight radiation, which belongs to the support area most affected by temperature and is the most unfavorable position affected by temperature. When the input steel temperature measuring point is located in the position that can accept the full radiation of the sun, the calculation and analysis results of temperature action according to the high temperature environment are too large, but it is beneficial to the structural safety from the engineering point of view.

Then, the finite element dynamic analysis is carried out under the vibration action of the earthwork truck after amplitude modulation. The total time of model calculation is 6 s, the step size is 0.005 s, and the sampling frequency of the field vibration sensor is 200 Hz. It is assumed that the temperature of steel is approximately in a stable state of 50.6 °C during the period when the earthwork truck passes through the trestle bridge. As shown in Figures 17–20 below, the monitoring data of the dynamic strain of typical position number DSS-06 (the same position as static strain SSTS-01), DSS-10, DSS12 and DSS-14 of T01 are compared with the calculation results at the same point of the model (the monitoring data of dynamic strain of other positions are in Supplementary Materials).

The following conclusions can be drawn from the data analysis of Figures 17–20: (1) Dynamic strain responses monitored, except for DSS-10, are less than the calculated results, including the dynamic strains of other measuring points not listed in this paper; (2) The maximum monitoring value of DSS-10 is 0.76 με, which is slightly larger than the calculated result of 0.46 με. Considering the influence of a tower crane near the measured support point, the larger monitoring value may be acceptable; (3) The maximum dynamic strain of measuring points is 1.4 με, which accounts for about 1% of the static strain of 146.44 με of steel assembly bracing under a high temperature of 50.6 °C. The sum of the stress caused by temperature rise and the maximum stress caused by vibration is 30.46 MPa, which is larger than that of the first (initial pre-axial force $P_{01}$ = 3200 kN, compressive stress 18.22 MPa) and the second (initial pre-axial force $P_{01}$ = 8000 kN, compressive stress 22.78 MPa) bracing initial prestress of T01 and less than the yield strength of steel.

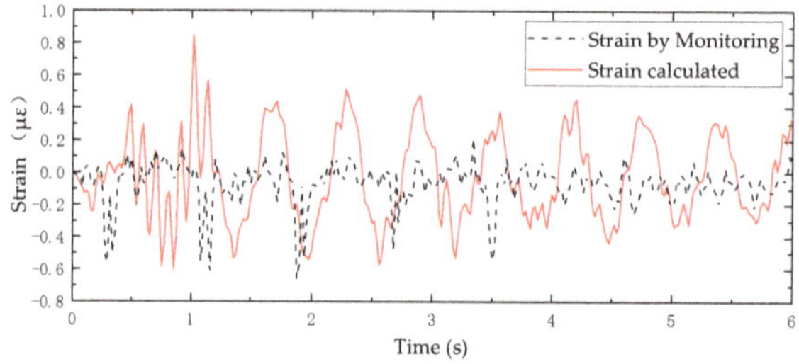


**Figure 17.** Comparison of dynamic strain between measurement and simulation results of DSS-06.

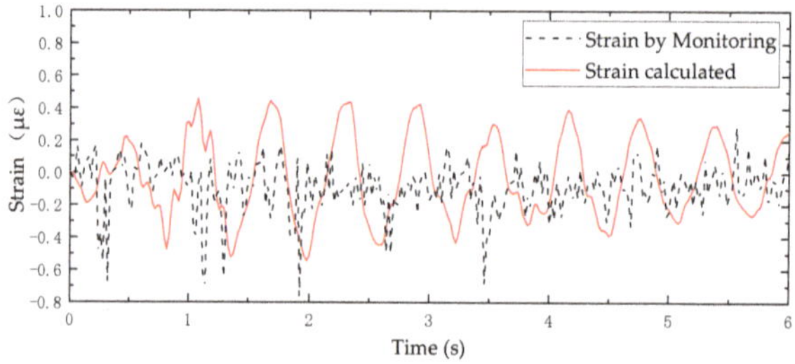


**Figure 18.** Comparison of dynamic strain between measurement and simulation results of DSS-10.

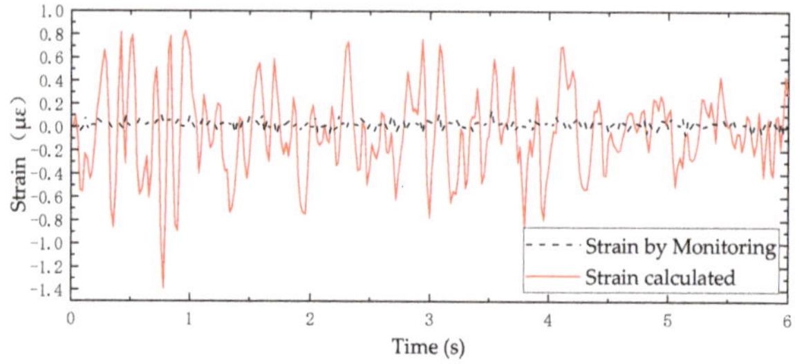


**Figure 19.** Comparison of dynamic strain between measurement and simulation results of DSS-12.

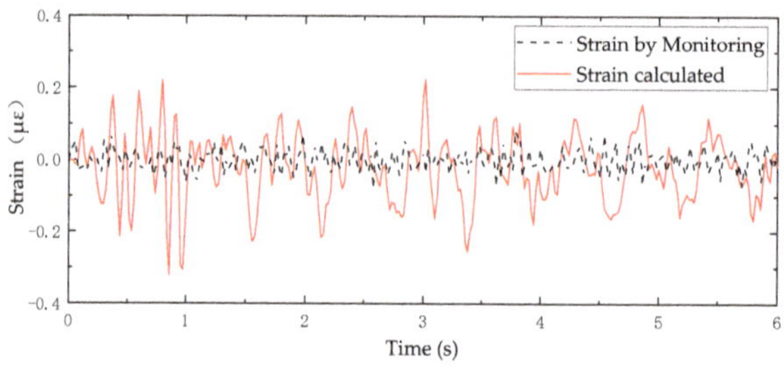


**Figure 20.** Comparison of dynamic strain between measurement and simulation results of DSS-14.

### 3.2. Acceleration of the Steel Assembly Bracing

Figure 21 shows the vertical calculated acceleration response and monitored acceleration response of measuring point 3AS-02 on steel assembly bracing. The acceleration response is less than the calculated results. It should be noted that the vertical vibration of T01 steel assembly bracing causes complaints from on-site construction technicians, which can be clearly felt during the construction. In order to understand the influence of the vibration of steel assembly bracing on the walking comfort of technicians, according to ISO 2631-1:1997, the vertical frequency-weighted vibration acceleration is calculated as $a_W$ = 79.30 dB for Figure 21. Obviously, the vertical frequency weighted acceleration $a_W$ is really large and it is reasonable for technicians to complain about the insecurity of bracing vibration [24].

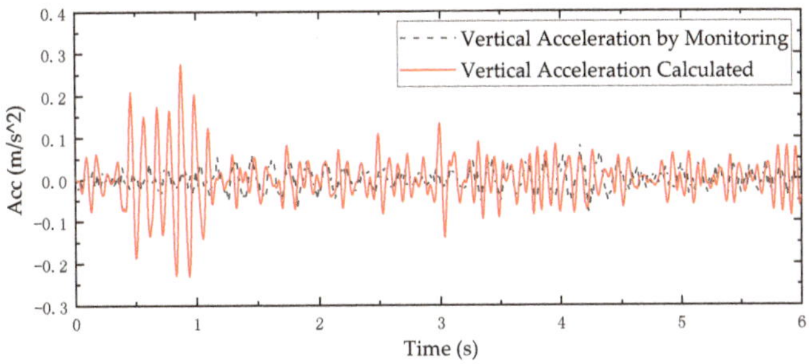


**Figure 21.** Comparison of vertical acceleration between measurement and simulation results of 3AS-02.

## 4. Conclusions

As a new type of bracing system for the construction of deep, large foundation pits, steel assembly bracing has been gradually applied in the civil engineering field. In this paper, the effects of vibration and temperature of the construction site on steel assembly bracing of the foundation pit is studied by virtue of the safety real-time monitoring system and FEM technology based on an engineering case. The following conclusions can be drawn:

(1) The influence of environmental temperature on steel assembly bracing is significant and cannot be ignored. The axial compressive stress of bracing caused by environmental temperature even exceeds the prestress. Under the most unfavorable conditions, the stress caused is less than the yield strength of material Q355b, and hence the steel assembly bracing is in a safe state.

(2) The steel assembly bracing is made up of hundreds of H-beams of different specifications connected by high-strength bolts, and the bolt joints are easy to loosen due to construction vibration. Therefore, the working state of key joints of the steel assembly bracing should be checked regularly during construction. Meanwhile, the steel assembly bracing is under axial compression, and the out-of-plane vertical deformation has great influence on its stability. Accordingly, it is necessary to regularly monitor the out-of-plane deformation of brace.

(3) The vertical frequency-weighted vibration acceleration $a_W$ is an important indicator to reflect the comfort related with vibration. In this engineering case, this value of the steel assembly bracing $a_W$ = 79.30 dB, which is large. The reality is that on-site technicians complained about the insecurity of bracing vibration which is consistent with our theoretical analysis.

(4) The safety real-time monitoring system can be used as an effective way to analyze the steel assembly bracing under complex working conditions. The results of the paper can provide guidance for the systematic design and further application of steel assembly bracing in engineering.

**Supplementary Materials:** The following supporting information can be downloaded at: https://www.mdpi.com/article/10.3390/buildings13020450/s1, Figures S1–S11 about Dynamic strain comparison between simulation and monitoring of DSS-01 to DSS-16.

**Author Contributions:** Conceptualization, Y.Y., B.Y. and Y.L.; methodology, Y.Y. and Y.L.; software, B.Y., H.Z. and X.L.; validation, Y.L.; formal analysis, Y.Y. and H.Z.; investigation, Y.Y. and H.Z.; resources, Y.L.; data curation, Y.Y. and H.Z.; writing—original draft preparation, Y.Y. and H.Z.; writing—review and editing, B.Y. and X.L.; visualization, X.L. and B.Y.; supervision, X.L. and B.Y.; project administration, Y.Y., B.Y. and Y.L.; funding acquisition, Y.Y. and Y.L. All authors have read and agreed to the published version of the manuscript.

**Funding:** This research was funded by the Natural Science Foundation of Zhejiang Province, grant number LGG20E080002.

**Data Availability Statement:** Not applicable.

**Conflicts of Interest:** The funders had no role in the design of the study; in the collection, analyses, or interpretation of data; in the writing of the manuscript; or in the decision to publish the results.

## References

1. Cai, W.; Lin, Y.; Wen, F. Analysis of Temperature Difference Effect of Long-Span Plate-Truss Composite Girder. *Bridge Constr.* **2019**, *49*, 41–46.
2. Zhu, Q.; Wang, H.; Mao, J.; Wan, H.; Zheng, W.; Zhang, Y. Investigation of Temperature Effects on Steel-Truss Bridge Based on Long-Term Monitoring Data: Case Study. *J. Bridge Eng.* **2020**, *25*, 05020007. [CrossRef]
3. Batten, M.; Powrie, W.; Boorman, R.; Yu, H.; Leiper, Q. Use of vibrating wire strain gauges to measure loads in tubular steel props supporting deep retaining walls. *Proc. Inst. Civil Eng.-Geotech. Eng.* **1999**, *137*, 3–13. [CrossRef]
4. Han, L.; Sun, M.; Huang, P.; Chen, H.; Wang, Q. Design analysis and engineering application research of new H-section steel support system. *Build. Struct.* **2021**, *51*, 95–102.
5. Venglar, M.; Lamperova, K. Effect of the temperature on the modal properties of a steel railroad bridge. *Slovak. J. Civ. Eng.* **2021**, *29*, 1–8. [CrossRef]
6. Luo, J.; Huang, M.; Lei, Y. Temperature Effect on Vibration Properties and Vibration-Based Damage Identification of Bridge Structures: A Literature Review. *Buildings* **2022**, *12*, 1209. [CrossRef]
7. de Battista, N.; Brownjohn, J.M.W.; Tan, H.P.; Koo, K.Y. Measuring and modelling the thermal performance of the Tamar Suspension Bridge using a wireless sensor network. *Struct. Infrastruct. Eng.* **2015**, *11*, 176–193. [CrossRef]
8. Li, Y.; Chen, D.; Liu, X.; Zhu, H.; Tang, D. Application of prestressed assemble steel support system in retaining structures for foundation pits in soft soils. *Chin. J. Geotech. Eng.* **2014**, *36*, 51–55.
9. Liu, X.; Tong, G.; Li, Y.; Hu, H.; Chen, D. Stability Analysis of Assembly Steel Struts in Deep Excavation. *Eng. Mech.* **2020**, *35*, 200–207.
10. Li, H.; Tang, Y.; Liao, S.; Shen, M. Structural Response and Preservation of Historic Buildings Adjacent to Oversized Deep Excavation. *J. Perform. Constr. Facil.* **2021**, *35*, 04021095. [CrossRef]
11. Cui, X.; Ye, M.; Zhuang, Y. Performance of a foundation pit supported by bored piles and steel struts: A case study. *Soils Found.* **2018**, *58*, 1016–1027. [CrossRef]
12. He, J.; Liao, S.; Sun, J.; Tang, J. Study on axial force interference of soft soil deep excavation under multi-point synchronous loading. *China Civil. Eng. J.* **2020**, *53*, 99–107.
13. Boone, S.J.; Crawford, A.M. Braced Excavations Temperature, Elastic Modulus, and Strut Loads. *J. Geotech. Geoenviron. Eng.* **2000**, *126*, 870–881. [CrossRef]
14. Hu, Q.; Zhang, K.; Song, J.; Fang, H.; Xie, J.; He, P.; Ran, L. Numerical Simulation and Analysis of Thermal Stress on Restressed Assembly Steel Struts System in Foundation Excavation. *IOP Conf. Ser. Earth Environ. Sci.* **2019**, *300*, 022162. [CrossRef]
15. Wu, M.; Du, C.; Yang, K.; Geng, X.; Liu, X.; Xia, T. A new empirical approach to estimate temperature effects on strut loads in braced excavation. *Tunn. Undergr. Space Technol.* **2019**, *94*, 103115. [CrossRef]
16. Wang, F.; Shi, G.; Zhai, W.; Li, B.; Zhang, C.; Fang, H. Internal Force on and Deformation of Steel Assembled Supporting Structure of Foundation Pit under Thermal Stress. *Appl. Sci.* **2021**, *11*, 2225. [CrossRef]
17. Li, B.; Qu, J.; Ge, S. Effect of Temperature on Supporting Structure of Metro Deep Foundation Pit. *J. Shenyang Jianzhu Univ (Nat. Sci.)* **2021**, *37*, 69–76.
18. Zhang, X.; Chen, J.; Wan, L.; Luo, W.; Xu, C. Field test and analysis of vehicle load and its influence on the retaining structure of adjacent foundation pits. *China Earthq. Eng. J.* **2020**, *42*, 490–497.
19. Wang, Z.; Guo, X.; Wang, C. Field Monitoring Analysis of Construction Process of Deep Foundation Pit at Subway Station. *Geotech. Geol. Eng.* **2019**, *37*, 549–559. [CrossRef]
20. Yang, Y.; Lu, J.; Huang, X.; Tu, X. Sensor monitoring of a newly designed foundation pit supporting structure. *J. Cent. South Univ.* **2013**, *20*, 1064–1070. [CrossRef]

21. Zhuang, S.; Zhang, J.; Zhang, C.; Li, S. On stiffness of prefabricated prestressed fish-belly steel support system. *China Civil. Eng. J.* **2021**, *54*, 18–25.
22. Tong, L.; Li, H.; Ha, S.; Liu, S. Lateral bearing performance and mechanism of piles in the transition zone due to pit-in-pit excavation. *Acta Geotech.* **2022**, *17*, 1935–1948. [CrossRef]
23. Shi, X.; Rong, C.; Cheng, H.; Cui, L.; Wang, B.; Sun, S. Analysis on deformation and stress characteristics of a multibraced pit-in-pit excavation in a subway transfer station. *Adv. Civ. Eng.* **2020**, *2020*, 8844461. [CrossRef]
24. *ISO 2631-1*; Mechanical Vibration and Shock Evaluation of Human Exposure to Whole-Body Vibration Part 1: General Requirements. International Organization Standardization: Geneva, Switzerland, 1997; p. 12.

*buildings*

MDPI

*Article*

# Automated Semantic Segmentation of Indoor Point Clouds from Close-Range Images with Three-Dimensional Deep Learning

**Chia-Sheng Hsieh * and Xiang-Jie Ruan**

Department of Civil Engineering, National Kaohsiung University of Science and Technology, Kaohsiung 807, Taiwan
* Correspondence: hsieh@nkust.edu.tw

**Abstract:** The creation of building information models requires acquiring real building conditions. The generation of a three-dimensional (3D) model from 3D point clouds involves classification, outline extraction, and boundary regularization for semantic segmentation. The number of 3D point clouds generated using close-range images is smaller and tends to be unevenly distributed, which is not conducive to automated modeling processing. In this paper, we propose an efficient solution for the semantic segmentation of indoor point clouds from close-range images. A 3D deep learning framework that achieves better results is further proposed. A dynamic graph convolutional neural network (DGCNN) 3D deep learning method is used in this study. This method was selected to learn point cloud semantic features. Moreover, more efficient operations can be designed to build a module for extracting point cloud features such that the problem of inadequate beam and column classification can be resolved. First, DGCNN is applied to learn and classify the indoor point cloud into five categories: columns, beams, walls, floors, and ceilings. Then, the proposed semantic segmentation and modeling method is utilized to obtain the geometric parameters of each object to be integrated into building information modeling software. The experimental results show that the overall accuracy rates of the three experimental sections of Area_1 in the Stanford 3D semantic dataset test results are 86.9%, 97.4%, and 92.5%. The segmentation accuracy of corridor 2F in a civil engineering building is 94.2%. In comparing the length with the actual on-site measurement, the root mean square error is found to be ±0.03 m. The proposed method is demonstrated to be capable of automatic semantic segmentation from 3D point clouds with indoor close-range images.

**Keywords:** building information model; 3D point cloud; semantic segmentation; deep learning

**Citation:** Hsieh, C.-S.; Ruan, X.-J. Automated Semantic Segmentation of Indoor Point Clouds from Close-Range Images with Three-Dimensional Deep Learning. *Buildings* **2023**, *13*, 468. https://doi.org/10.3390/buildings13020468

Academic Editor: Ahmed Senouci

Received: 13 January 2023
Revised: 4 February 2023
Accepted: 6 February 2023
Published: 9 February 2023



## 1. Introduction

With the rapid development of laser scanners and digital images in recent years, spatial three-dimensional (3D) point cloud data have been widely used in many fields. Point clouds are convenient for spatial measurements and can show object shapes. However, point clouds only have 3D coordinates and color information; moreover, they do not contain attribute information. Extraction of accuracy objects from a 3D point cloud is a challenge [1]. Therefore, this issue is currently a hot research topic [2–4]. The main purpose of this study is to expand the application of point clouds through 3D point cloud classification and segmentation technology. The 3D information of point clouds can be widely applied to different fields for the visual display and management of engineering information.

With the recent development of laser technology and digital photogrammetry, the real appearance of an object can be restored through a 3D point cloud model. Point clouds are easy to visualize; they are simply point clusters without attribute information. Consequently, designers find them difficult to use in drawings. If the point cloud can be segmented, errors in drawings can be reduced [5]. Moreover, point cloud attributes can enable semi-automatic or fully automatic modeling.

The early development of artificial intelligence in the field of computer vision was intended for the classification, detection, and semantic segmentation of two-dimensional (2D) images. The advances in deep learning have indirectly promoted the development of the field of combining deep neural networks with 3D information [1]. With 3D point clouds, supervised learning or unsupervised methods can be used for feature learning so that the neural network can recognize geometric shapes. Because point clouds do not contain attribute information, the attributes of an object can be obtained from a segmented point cloud. Then, the rules of 3D modeling can be formulated, enabling the use of point clouds in automatic modeling.

In recent years, the development of deep learning networks has been effective in the semantic segmentation processing of 3D point clouds [4,6–8]. Using the segmentation results, the point cloud can be assigned to a corresponding object label. Accordingly, this study uses a deep learning network to segment 3D point clouds, improving the efficiency and accuracy of artificial segmentation.

This research aims to segment the 3D point cloud of an indoor space using a deep learning network, develop a set of point cloud feature extraction procedures, and complete the automatic modeling of parametric components [9–12]. The dynamic graph convolutional neural network (DGCNN) proposed by Wang et al. (2018) was used to perform indoor point cloud segmentation [13]. After segmentation, feature extraction technology is applied to derive the endpoints of components. Finally, the endpoint coordinates are imported into the automatic modeling rules to generate parametric components. To ensure the correctness of model reconstruction, the difference between a model and an object (i.e., between a 3D model and a real 3D housing condition, respectively) is evaluated.

This paper presents a framework for automated building component recognition based on close-range images. The proposed approach consists of three main steps: (1) grouping 3D point clouds into five categories using a deep learning classification model; (2) extracting outlines from the five categories of building structure point clouds; and (3) identifying boundary regulation and parametric components. The reason for choosing "columns, beams, walls, ceilings, and floors" as the segmentation target is that these five types of data represent the basic structure and layout of the house that cannot be easily changed, and the simple geometry is also conducive to feature extraction and automatic modeling. The proposed method automatically reconstructs the complete geometry of columns, beams, walls, ceilings, and floors from 3D point clouds using close-range images. Moreover, the material properties of components are included, thus allowing the generation of building information models (BIMs). The proposed approach is then field-validated using an actual building on campus.

## 2. Related Work

With the development of laser scanner technology and digital photogrammetry in recent years, 3D point cloud models are typically employed to represent the surfaces of objects. Point clouds have spatial coordinates that provide measurement information. In addition, a colored point cloud can be used as a basis for browsing the housing environment and querying the relative positions of components.

The generation of 3D graphs from 3D point clouds of indoor spaces is a current research focus. The early method for this purpose was to construct a point cloud into 3D elements using artificial methods [2,14,15]. For example, based on the geometric shape and edge features of the point cloud distribution, the centerline of the object, the boundary of the structure, and other details are used to build a model. However, after 3D reconstruction, the point cloud becomes non-attribute data. If the 3D point cloud can be effectively segmented and provided with attribute data, the results can aid in the development of automatic modeling. Accordingly, 3D point cloud segmentation has become an important research topic [3,14,16,17]. There are some review-type articles that organize and analyze the progress of relevant research [1,14,15,18].

### 2.1. Three-Dimensional Point Cloud Classification

A point cloud does not contain geometric information. In contrast, a segmented point cloud contains attribute data to which the rules of 3D modeling can be applied. Hence, point clouds can benefit from automatic modeling.

The current research on 3D data combined with deep learning can be broadly classified into categories, such as RGB-D (red–green–blue depth), volumetric approach, multiview convolutional neural network (CNN), and unordered point set processing [1]. The first three data items are regularly structured data with clear connection information; they yield acceptable results in both object detection and segmentation. However, with automated processing, the direct processing of an out-of-sequence point cloud to achieve point-to-point classification, part segmentation, or semantic segmentation can be implemented. Moreover, the use of a voxel grid or other conversion methods to reduce the risk of potential loss of 3D point cloud data during the conversion process is not necessary.

In recent years, the classification and segmentation technology of point clouds for 3D processing has been investigated [6,19–23]. In 2017, Qi et al. proposed the PointNet method for 3D point cloud processing based on deep learning. The overall semantic segmentation accuracy of the indoor scene point cloud in the mixed test of the S3DIS dataset can reach 78.5% [24].

By ignoring related details on geometry among the points, some local features are lost. After discovering this problem, Qi et al. proposed an improved method. In the improved version, i.e., PointNet++, a 2D CNN processing mechanism is added to the original architecture of the method. The overall semantic segmentation accuracy of PointNet++ in the S3DIS dataset hybrid test is 81.0% [25].

To improve segmentation accuracy, Wang et al. proposed the DGCNN method in 2019 [13]. In addition to obtaining local features, the feature information of the overall scene can be extracted through repeated stacking. The overall accuracy of the point cloud semantic segmentation in indoor scenes reached 84.1%.

Presently, the development of deep learning in the field of computer vision has shifted from a mature 2D platform to 3D space. Since Qi et al. proposed PointNet, breakthroughs have been made in object classification and semantic segmentation applied to 3D point clouds by learning their features [24].

With the introduction of DGCNN, more accurate semantic segmentation of indoor scenes can be achieved. In this study, after referring to relevant research on 3D point clouds [14], the DGCNN with improved performance and a simple operational process is selected for testing.

### 2.2. Semantic Segmentation and Modeling

A 3D point cloud can be provided with attribute data after semantic segmentation using a deep neural network; for example, certain points can represent columns. For 3D modeling, a 3D point cloud with attributes can be used to extract the feature information of a corresponding target using a feature extraction algorithm [26–29].

The study focuses on columns, beams, walls, ceilings, and floors in interior space. These objects have clear corners, edges, and other characteristic information in the expression of geometric shapes.

In general, the procedure for generating a building footprint involves three steps using point cloud data: (1) segmentation; (2) extraction of building outlines; and (3) regularization or generalization of boundaries. The first step classifies the points of the building from a point cloud dataset. The second step involves the extraction of building boundaries and the generation of a preliminary polygon. Finally, the third step involves the adjustment of the generated boundary and the retrieval of simple and regular polygons [27,30].

According to Awrangjeb (2016), the methods for extracting building outlines can be divided into two types: direct and indirect [30]. Direct methods extract building outlines based on the points. However, these methods are sensitive to the selection of parameters (such as neighborhood radius) and are easily affected by noise in point cloud data.

The indirect method uses image processing technology to extract edge features from 2D images and then matches them to point cloud data to extract 3D edge features. Wang et al. (2013) pointed out that this method includes detecting 2D edge information from a 2D image corresponding to the point cloud [31]. The image depth is generated from the point cloud, and then it is matched with the original 3D point cloud data. Finally, multiple groups of edge points are merged and used as detected 3D point cloud edges.

When edge features are extracted using the indirect method, spatial information can easily be lost during dimension conversion; thus, 3D edge feature information may be missed. In addition, the actual semantic segmentation results may be less than the number of point clouds before processing due to the parameter setting of deep learning in the point cloud sampling process. Accordingly, this study adopts the direct method for feature extraction.

The direct method extracts edge information directly from a 3D point cloud. For example, Borges (2010) first divided the point cloud and then detected the intersection of the segmentation surface and depth discontinuity edge [32]. In addition, Sampath and Shan (2007) proposed the use of a convex hull algorithm to establish the plane point information of a roof [33]. Then, the same algorithm is used to obtain the edge lines and finally perform boundary regularization.

## 3. Methodology

The main processing step in this study is to automatically generate the parametric components of BIM from the close-range images. A series of processes in this study can be referred to in Figure 1.

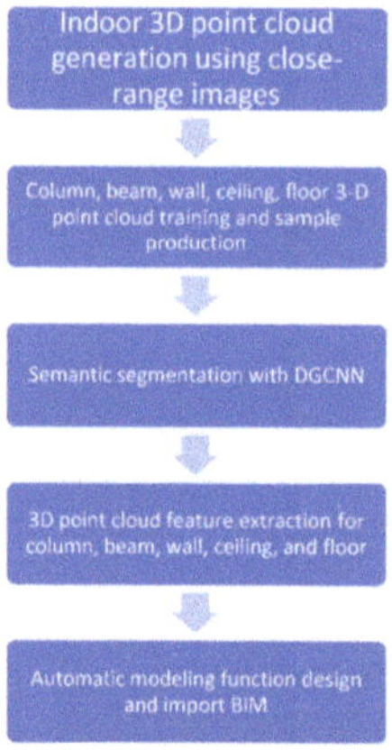


**Figure 1.** The overall process of the proposed method in this study.

### 3.1. *Three-Dimensional Point Cloud Classification*

#### 3.1.1. Sample Data

A 3D point cloud can be applied to surveying and mapping, unmanned driving, robotics, reverse engineering, and other fields. This is because it has visualization characteristics, and each point contains coordinate information. In addition to the complete and accurate preservation of the actual size of a target object, 3D point clouds present the characteristics of irregular surface changes and image space information. To understand the current geometric environment, the construction plans can be immediately viewed, improved, and modified. Accurate measurements of indoor spaces can also be obtained.

Several methods for obtaining 3D indoor point clouds are available, including laser scanning and close-range photogrammetry. The point cloud properties obtained using these methods vary. After a point cloud is obtained, determining how to classify it is typically required to obtain useful information. Therefore, point cloud segmentation technology is necessary for many applications. Consider BIMs in civil engineering as an example. To facilitate the subsequent surface reconstruction and boundary extraction, the segmentation of the different surfaces of building components is necessary.

In the 3D point clouds of existing buildings, multiple attribute categories are typically present. For example, these categories are found in the 3D point clouds of columns, beams, walls, and panels of structures, pipelines, lamps, desks, and firefighting appliances for non-structural objects. However, existing point cloud segmentation algorithms are mainly intended for specific shapes. For spatial regions with complex environments, manually preprocessing the point cloud first before using segmentation algorithms may be necessary.

Accordingly, this study attempts to use the DGCNN with a deep neural network to apply semantic segmentation to 3D point clouds and maintain the neighborhood relationship among point clouds through edge convolution. Consequently, the semantic segmentation of 3D point clouds of columns, beams, walls, ceilings, and floors can be achieved.

1. S3DIS Dataset

The deep learning process typically relies on numerous samples for training and requires relevant benchmark data to evaluate the prediction results of deep neural networks. The S3DIS dataset (more completely described as the Stanford large-scale 3D indoor space dataset) is used in this study. The dataset is built by capturing RGB-D images with a Matterport camera to create a grid and then generating an indoor point cloud through grid sampling. This dataset has approximately 700 million point clouds, and ground truth has also been established [11]. We use five types of samples from the S3DIS dataset, such as columns, beams, walls, floors, and ceilings, to increase the number of training samples to obtain better overall accuracy and to verify our training results.

2. Close-range Images

The main point cloud acquisition methods can be classified into two categories: laser scanning and close-range photogrammetry methods. Close-range photogrammetry has the advantage of capturing images from multiple perspectives using a general, non-measuring digital camera or mobile phone. It can also produce point clouds through SFM technology, significantly reducing the production time of 3D point clouds and improving the convenience of point cloud acquisition.

Because close-range photogrammetry is characterized by low cost, high mobility, and high precision, it can obtain an indoor 3D point cloud in a more economical, convenient, and reliable manner.

In view of the foregoing, this study adopts close-range photogrammetry to capture indoor images and SFM technology to produce 3D point clouds. SFM technology can produce high-precision 3D point clouds quickly and massively. It is a common technology for generating 3D point clouds from close-range photogrammetry images [34,35]. The precision of the 3D point cloud is within ±6 cm for the control point and ±3 cm for the check point. Consequently, 3D point clouds with sufficient precision and quantity can be generated as deep learning samples.

#### 3.1.2. Sample Training

In this study, the DGCNN is employed to classify 3D point clouds for training using supervised learning. Therefore, to evaluate the correctness of the training results, ground truth samples are required. The ground truth samples in this experiment include those of columns, beams, walls, floors, and ceilings. In the S3DIS dataset, indoor 3D point cloud data are established to complete the ground truth samples for each category. Hence, the ground truth data of close-range images are generated by artificially segmenting the 3D point cloud to train the discriminative parameters of the deep learning model. By manually segmenting the point clouds with this accuracy, we can ensure the accuracy of the ground truth data.

*3.2. Semantic Segmentation and Modeling*

Each point can be assigned a corresponding attribute value using the trained model to segment the 3D point cloud. For example, the semantic segmentation result of a certain point in space is "column" or "beam." However, only the segmented point cloud can achieve the purpose of automatically creating 3D model components.

To achieve automatic modeling, the segmentation result must be preprocessed. The features in the 3D point cloud must be extracted, and the extraction rules must be established to convert the point cloud with attributes into model components. This section describes the feature extraction and automatic modeling rules. After automatic processing, the point cloud results can be automatically converted into parametric components.

Sampath and Shan (2007) reported that in their study of the normalized edge results of roof edge extraction, the object model had distinct endpoint features [33]. If the endpoint coordinates can be effectively extracted, they can be used in formulating the size of the parameterized element. Endpoint coordinates can also be based on the defined 3D coordinates where the components are to be placed. To extract endpoint coordinates from the segmented point cloud, data preprocessing must be performed. Without preprocessing, false edges and connection problems can occur because point clouds typically contain noise, errors, and edge irregularities.

In addition, previous related research shows that columns and beams are not consistently considered when classifying 3D point clouds; nevertheless, in such cases, classification is extremely inadequate. After the analysis, the column and beam characteristics are as follows: (1) The point clouds are few, small in size, and difficult to classify. (2) The columns and beams overlap with other structural components of the building. To resolve this problem, this study focuses on the characteristics of columns and beams. The following processing is proposed. (1) First, each category is extracted from the results of DGCNN classification. (2) The point cloud is classified because the columns and beams overlap with other categories; thus, the components are simplified. (3) Outlier points are removed. The characteristics of a category are used to remove incorrect points, avoiding the lines and results of the model. (4) Feature extraction is performed on the point cloud of the confirmed category to extract the outlines of the model. (5) The appearance of the previous model is integrated, and the correct model components are built.

3.2.1. Category Extraction

The 3D point cloud processed by the DGCNN is divided into five categories, which can be extracted separately; colors are assigned to indicate different categories.

This study considers five types of data: "columns," "beams," "walls," "floors," and "ceilings." These categories can be distinguished by RGB colors: columns are pink, beams are yellow, walls are light blue, floors are dark blue, and ceilings are green.

The semantic segmentation results have RGB band values; therefore, they are used as classification indicator references. The results are shown in Figure 2.

(a) (b)

**Figure 2.** (**a**) Classification results for each category (from left to right: column, beam, wall, floor, ceiling). (**b**) Combination of all classification results.

3.2.2. Labeled Category

In the classified data, "column" and "beam" are repeatedly found in the same space with other categories. These data must first be divided and processed into a single point cloud component, which is beneficial for subsequent feature extraction and component construction. This study adopts the minimum distance segmentation method in Euclidean space for division. After analyzing the actual building, the minimum segmentation distance for the columns and beams was determined to be 60 cm. After checking the minimum distance, effectively preserving the point clouds of beams and columns was found to be possible.

### 3.2.3. Removal of Outliers

The overall accuracy rates of segmenting beams and columns are low. This may cause the subsequent automatic modeling of components to be inconsistent with the current situation. To solve this problem, this study refers to the method of Torr and Zisseee [36]. The use of building structure characteristics to filter out erroneous point clouds is proposed, and boundary errors are avoided. Three factors are considered: (1) point-to-plane distance; (2) plane normal vector; and (3) maximum angular distance, which uses the range interval and the directionality of the point cloud to remove the erroneous point cloud. The method first identifies the farthest point. Then, this point is used as the center to find the point cloud within a certain radius. This point is evaluated using the maximum and minimum values of the plane coordinates of the point cloud within the radius. Error points are filtered out.

In the category of beams and columns that are difficult to classify, a method for direction evaluation is added to filter out erroneous points and thus improve the accuracy of beam and column models. After this preliminary filtering of error points, the outline of the component becomes visible. However, if the coordinates of the maximum endpoint value of the point cloud are used directly as a component range, the appearance of the model may differ from the real situation.

Therefore, in extracting the boundaries of elements such as columns and beams in this study, the vertical axis (Z axis) is used as the normal vector for the column, and the horizontal axes (X and Y) are used as normal vectors for the beam element. Then, the erroneous points of the 3D point cloud with the maximum angle are removed. This method considers the allowable value of the angle between the point cloud and the normal vector. The point cloud within the allowable range is retained; otherwise, it is eliminated.

### 3.2.4. Feature Extraction

The endpoint of the component must be extracted from the cross-section of the point cloud. If the cross-sectional information of a certain plane is directly obtained, it can readily result in insufficient information (Figure 3a). To resolve this, the method used in this research projects the segmented point cloud data to the minimum value of the reference direction according to the three axes (XYZ). The planarized point cloud has dense point clouds (Figure 3b).

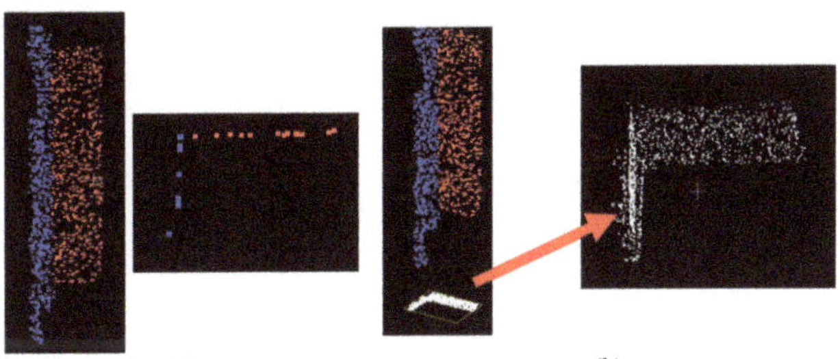

(a) (b)

**Figure 3.** (**a**) 3D point clouds are projected onto a certain section; point clouds are few. (**b**) Planarized point cloud from proposed method has dense point clouds.

An endpoint feature extraction process can then be performed. The planarization results in the three dimensions of XYZ are considered separately during the extraction operation. The coordinate values of components in three dimensions are extracted to achieve feature extraction.

3.2.5. Three-Dimensional Modeling

After completing the point cloud feature extraction, the attribute information can be used for the automatic modeling rules [17,18,37]. In the BIM operating platform, this study uses the Revit software produced by Autodesk for overall modeling display, component information viewing, conflict analysis, and other functional purposes. Design using Dynamo (a visual plug-in program for Revit) in the automatic modeling stage is implemented. Automatic modeling processes five parts: column, beam, wall, floor, and ceiling. Because the research objects are mainly preset components in the parameter library of Revit, several preset family types are directly selected as the modeling types in this study.

## 4. Experiments and Analyses

### *4.1. Test Area: Civil Engineering Building*

The preliminary planning of the experimental area considered different indoor spaces as research targets; control points were set indoors; and coordinates were obtained for scale constraints and precision analysis. The number of training samples was increased using data amplification methods.

Four experimental areas, 2F, 4F, 6F, and the basement of the civil engineering building of our school, were selected as 3D reconstruction targets (Figure 4). The common characteristics of the four areas are as follows: they have distinct "columns," "beams," "walls," "ceilings," and "floors," and a square layout.

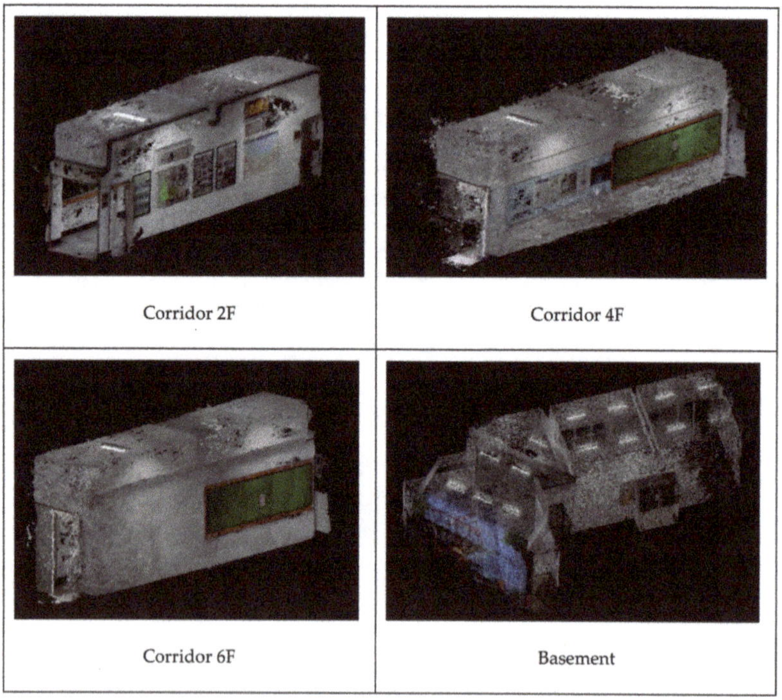


**Figure 4.** Three-dimensional point cloud results of test area. Four experimental areas, the corridor of 2F, 4F, 6F, and the basement of the civil engineering building.

## 4.2. Three-Dimensional Point Cloud Classification

In this study, the DGCNN is used to train and segment the S3DIS dataset and the civil engineering hall of our school. This section presents the evaluation of the results of using the S3DIS dataset.

### 4.2.1. S3DIS Dataset

The S3DIS dataset has six areas: Area1, Area2, Area3, Area4, Area5, and Area6. In this study, Area2, Area3, Area4, Area5, and Area6 are used as training samples and Area1 is used as the test area. The data in Area1 include 13 categories of objects, such as tables and chairs. To explore the internal structure of the building, this study only retains the point cloud sample data of columns, beams, walls, floors, and ceilings for training and testing.

The parameter setting before training affects the subsequent semantic segmentation results; therefore, the parameters of the DGCNN model can be adjusted before training. After training, the training parameters set by S3DIS were as follows: batch size = 3, decay rate = 0.5, decay step = 300,000, learning rate = 0.001, momentum = 0.9, num point = 4096, and epoch = 40.

Each iteration of the training process lasted approximately 33 min, and the training accuracy started to flatten upon reaching 0.96. In the 40th iteration, the training loss rate was 0.019 and the training accuracy rate was 0.993; overfitting was not observed.

Based on the training results, this study selects the 40th iteration model for the segmentation test of the Area1 indoor area. Three small areas in Area1 were randomly selected for comparative analysis: Conference_Room2, Office_2, and Office_6; the overall segmentation accuracy rates are 86.90%, 97.49%, and 92.47%, respectively. Overall accuracy is calculated as the sum of correctly classified pixels divided by the total number of pixels. Tables 1–3 are the confusion matrices.

**Table 1.** Conference_Room2 confusion matrix. The overall accuracy is 86.9%.

| **Conference_Room2** | | **Predicted Class** | | | | | |
|---|---|---|---|---|---|---|---|
| | | **Beam** | **Ceiling** | **Column** | **Floor** | **Wall** | % |
| True Class | beam | 8214 | 287 | | | 3806 | 66.7 |
| | ceiling | 99 | 61,073 | 24 | | 56 | 99.7 |
| | column | 276 | 7 | 3704 | | 4084 | 45.9 |
| | floor | | 17 | 425 | 47,700 | 1204 | 96.7 |
| | wall | | 2627 | 9476 | 152 | 28,801 | 70.2 |
| | % | 95.6 | 95.4 | 27.2 | 99.7 | 75.9 | 86.9 |

**Table 2.** Office_2 confusion matrix. The overall accuracy is 97.5%.

| **Office2** | | **Predicted Class** | | | | | |
|---|---|---|---|---|---|---|---|
| | | **Beam** | **Ceiling** | **Column** | **Floor** | **Wall** | % |
| True Class | beam | 5509 | 10 | | | 28 | 99.3 |
| | ceiling | 44 | 21,155 | 3 | | 8 | 99.7 |
| | column | | 22 | 1179 | | 3 | 97.9 |
| | floor | | | 9 | 15,211 | 130 | 99.1 |
| | wall | 61 | 356 | 1177 | 3 | 28,820 | 94.7 |
| | % | 98.1 | 98.2 | 49.8 | 100.0 | 99.4 | 97.5 |

**Table 3.** Office_6 confusion matrix. The overall accuracy is 92.5%.

| **Office6** | | **Predicted Class** | | | | | |
|---|---|---|---|---|---|---|---|
| | | **Beam** | **Ceiling** | **Column** | **Floor** | **Wall** | **%** |
| True Class | beam | 4125 | 2693 | | | 130 | 59.4 |
| | ceiling | 19 | 16,823 | 2 | | 47 | 99.6 |
| | column | | | 1176 | | 5 | 99.6 |
| | floor | | | | 14,390 | 353 | 97.6 |
| | wall | 23 | 97 | 1235 | 22 | 20,300 | 93.6 |
| | % | 99.0 | 85.8 | 48.7 | 99.8 | 97.4 | 92.5 |

After analyzing the five structures, the segmentation results of the ceiling, floor, and wall were all found to be excellent; however, the classification results of the beam and column were inadequate.

Some wall point clouds were misclassified as columns, and beams were misjudged as walls and ceilings. The ground truth and segmented results are summarized in Table 4.

**Table 4.** Categories of ground truth and segmented results for S3DIS dataset.

| | **Ground Truth** | **Segmented Results** |
|---|---|---|
| Conference _Room2 |  |  |
| Office_2 |  | 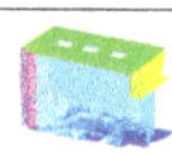 |
| Office_6 |  |  |

4.2.2. Civil Engineering Building

The experimental area of the civil engineering building in our school has four sections: corridors 2F, 4F, and 6F, and the basement.

The information obtained from corridors 4F and 6F as well as the basement is used as a training sample, and that from corridor 2F is used as a test sample.

In this study, the data augmentation method is used to increase the number of samples effectively. The training samples were sequentially divided at 10° intervals, and samples from 10° to 90° were also added. After adding the samples, the parameters obtained through training were as follows: batch size = 5, decay rate = 0.5, decay step = 300,000, learning rate = 0.001, momentum = 0.9, number of points = 4096, epoch = 40, and dropout = 0.4–0.7. In the training results, determining whether overfitting occurs was necessary. The tests for loss and accuracy of calculations using the sampling model was performed. No overfitting occurred during the S3DIS sample training, but overfitting started in round 34 after adding the close-range image sample data. It is assumed that the overfitting problem occurred because the training sample number of close-range images was small.

In the analysis, the lowest loss and highest accuracy rates occurred in the 33rd iteration; these were 0.182% and 94.2%, respectively. Subsequently, at the 34th iteration, the loss rate started to increase and the accuracy rate started to decrease. Accordingly, 0.7 was selected as the dropout point. The 33rd iteration yielded the best segmentation result after adding the samples.

The classification test results for the point cloud of corridor 2F are listed in Table 5.

**Table 5.** Confusion matrix for corridor 2F. The overall accuracy is 94.2%.

| Corridor 2F | | Predicted Class | | | | | |
|---|---|---|---|---|---|---|---|
| | | **Beam** | **Ceiling** | **Column** | **Floor** | **Wall** | **%** |
| True Class | beam | 4642 | 51 | 538 | | 818 | 76.7 |
| | ceiling | 93 | 23,378 | 43 | | 71 | 99.1 |
| | column | 8 | 2 | 5972 | | 398 | 93.6 |
| | floor | | | 80 | 41,111 | 1345 | 96.6 |
| | wall | 585 | 1528 | 2300 | | 52,205 | 92.2 |
| | % | 87.1 | 93.7 | 66.9 | 100 | 95.2 | 94.2 |

In the confusion matrix in Table 4, the production accuracy rates were 99.1%, 93.6%, 96.6%, and 92.2% for the ceiling, column, floor, and wall, respectively. The beam achieved an accuracy rate of 76.7%. With regard to user accuracy, the beam, ceiling, floor, and wall reached accuracy rates of 87.1%, 93.7%, 100%, and 95.2%, respectively. The column attained an accuracy rate of 66.9%. The ground truth data of the point cloud and the visualization of the segmentation results are summarized in Table 6.

**Table 6.** Categories of ground truth and segmented results for corridor 2F of civil engineering building.

| | **Ground Truth** | **Segmented Result** |
|---|---|---|
| Corridor 2F of Civil Engineering Building |  |  |

Two types of sample data are used in this study: the S3DIS indoor dataset and the self-constructed point cloud sample of the civil engineering building. In the training process, owing to the sufficient training samples in S3DIS, the trend graphs of test and training accuracy rates were parallel; overfitting did not occur. In the 40th iteration of training, the overall segmentation accuracy rates of Area1_ConferenceRoom2, Area1_Office 2, and Area1_Office 6 reached 86.90%, 97.49%, and 92.47%, respectively.

However, in the training results of the civil engineering gymnasium, owing to the small number of original samples, sample training was performed in the form of data augmentation. The test sample was segmented using the training results of the 33rd iteration; the overall accuracy was 94.2%. After the analysis, the accuracy of beams and columns remained low.

#### 4.2.3. Discussion of Classification Results

In the S3DIS dataset, the segmentation results of the ceiling, floor, and wall were all found to be excellent; however, the classification results of the beam and column were inadequate. Some wall point clouds were misclassified as columns, and beams were misjudged as walls and ceilings. The overabundance of these two types of components was due to the small size and number of point cloud samples; hence, this type of error was expected.

In the civil engineering building dataset, the segmentation accuracy of columns and beams is lower than that of walls, floors, and ceilings because of two possible reasons.

1. Number of point clouds

In a single indoor space, the areas of walls, floors, and ceilings are larger than those of columns and beams. The original sample training results of the hall in the civil engineering building indicate that the segmentation results of columns and beams are lower than those of the walls, floors, and ceilings. Segmentation can be improved by increasing the number of training samples.

2. Geometric distribution of point cloud

The point cloud data of walls, floors, and ceilings has a flat and wide distribution. In contrast, the point cloud geometry of columns and beams has a 90° corner.

### 4.3. Semantic Segmentation and Modeling

After completing segmentation, the result of each test area was imported into the feature extraction operation. Moreover, parametric modeling was performed using automatic modeling rules.

#### 4.3.1. S3DIS Modeling

In the test sample of S3DIS, three sections of Area_1 were selected for the analysis of segmentation results: Conference_Room2, Office_2, and Office_6. Feature extraction and parametric modeling of the three areas were performed sequentially. The modeling results are shown in Figure 5 to verify the feasibility of the automatic modeling rule design. To facilitate the visualization of internal modeling, the ceiling is removed, as shown in Figure 5.

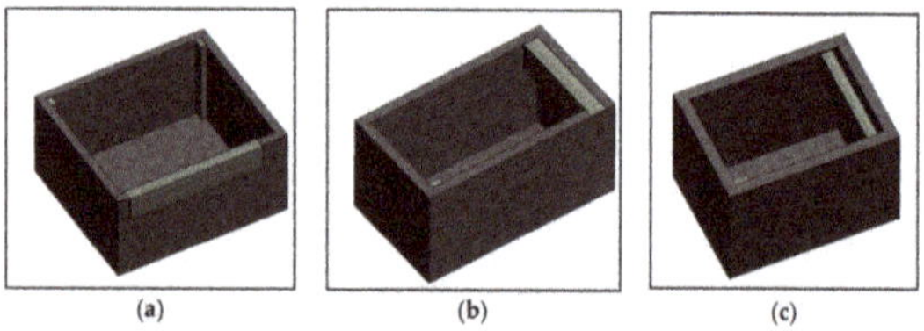


**Figure 5.** Automatic modeling results of (**a**) Conference_Room2, (**b**) Office_2, and (**c**) Office_6 without ceiling.

#### 4.3.2. 2F Corridor of Civil Engineering Building

After the segmentation of the point cloud in corridor 2F, feature extraction and automatic modeling were performed sequentially. The modeling results are shown in Figure 6. The ceiling is also removed to visualize the interior.

**Figure 6.** Automated modeling results of corridor 2F (without ceiling) in civil engineering building.

### 4.4. Evaluation of 3D Model

As presented in this section, the modeling results of corridor 2F in the civil engineering building of our school were selected for testing. This is because corridor 2F is more convenient to measure on site than the other areas. There were two columns, four walls, two beams, one floor, and a ceiling in the area. This study analyzes the top view and sectional view, as shown in Figure 7.

To verify whether the modeling result is consistent with the actual length of the selected area, a total station was used for measuring the points, as shown in Figure 8. The comparison results are listed in Table 7. The root mean square error (RMSE) of the line length given by BIM compared with the actual length is ±0.03 m.

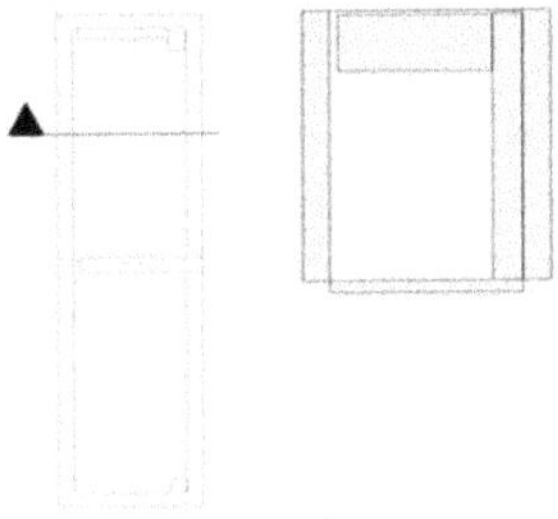

(a) (b)

**Figure 7.** (**a**) Top view and (**b**) sectional view of corridor 2F.

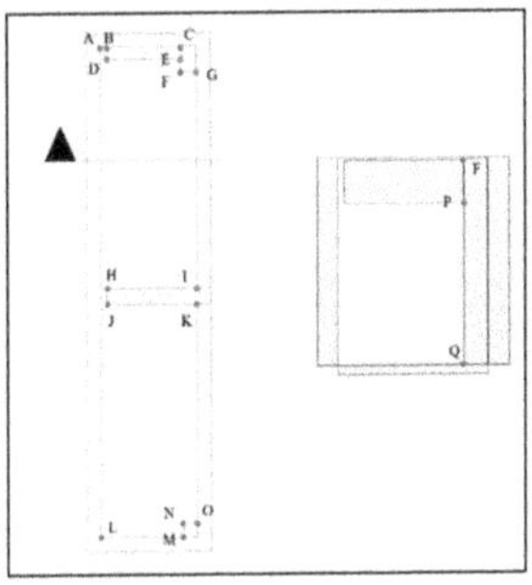

**Figure 8.** Line position map for comparing 3D model and in situ field. The letters A–Q indicate the location of the points, and the length of the line segment connecting the two points is measured and compared in this study.

**Table 7.** Three-dimensional model and actual line length difference (unit: m).

| Line | Model Length | Actual Length | Difference | Line | Model Length | Actual Length | Difference |
|---|---|---|---|---|---|---|---|
| AB | 0.094 | 0.09 | 0.004 | BD | 0.259 | 0.232 | −0.027 |
| BC | 1.715 | 1.721 | −0.006 | EF | 0.286 | 0.269 | 0.017 |
| DE | 1.715 | 1.721 | −0.006 | HJ | 0.347 | 0.321 | 0.026 |
| FG | 0.337 | 0.322 | 0.015 | IK | 0.347 | 0.324 | 0.023 |
| HI | 2.032 | 2.05 | −0.018 | GI | 4.671 | 4.643 | 0.028 |
| JK | 2.032 | 2.052 | −0.02 | NM | 0.360 | 0.304 | 0.056 |
| LM | 1.778 | 1.784 | −0.006 | KO | 4.631 | 4.716 | −0.085 |
| NO | 0.334 | 0.325 | 0.009 | FP | 0.749 | 0.781 | −0.032 |
| AL | 10.598 | 10.609 | 0.011 | FQ | 3.455 | 3.430 | 0.025 |

The feature extraction process developed in this study derives the features of endpoints from the point cloud segmentation results. Then, the parametric elements of columns, beams, walls, floors, and ceilings were automatically modeled based on the attribute information. The experimental results indicate that Area1_ConferenceRoom2, Area1_Office 2, Area1_Office 6 of S3DIS, and corridor 2F in the civil engineering building can be used to create the 3D model data of indoor components automatically. Each component has attribute information, such as material, length, volume, and quantity. The overall process not only reduces the time cost of manual model construction but also serves as follow-up application management. It is a rapid BIM method for reconstructing existing indoor spaces.

In comparing the actual length (obtained by inspection) of corridor 2F with the indoor measurement yielded by automatic modeling, the RMSE is found to be ±0.03 m; hence, the accuracy is acceptable.

## 5. Conclusions and Future Work

In this study, a DGCNN was used to learn indoor 3D point cloud features. Five items, i.e., columns, beams, walls, floors, and ceilings, were considered the research objects. Training and learning were conducted using two different sources of point cloud data: S3DIS and photogrammetry. Area_1 data in the S3DIS dataset and those of corridor 2F in the civil engineering building were used as test samples for segmentation.

The endpoint feature extraction program developed in this study was applied to process the segmented results. Other details, such as endpoint coordinates, quantity, and length of the research object, are derived. Finally, feature information is imported into an automatic modeling program for parametric element modeling.

This study uses a DGCNN to learn the features of indoor point clouds and segment the point clouds of columns, beams, walls, floors, and ceilings automatically. The overall accuracies using the S3DIS indoor dataset and civil engineering building information were 86.9% and 94.2%, respectively.

An endpoint feature extraction method that overcomes the errors caused by irregular line segments is proposed in this paper. In addition, for columns and beams with low semantic segmentation accuracy, a range processing method is devised to reduce semantic segmentation errors.

The method can be employed to calculate the number of components, boundary length and size, and relative information from the extracted endpoint. In comparing the inspected size of corridor 2F in the civil engineering building with the measurement yielded by automatic modeling, the RMSE is found to be approximately $\pm 0.03$ m. Because the point clouds are constrained by control points, the model is similar to the building.

The results of this study demonstrate that indoor 3D point clouds produced by close-range images can be segmented using a trained 3D deep learning network. The automatic feature point extraction method proposed in this study is employed to derive the feature point information of components. Using this information, the point cloud can be imported into an automatic modeling system to generate BIM parametric components and create indoor drawings.

There are a lot of objects in the room, and this study only sets out to study and discuss five categories of structural objects. In the future, we intend to increase the number of samples, increase the types of objects, reduce the noise, and explore ways to improve accuracy.

**Author Contributions:** Conceptualization, funding acquisition, methodology, writing—original draft preparation, C.-S.H. and X.-J.R.; data curation, software (data experiments), writing—review and editing, visualization, C.-S.H. and X.-J.R.; formal analysis, investigation (physical experiments), validation, X.-J.R. All authors have read and agreed to the published version of the manuscript.

**Funding:** This research received no external funding.

**Data Availability Statement:** This research did not use publicly archived datasets.

**Conflicts of Interest:** The authors declare no conflict of interest.

## References

1. Griffiths, D.; Boehm, J. A Review on Deep Learning Techniques for 3D Sensed Data Classification. *Remote Sens.* **2019**, *11*, 1499. [CrossRef]
2. Wang, Y.; Ahsan, U.; Li, H.; Hagen, M. A Comprehensive Review of Modern Object Segmentation Approaches. *Found. Trends Comput. Graph. Vis.* **2022**, *13*, 111–283. [CrossRef]
3. Guo, Y.-C.; Weng, T.-H.; Fischer, R.; Fu, L.-C. 3D semantic segmentation based on spatial-aware convolution and shape completion for augmented reality applications. *Comput. Vis. Image Underst.* **2022**, *224*, 103550. [CrossRef]
4. Chen, H.; Liang, M.; Liu, W.; Wang, W.; Liu, P.X. An approach to boundary detection for 3D point clouds based on DBSCAN clustering. *Pattern Recognit.* **2022**, *124*, 108431. [CrossRef]
5. Biagini, C.; Capone, P.; Donato, V.; Facchini, N. Towards the BIM implementation for historical building restoration sites. *Autom. Constr.* **2016**, *71*, 74–86. [CrossRef]
6. Bello, S.A.; Yu, S.; Wang, C.; Adam, J.M.; Li, J. Review: Deep Learning on 3D Point Clouds. *Remote. Sens.* **2020**, *12*, 1729. [CrossRef]

7. Camuffo, E.; Mari, D.; Milani, S. Recent Advancements in Learning Algorithms for Point Clouds: An Updated Overview. *Sensors* **2022**, *22*, 1357. [CrossRef]
8. Diab, A.; Kashef, R.; Shaker, A. Deep Learning for LiDAR Point Cloud Classification in Remote Sensing. *Sensors* **2022**, *22*, 7868. [CrossRef]
9. Sun, Y.; Paik, J.; Koschan, A.; Abidi, M. 3D reconstruction of indoor and outdoor scenes using a mobile range scanner. In Proceedings of the 2002 International Conference on Pattern Recognition, IEEE, Quebec City, QC, Canada, 11–15 August 2002.
10. Wu, H.; Yue, H.; Han, J.; Zhang, S. Indoor Map Boundary Correction Based on Normalized Total Least Squares of Condition Equation. *J. Surv. Eng.* **2021**, *147*, 04021015. [CrossRef]
11. Armeni, I.; Sener, O.; Zamir, A.R.; Jiang, H.; Brilakis, I.; Fischer, M.; Savarese, S. 3D semantic parsing of large-scale indoor spaces. In Proceedings of the IEEE Conference on Computer Vision and Pattern Recognition, Las Vegas, NV, USA, 27–30 June 2016.
12. Wang, T.; Wang, Q.; Ai, H.; Zhang, L. Semantics-and-Primitives-Guided Indoor 3D Reconstruction from Point Clouds. *Remote. Sens.* **2022**, *14*, 4820. [CrossRef]
13. Wang, Y.; Sun, Y.; Liu, Z.; Sarma, S.E.; Bronstein, M.M.; Solomon, J.M. Dynamic Graph CNN for Learning on Point Clouds. *ACM Trans. Graph.* **2019**, *38*, 1–12. [CrossRef]
14. Guo, Y.; Wang, H.; Hu, Q.; Liu, H.; Liu, L.; Bennamoun, M. Deep Learning for 3D Point Clouds: A Survey. *IEEE Trans. Pattern Anal. Mach. Intell.* **2021**, *43*, 4338–4364. [CrossRef]
15. Liu, W.; Sun, J.; Li, W.; Hu, T.; Wang, P. Deep Learning on Point Clouds and Its Application: A Survey. *Sensors* **2019**, *19*, 4188. [CrossRef]
16. Yang, J.; Zou, B.; Qiu, H.; Li, Z. MLFNet-Point Cloud Semantic Segmentation Convolution Network Based on Multi-Scale Feature Fusion. *IEEE Access* **2021**, *9*, 44950–44962. [CrossRef]
17. Chen, P.; Luo, Z.; Shi, W. Hysteretic mapping and corridor semantic modeling using mobile LiDAR systems. *ISPRS J. Photogramm. Remote. Sens.* **2022**, *186*, 267–284. [CrossRef]
18. Ochmann, S.; Vock, R.; Klein, R. Automatic reconstruction of fully volumetric 3D building models from oriented point clouds. *ISPRS J. Photogramm. Remote. Sens.* **2019**, *151*, 251–262. [CrossRef]
19. Xie, Y.; Tian, J.; Zhu, X.X. Linking Points with Labels in 3D: A Review of Point Cloud Semantic Segmentation. *IEEE Geosci. Remote Sens. Mag.* **2020**, *8*, 38–59. [CrossRef]
20. Rashdi, R.; Martínez-Sánchez, J.; Arias, P.; Qiu, Z. Scanning Technologies to Building Information Modelling: A Review. *Infrastructures* **2022**, *7*, 49. [CrossRef]
21. Jhaldiyal, A.; Chaudhary, N. Semantic segmentation of 3D LiDAR data using deep learning: A review of projection-based methods. *Appl. Intell.* **2022**, 1–12. [CrossRef]
22. Lin, H.; Wu, S.; Chen, Y.; Li, W.; Luo, Z.; Guo, Y.; Wang, C.; Li, J. Semantic segmentation of 3D indoor LiDAR point clouds through feature pyramid architecture search. *ISPRS J. Photogramm. Remote. Sens.* **2021**, *177*, 279–290. [CrossRef]
23. Jeon, J.; Jung, J.; Kim, J.; Lee, S. Semantic Reconstruction: Reconstruction of Semantically Segmented 3D Meshes via Volumetric Semantic Fusion. *Comput. Graph. Forum* **2018**, *37*, 25–35. [CrossRef]
24. Qi, C.R.; Su, H.; Mo, K.; Guibas, L.J. PointNet: Deep learning on point sets for 3D classification and segmentation. In Proceedings of the 30th IEEE Conference on Computer Vision and Pattern Recognition, CVPR 2017, Honolulu, HI, USA, 21–26 July 2017.
25. Qi, C.R.; Yi, L.; Su, H.; Guibas, L.J. PointNet++: Deep hierarchical feature learning on point sets in a metric space. *Adv. Neural Inf. Process. Syst.* **2017**, *30*, 5105–5114.
26. Mitropoulou, A.; Georgopoulos, A. An Automated Process to Detect Edges in Unorganized Point Clouds. *ISPRS Ann. Photogramm. Remote. Sens. Spat. Inf. Sci.* **2019**, *IV-2/W6*, 99–105. [CrossRef]
27. Kong, G.; Fan, H.; Lobaccaro, G. Automatic building outline extraction from ALS point cloud data using generative adversarial network. *Geocarto Int.* **2022**, 1–18. [CrossRef]
28. Mahphood, A.; Arefi, H. Grid-based building outline extraction from ready-made building points. *Autom. Constr.* **2022**, *139*, 104321. [CrossRef]
29. Chen, X.; Yu, K. Feature Line Generation and Regularization from Point Clouds. *IEEE Trans. Geosci. Remote Sens.* **2019**, *57*, 9779–9790. [CrossRef]
30. Awrangjeb, M. Using point cloud data to identify, trace, and regularize the outlines of buildings. *Int. J. Remote Sens.* **2016**, *37*, 551–579. [CrossRef]
31. Wang, Y.; Jeschke, S. Edge extraction by merging 3D point cloud and 2D image data. In Proceedings of the 2013 10th International Conference and Expo on Emerging Technologies for a Smarter World (CEWIT), IEEE, Melville, NY, USA, 21–22 October 2013.
32. Borges, P.; Zlot, R.; Bosse, M.; Nushke, S.; Tews, A. Vision-based localization using an edge map extracted from 3D laser range data. In Proceedings of the 2010 IEEE International Conference on Robotics and Automation, IEEE, Anchorage, AK, USA, 3–7 May 2010.
33. Sampath, A.; Shan, J. Building Boundary Tracing and Regularization from Airborne Lidar Point Clouds. *Photogramm. Eng. Remote Sens.* **2007**, *73*, 805–812. [CrossRef]
34. Jiang, S.; Jiang, C.; Jiang, W. Efficient structure from motion for large-scale UAV images: A review and a comparison of SfM tools. *ISPRS J. Photogramm. Remote Sens.* **2020**, *167*, 230–251. [CrossRef]
35. Braun, A.; Tuttas, S.; Borrmann, A.; Stilla, U. Improving progress monitoring by fusing point clouds, semantic data and computer vision. *Autom. Constr.* **2020**, *116*, 103210. [CrossRef]

36. Torr, P.H.S.; Zisserman, A. MLESAC: A New Robust Estimator with Application to Estimating Image Geometry. *Comput. Vis. Image Underst.* **2000**, *78*, 138–156. [CrossRef]
37. Macher, H.; Landes, T.; Grussenmeyer, P. From Point Clouds to Building Information Models: 3D Semi-Automatic Reconstruction of Indoors of Existing Buildings. *Appl. Sci.* **2017**, *7*, 1030. [CrossRef]

*buildings*

MDPI

*Article*

# Determinants of Data Quality Dimensions for Assessing Highway Infrastructure Data Using Semiotic Framework

**Chenchu Murali Krishna [1,*], Kirti Ruikar [2] and Kumar Neeraj Jha [1]**

1 Department of Civil Engineering, Indian Institute of Technology Delhi, Delhi 110016, India
2 School of Architecture, Building and Civil Engineering, Loughborough University, Epinal Way, Loughborough LE113TU, Leicestershire, UK
* Correspondence: cez198622@iitd.ac.in

**Abstract:** The rapid accumulation of highway infrastructure data and their widespread reuse in decision-making poses data quality issues. To address the data quality issue, it is necessary to comprehend data quality, followed by approaches for enhancing data quality and decision-making based on data quality information. This research aimed to identify the critical data quality dimensions that affect the decision-making process of highway projects. Firstly, a state-of-the-art review of data quality frameworks applied in various fields was conducted to identify suitable frameworks for highway infrastructure data. Data quality dimensions of the semiotic framework were identified from the literature, and an interview was conducted with the highway infrastructure stakeholders to finalise the data quality dimension. Then, a questionnaire survey identified the critical data quality dimensions for decision-making. Along with the critical dimensions, their level of importance was also identified at each highway infrastructure project's decision-making levels. The semiotic data quality framework provided a theoretical foundation for developing data quality dimensions to assess subjective data quality. Further research is required to find effective ways to assess current data quality satisfaction at the decision-making levels.

**Keywords:** highway data quality assessment; data quality dimensions; semiotic framework; decision-making

**Citation:** Krishna, C.M.; Ruikar, K.; Jha, K.N. Determinants of Data Quality Dimensions for Assessing Highway Infrastructure Data Using Semiotic Framework. *Buildings* **2023**, *13*, 944. https://doi.org/10.3390/buildings13040944

Academic Editors: Ming-Hung Hsu, Osama Abudayyeh, Zheng-Yun Zhuang and Ying-Wu Yang

Received: 30 December 2022
Revised: 27 March 2023
Accepted: 29 March 2023
Published: 2 April 2023



## 1. Introduction

Highway agencies devote significant resources to collecting, storing, and maintaining many forms of data, ranging from preliminary survey data to pavement condition data, throughout the life cycle of a highway project. For instance, the National Highway Authority of India launched Data Lake, a project monitoring tool to track and monitor the progress of projects and to act as the central repository of documents across the project life cycle [1]. According to the FMI's (2019) report titled "Big Data Equals Big Questions for the Engineering and Construction Industry," some of the most significant infrastructure projects require an average of 130 million emails, 55 million documents, and 12 million workflows. At the same time, 95.5% of all data collected in the engineering and construction industry is unutilised because many firms cannot manage and process vast amounts of data for decision-making [2]. According to a 2018 industry report titled "Construction Disconnected" by FMI, 48% of all reworks in infrastructure projects in the United States are caused by poor data and miscommunication, resulting in an annual cost of over USD 31.3 billion. Globally, an average of 52% of rework was caused by poor data and communication, amounting to USD 280 billion. The primary cause of poor data and information was that 34.4 percent of reworks were caused by incorrect project data, meaning it was out-of-date or otherwise flawed data, while 28.8 percent of reworks were caused by difficulty gaining access to necessary project data [3]. Despite the significant investment, data utilisation to users' needs for extracting information, knowledge, and support decisions has become debatable [4]. Data collection is becoming an increasingly significant asset for today's highway arena within highway management and operation. Several systems and technologies have created significant infrastructure data in recent years [5].

Data have been widely used to manage system operations and provide information on highway conditions. However, public and private users discovered that utilising and operating the data is becoming increasingly complex. Data are collected with varying degrees of precision and resolution, and data formats are often incompatible [6].

Technological advancements in data collection result in the real-time monitoring of data and a massive volume of data, Such as data collected in the structural health monitoring of a bridge [7] and data collected during the degradation process of concrete material [8]. In addition, the issue intensifies as the volume of data continues to increase [9–15]. Ghasemaghaei and Calic [16] discussed the role of data quality and diagnosticity in the firm's decision-making, considering the effect of big data processing. However, there is substantial evidence that data quality issues are pervasive in practice and that relying on poor or uncertain data results in less effective decision-making. It also increases the cost of correcting the data in the decision-making process of highway projects [17,18].

Data quality has been extensively studied in various disciplines for several decades [19]. It has become a professional field, emphasising organisational strategy and effective decision-making [20,21]. In addition, data quality is considered a multi-dimensional concept in the literature [22–24]. In the last two decades, scholars and practitioners have proposed several classifications of data quality dimensions, many of which have overlapping and occasionally contradictory meanings concerning respective disciplines (e.g., [14,24–26]). Despite the different classifications, few investigations have attempted to integrate these perspectives of data quality dimensions to assess the quality of highway data for effective decision-making. For instance, Coleman [27] gave an insightful examination of the various current classifications of data quality dimensions and identified sixteen mutually incompatible dimensions.

Although numerous studies have found the significance of data quality for decision-making based on various frameworks and methodologies, not much focus has been given to assessing data quality at different decision-making levels of highway projects [5,28,29]. Samitsch et al. [30] provided a guide for companies seeking to improve organisational performance by improving data quality, with a combination of 16 dimensions. Addressing this issue necessitates a method for comprehending data quality, followed by methods for enhancing data quality and decision-making based on data quality information. This research proposes a semiotic-based framework for comprehending highway infrastructure data quality, consisting of four levels: syntactic (form), empiric (connection), semantic (meaning), and pragmatic (use) [29]. The semiotic-based framework assesses and understands data quality based on the semiotic theory's application. Semiotic theory concerns using signs and symbols to convey data, information, and meaning [31]. A review of data quality frameworks applied in various fields was also carried out. Such as the semiotics framework, AIMQ methodology, data quality assessment (DQA), the observe-orient-decide-act methodology (OODA DQ), and the Canadian Institute for health information methodology (CIHI) framework are used in the healthcare industry for data quality assessment [32–36], while the total data quality management (TDQM) framework, comprehensive methodology for data quality management (CDQ), data quality practical approach (DQPA), task-based data quality method (TBDQ), and data quality assignment framework (DQAF) are used in the IT industry to deliver high-quality information products (IP) to information consumers [9,37–41]. A DQMos model and DQMes methodology are used for evaluating data quality in software engineering experiments data [42]. A questionnaire survey identified the critical data quality dimensions of the proposed semiotic framework levels from the National Highway stakeholders for decision-making. The survey helps the National highway stakeholders understand the parameters or dimensions of data quality to assess the quality of data stored in the data lake. The study investigated identifying the framework for analysing data quality and determining the appropriate framework for assessing highway infrastructure data. Currently, there are no specialised studies of data quality dimensions for evaluating highway infrastructure data.

A literature review was conducted first for the study, followed by identifying data quality frameworks. The second step identified data quality dimensions within the four levels of the semiotic data quality framework. In the third step, an interview and questionnaire were conducted in two stages. Initially, an interview survey was undertaken to develop a list of data quality characteristics that reflect the opinions of data consumers regarding data quality. For the second stage, a questionnaire was developed from the identified dimensions through an interview study. The questionnaire survey was conducted to gather information on the importance of each of these dimensions to data consumers at the individual level of decision-making, followed by a ranking of the dimensions within the categories of semiotic frameworks to comprehend stakeholders' priorities for each characteristic data quality.

The paper is structured as follows: The next section focuses on the literature review of frameworks and data quality dimensions and identifies the most effective framework for evaluating highway infrastructure data. The subsequent section addresses the research methodology, the findings, and an analysis of the findings. Finally, conclusions and future work scope are presented.

## 2. Objectives

The study's main objective was to investigate highway infrastructure data quality dimensions and the framework for assessing data quality. According to the 2018 FMI report, the cost of reworks caused by poor data quality and accessibility of data in the United States was USD 31.3 billion, while in Australia and New Zealand, it was USD 8.4 billion, and in the United Kingdom, it was USD 10.2 billion [3]. The literature shows that poor data quality negatively impacts the time and cost to make a decision and decision-making performance in the highway infrastructure project lifecycle. Hence, assessing data quality is critical for organisations and creates importance for identifying the dimensions to define data quality. The objective of the study was divided into three key research objectives as follows:

- To establish the data quality dimensions necessary for determining the data quality of highway infrastructure data to facilitate effective decision-making.
- To determine the importance of data quality dimensions at each level of decision-making.
- To determine the priority of dimensions within the semiotic framework categories.

## 3. Literature Review

### 3.1. Data and Data Quality

Before going to the concept of data quality dimensions, let us review the first-order questions that arise from the history of the data quality domain. What is data, and what is data quality? Liebenau and Backhouse [43] defined data as "linguistic, mathematical or other symbolic representation that is universally accepted to represent people, things, events, and ideas." Data represent objects or processes in the actual world in their most basic form. Thus, while addressing data quality, we may argue that poor data quality results from an inaccurate depiction of the real world [44]. Abedjan et al. [45] addressed the tools used for detecting data errors. The study of data quality assessment began in the 1950s, particularly regarding the quality of products and services. Several researchers published several definitions, though no universally accepted definition of data quality exists. Wang and Strong [46] defined data quality as information usable by data consumers, and Crosby [47] defined it as "conformance to requirements." The General Administration of Quality Supervision, 2008, defined data quality as "the degree to which a set of inherent characteristics fulfil the requirements" [15]. At the same time, Fu and Easton [48] explained that data quality is commonly referred to as a collection of "characteristics" of data, such as precision, exhaustiveness, consistency, and timeliness. Most of these characteristics dictate the various dimensions along which data quality may be represented. A low degree of data quality can significantly influence the overall effectiveness of the associated data applications [49].

### 3.2. Data Quality Assessment Framework

Researchers define various frameworks and approaches for data quality assessment. For example, Madnick and Zhu [50], English [51], and Redman [52] explored strategies for increasing data quality, Batini et al. [53] provided a thorough and comparative description of data quality techniques for assessing and improving data quality, Gao et al. [54] proposed a fussing attributes approach for improving uncertain data quality, and Madnick et al. [55] reviewed current practices and research in the field. The research literature describes or defines data quality from simple lists of data quality dimensions to comprehensive frameworks (for example, [24,25,29,56]).

Hassenstein and Vanella [57] presented a data quality encyclopedia for the data life cycle. It describes the data quality dimensions, the data quality evaluation procedure, and the data quality context and practices in various fields. At the same time, Gabr et al. [58] comprehensively defined each traditional and big data quality dimension, metrics, and handling approach with specific definitions. They examined the metrics and methodologies used to monitor and manage each dimension and how they are monitored and managed. The study also examined the most-used data quality dimensions of traditional and large data sets.

Svetlana [59] presented the findings of an expert survey on data quality concerns to demonstrate that it is not required to employ all the numerous dimensions of data quality provided by researchers. However, the essential data quality criteria may be blended for a particular application. The study equips data users and producers with the knowledge necessary to effectively address application-specific data quality issues. In addition to the Svetlana findings, Eliza et al. [60] provided a methodology that allows users to manage data quality and make decisions based on data quality. It eliminates the requirement to fully integrate insufficient data by considering the operational context of the user to enhance a specific element of data quality.

Different approaches from the literature review were summarised to review the well-known and established frameworks for assessing and improving data quality for different data types. Table 1 lists fourteen data quality frameworks identified from the literature.

**Table 1.** Frameworks identified from the literature review.

| S. No. | Framework | Dimensions | References |
|---|---|---|---|
| 1 | TDQM: Total Data Quality Management | Accuracy, objectivity, believability, reputation, access, security, relevance, value-added, timeliness, completeness, amount of data, interpretability, ease of understanding, concise representation, and consistent representation. | [38] |
| 2 | TIQM: Total Information Quality Management | Definition conformance, completeness, validity, accuracy, precision, non-duplication, the equivalence of redundant or distributed data, accessibility, timeliness, contextual clarity, derivation integrity, usability, usability, and rightness. | [51] |
| 3 | COLDQ: Cost-effect of Low Data Quality | *Data model:* Clarity of definition, comprehensiveness, flexibility, robustness, essentialness, attribute granularity, the precision of domains, homogeneity, naturalness, identifiability, obtainability, relevance, simplicity, and semantic and structural consistency. *Data values:* Accuracy, completeness, consistency, currency, null values, and timeliness. Information Policy: Accessibility, metadata, privacy, redundancy, security, and unit cost. *Presentation:* Appropriateness, correct interpretation, flexibility, format precision, portability, consistent representation, representation of null value, and use of storage. | [61] |

**Table 1.** *Cont.*

| S. No. | Framework | Dimensions | References |
| --- | --- | --- | --- |
| 4 | AIMQ: A Methodology for Information Quality Assessment | Accessibility, appropriate amount, believability, completeness, concise representation, consistent representation, ease of operation, free-of-error, interpretability, objectivity, relevancy, reputation, security, timeliness, and understandability. | [34] |
| 5 | DQA: Data Quality Assessment | Accessibility, appropriate data, objectivity, believability, reputation, security, relevancy, value-added, timeliness, completeness, interpretability, ease of manipulation, understandability, concise representation, consistent representation, and free-of-error. | [35] |
| 6 | HIQM: Hybrid Information Quality Management | Accuracy, completeness, consistency, and timeliness. | [62] |
| 7 | CDQ: Comprehensive Methodology for Data Quality Management | Accuracy, completeness, and currency, Unstructured: Currency, relevance, and reliability. | [63] |
| 8 | DQPA: A Data Quality Assessment Framework | Accuracy, completeness, consistency, timeliness, uniqueness, and volatility. | [39] |
| 9 | SPDQM: Square-Aligned Portal Data Quality Model | Accuracy, traceability, correctness, expiration, completeness, consistency, accessibility, compliance, confidentiality, efficiency, precision, and understandability. Availability, accessibility, verifiability, confidentiality, portability, and recoverability. Validity, value-added, relevancy, specialisation, usefulness, efficiency, effectiveness, traceability, compliance, precision, concise representation, consistent representation, attractiveness, and readability. | [64] |
| 10 | HDQM: A Data Quality Methodology for Heterogeneous Data | Accuracy and currency. | [65] |
| 11 | DQAF: Data Quality Assessment Framework | Completeness, timeliness, validity, consistency, and integrity. | [40] |
| 12 | TBDQ: Task-Based Data Quality Method | Accuracy, completeness, consistency, and timeliness. | [41] |
| 13 | OODADQ: The Observe-Orient-Decide-Act Methodology | Speed and volume. | [36] |
| 14 | Semiotic Approach Data Quality-SESP model | Accuracy, consistency representation, unbiased, accessibility, up-to-date, traceability, security, believability, interpretability, ease of manipulation, understandability, completeness, appropriate amount of information, relevancy, concise representation, value-added, and reputation. | [32] |

According to the analysis of the frameworks listed in Table 1, the data quality dimensions considered by each framework vary considerably. Some data quality dimensions are recognised by only one framework, whereas specific dimensions appear frequently. For example, the HDQM and OODADQ frameworks considered only two dimensions for assessment, while the frameworks DQA and HIQM considered more than four dimensions. The dimensions varied according to the field of applications and perspective of the application, such as the health care industry, information technology, and business management. For example, let us consider how the accuracy dimension has been used in the HDQM and HIQM frameworks. In the HDQM framework in the IT industry, dimension accuracy is defined as the proximity between a value "v" and another value "v." of the domain D in the user interface development. This is regarded as the correct representation of the real-world phenomenon value "v" seeks to represent. At the same time, the HIQM framework in the business management sector defines accuracy as the value difference between two databases containing the same value as the correct representation of the real-world value. To understand the most critical dimensions applied in the various fields, the frequency of usage by different data quality dimensions was considered and is shown in Figure 1. Only dimensions used more than once are considered in the figure. The study of Figure 1 helps finalise the dimensions from the literature review perspective to be identified in the data quality in the semiotic framework for assessing highway infrastructure data. The semiotic approach data quality framework is the most applicable of the 14 frameworks mentioned above for evaluating highway infrastructure data. The reason for the selection is explained in the semiotic framework section.

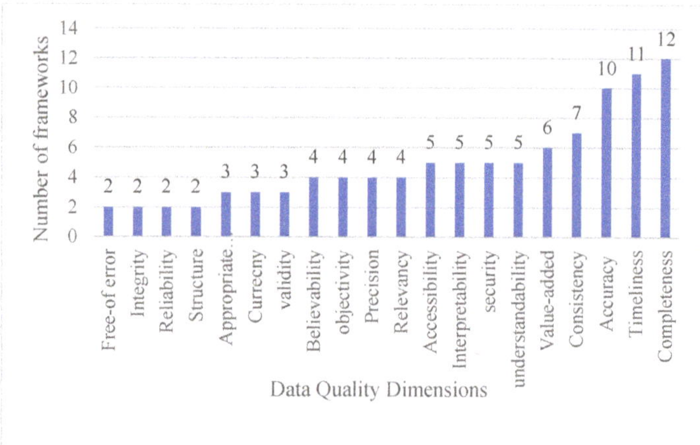


**Figure 1.** Number of frameworks that used specific data quality dimensions.

### 3.3. *Semiotic Framework*

Semiotics is the study of signs and symbols used to convey meaning to various users. Data quality researchers have also adopted the semiotic perspective of data; for instance, Price and Shanks [29] identified three data quality levels: syntactic, semantic, and pragmatic. Semiotic theory concerns using symbols to convert knowledge and define levels in the framework for analysing structure, physical form, meaning, and data usage. A thorough examination of the various levels of semiotics would reveal that the pragmatic level is associated with knowledge, the semantic level with information, and only the syntactic level with data. In other words, the dimensions operating at the pragmatic, semantic, and syntactic levels pertain to the quality of knowledge, information, and data.

According to Falkenberg et al. [66], data are meaningful symbolic creations consisting of a limited arrangement of signs and symbols. Thus, the semiotic framework was used in this study to define data quality dimensions. The semiotic framework consists of four levels: empiric, syntactic, pragmatic, and semantic. Each level of the semiotic framework facilitates data quality evaluation from several perspectives, including structure, data, information, and knowledge for assessing highway infrastructure data for decision-making at various levels of the highway decision-making hierarchy, for instance, while selecting a treatment technique for damaged pavement in a highway construction project.

Each decision-making level bases its decisions on the raw data, information, and knowledge available at that level. The strategic level is the top level of an organisation and is responsible for strategic planning. This involves making long-term, big-picture decisions and establishing policies that impact the organisation. For the decision of treatment technique, the system performance (policymaking) policies are established, requiring knowledge to make policies. Similarly, at the network level, the fund distribution (planning) decisions are made, i.e., allocating funds according to project requirements. At the program level, the decision of pavement evaluation and prioritisation is considered for each project.

At the project selection level, the project is selected according to the prioritisation made at the program level, and treatment selection is made at the project level.

Kahn et al. [67] addressed the relationships between semiotic levels, the data-information-knowledge (DIK) hierarchy, and associated data-quality issues, as shown in Figure 2. The relationship between semiotic levels and structure, data, information, and knowledge facilitate the identification of unique data quality issues that may necessitate the application of specialised skills to resolve. Knoke and Yang [68] claimed that information originates with data and is transferred to knowledge in the DIK hierarchy. Depending on how data's meaning, structure, and operation are communicated at different semiotic levels of the DIK hierarchy, such transference could increase or decrease data's meaningfulness, transferability, and applicability.

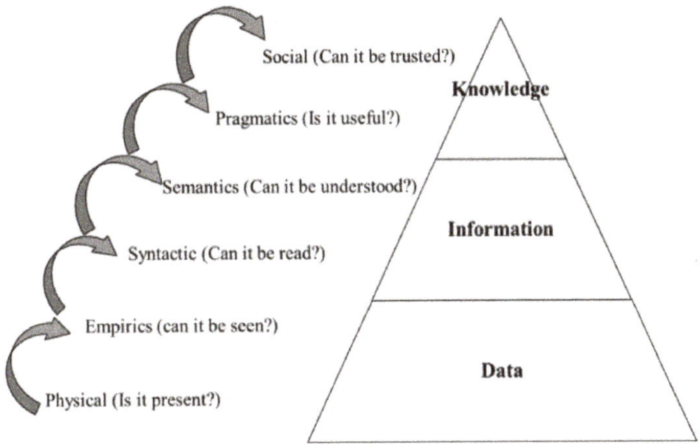


**Figure 2.** Semiotic levels and the data-information-knowledge hierarchy, adapted from [32]. 2018, Huang.

The empiric level focuses on the quality aspect of data access and the means of communication. It considers how much and in what way raw data are available for stakeholders for decision-making. In highway projects, decision-makers at each project phase, such as preconstruction, construction, and post-construction phases, consider data availability essential for effective decision-making. At the empirical level, accessibility, security, and timeliness (currentness) are considered to evaluate the data communication and access perspective of the raw data stored in the data lake [68]. For example, the dimension accessibility of highway data could be the availability of real-time traffic data on a particular highway. If the data are easily accessible through an open data portal such as a data lake, API, or mobile app, they would have a high level of accessibility. On the other hand, if the data are only available through a difficult-to-navigate website or requires complex technical skills, they would have a low level of accessibility.

On the other hand, the syntactic level concentrates on the forms and structure of data, or, more accurately, their physical form instead of their content. After assessing the accessibility criteria of raw data, the second crucial limitation for decision-makers is the kind and format of accessible data. To quantify the structure of raw data stored in a data lake, the syntactic level considers accuracy, concise presentation, ease of operation, consistency, integrity, and completeness as data quality dimensions [49]. For instance, the accuracy dimension in highway infrastructure data could be the precision of the measurements taken for the width of a particular road lane. Inaccurate measurements could lead to too narrow lanes, potentially causing safety issues or impeding traffic flow.

The semantic level of data quality is concerned with the meaning of data for information generation rather than the data [69]. The decision-makers at the program and project selection decision-making levels require information regarding project performance for decisions such as budget allocation and project prioritisation. The dimensions at the semantic level are credibility, interpretability, and understandability for assessing the interpretation of data that provides meaning. For example, dimension interpretability refers to the ease with which stakeholders can understand and use data. In the context of highway data, interpretability could be the use of visualisations or dashboards that make it easier for stakeholders to understand complex data sets. This could include interactive maps or charts that allow users to explore different aspects of highway infrastructure data, such as traffic volume or accident rates.

The pragmatic level is concerned with the relationship between data, information, and behaviour in a specific context of decision-making [69]. The generation of knowledge from the available data and information for making the policies and planning at the strategic and network levels of decision-making of highway infrastructure projects requires data utilisation quality. Dimensions of data quality associated with the pragmatic level include appropriateness, value-addition, reputation, relevancy, and usefulness [68]. Contextual features of pragmatic concerns are related to dimensions of relevance and utility of data and information for making decisions. As a dimension, reputation focuses on the user's expectations of data utility. The value-addition dimension aims to comprehend the user intent. These facets concern the data's compatibility with the challenging job. Related data quality dimensions are concerned with the intended application, i.e., how data would be utilised in connection to the current issue [70], for instance, value addition as a data quality dimension that refers to the extent to which data are valuable and add value to the organisation or individual stakeholders using it. In the context of highway data, it could use data analytics and machine learning algorithms to identify patterns and trends in data that are not immediately apparent. This could help highway agencies to identify areas of the highway system that require additional investment or maintenance and to prioritise their efforts accordingly.

Consequently, each semiotic level handles certain data quality and communication concerns. Understanding the overall data utilisation of highway infrastructure data stored in the data lake for making decisions at each decision-making level depends on the quality dimensions of the semiotic levels [32]. Within each semiotic level, it is crucial to identify the data quality requirements of decision-makers at their respective decision-making levels. For instance, strategic-level decision-makers focus on the utility of data and information for making effective policies throughout the organisation. Similarly, the other decision-making levels also required their specific data quality according to the requirement of decision-makers. Table 2 shows the data quality dimensions and the perspectives of dimensions along with the semiotic framework categories.

Applying a semiotic framework can be considered one of the philosophical approaches to studying data and its quality. In a semiotic framework, a top-down approach involves starting with high-level concepts or theories and breaking them into their constituent parts to understand how they work. In terms of the decision-making hierarchy, NHAI also follows a top-down approach. The higher officials make the authority's decisions at the top of the organisational structure and then communicate to the lower-level employees for implementation. Overall, by using a semiotic framework for data quality assessment, NHAI can ensure that its decision-making processes are informed by high-quality data that are relevant, accurate, and consistent. This can help to improve the efficiency and effectiveness of NHAI's operations and ensure that its highway and road networks are developed and maintained to the highest standard. However, the semiotic perspective has not become popular among researchers and practitioners to date [71]. The present study uses semiotic categories to describe the highway infrastructure data quality, specifically to identify the data quality dimensions to assess the data quality for effective decision-making [29]. Presently, no research has been reported to comprehend the link between data quality dimensions and highway infrastructure data about the semiotic levels that represent them.

**Table 2.** Data quality (DQ) dimensions and perspectives as per the semiotic framework levels.

| Semiotic Levels | DQ Dimensions | DQ Dimensions Perspective |
|---|---|---|
| Empiric = It addresses issues that arise when data are utilised repeatedly. This level focuses on developing means of communication and data handling. | Accessibility | Accessibility implies that data must be accessible, obtainable, or retrievable when necessary for data to be accessible. |
| | Timeliness | Timeliness is concerned with the age of data and whether data are current. It is achieved if the recorded value is not out of date. |
| | Security | As a dimension, security involves securing data and limiting access to it. |
| Syntactic = It focuses on the structures and formats of data. It deals with the physical form of data rather than their content. | Accuracy | The accuracy dimension is concerned with the conformity of the recorded value with the actual value. It implies that data are accurate, flawless, trustworthy, and error-free. |
| | Completeness | Completeness concerns capturing all values for a specific variable and preventing data loss. It implies that the data must have adequate breadth, depth, and scope for the given task. |
| | Conciseness | Conciseness is a well-organised, concise, and condensed representation of data. |
| | Consistency | Consistency is achieved when data are represented in the same format, are compatible with previous data, and are represented consistently. |
| | Ease of operation | Ease of operation implies that data are manipulatable, integrated, customised, and utilised for multiple purposes. It is similar to flexibility. |
| | Integrity | Integrity measures correctness and consists of semantic and physical integrity. Semantic integrity measures consistency and completeness concerning the rule of the description language. Physical integrity measures the correctness of implementation details. |
| | Structure | Format or structure implies that data are in the correct format and structure. |
| Semantic = At the semantic level, dimensions are connected with information rather than data. Information is selected data to which meaning has been assigned in a particular context. It is concerned with meaning. | Ambiguity | Ambiguity arises due to improper representation and is when data can be interpreted in more than one way. |
| | Believability | Believability is concerned with whether data can be believed or regarded as credible. |
| | Interpretability | Interpretability means that data should be interpreted; that is, it should be defined clearly and represented appropriately. |
| | Definition | Meaningfulness or definition is concerned with the interpretation of data. The failure of this dimension results in meaningless data. |
| | Reliability | Reliability in terms of concepts drawn from the field of quality control. |
| | Understandability | Understandability concerns whether data are clear, readable, unambiguous, and easily comprehendible. |
| | Validity | Data are valid when verified as genuine and satisfying appropriate standards related to other dimensions. |
| Pragmatic = It focuses on how individuals use information. It concerns the relationship between data, information, and behaviour in each context. | Appropriateness | Appropriateness as a data quality dimension means that data must be appropriate to the task at hand. |
| | Relevant | Relevancy is concerned with the applicability of data to the task at hand. It is a crucial dimension if the data do not address the customer's needs and when the customer finds the data inadequate. |
| | Value | Value is added as a dimension that addresses the benefits and advantages of using data. |

## 4. Methodology

In order to meet the research objectives, this study was carried out in three steps. The first step was to identify the data quality dimensions of highway infrastructure using the semiotic framework. Most appropriate dimensions that were applicable to the highway infrastructure project were identified. In the second step, the questionnaire was prepared to the selected data quality dimensions finalised in step one. The responses were collected for the questionnaire from the highway infrastructure stakeholders. Finally, the responses were analysed in the third step to identify the critical dimensions and to rank them according to their mean value. These steps are described in detail in the following sub-sections.

Step 1: Identification of data quality dimensions of highway infrastructure data using the semiotic framework.

The semiotic framework consists of 43 data quality dimensions, as defined by Tejay. G. et al. [72]. These data quality dimensions are defined in the context of information system security. For the study of highway infrastructure projects' data quality, the dimensions were reduced to 20 out of 43 data quality dimensions, according to the relevant literature sources. A few dimensions have synonyms dimensions, and those were combined and considered a single dimension. The dimension accessibility, portability, and locatability have a similar meaning in the context of data quality; thus, we considered accessibility the primary dimension for assessing data quality. The established data quality dimensions were used to determine the data quality of highway infrastructure data. The 20 dimensions were personally reviewed with the three highway stakeholders; one chief general manager from the headquarters office responsible for network-level decision-making, one regional officer from the regional office responsible for the program and project selection level decision-making, and one project director from the project-implementing unit responsible for project-level decision-making were selected to verify the exhaustiveness/comprehensiveness of the selected data quality dimensions. Among the professionals, the chief general manager had more than ten years of experience, the project director had eight years of experience, and the regional officer had six years of experience in highway construction projects. The responses were not uniform, and the experience of the stakeholders was considered a limitation. Hence, all 20 dimensions were considered for the questionnaire survey for a comprehensive understanding of highway data quality for the effective use of data for effective decision-making.

Step 2: Data Collection

The questionnaire was designed based on the 20 data quality dimensions identified in Step 1. The survey targeted the National Highway of India decision-makers who utilised these data in decision-making. A pilot study was undertaken with 40 responses to test the language and understanding of the questionnaire. The responses are from the site engineers, deputy engineers, and managers from the project implementing units and regional offices. According to the suggestions from the pilot study, some significant changes were made to the questionnaire to make it more understandable for the stakeholders. The questionnaire was then shared via google forms with the 220 stakeholders. The stakeholders included the members, chief general manager, managers, regional officers, deputy general managers, and project directors. A total of 105 experts participated in the survey, which is a 48% response rate. The stakeholders with significant experience deal with the critical decisions from the National Highway Authority of India (NHAI), representing the strategic, network, program, project selection, and project levels, respectively. The questionnaire consists of three parts. Part 1 deals with the basic contact details, role, responsibility, and decision-making level in the decision-making hierarchy. The second part evaluates each attribute's importance at each decision-making level for the available data. The third part deals with ranking the dimensions, which states the priority of dimensions required in decision-making within the category of the semiotic framework.

A five-point Likert scale of 1 to 5 was used to record the decision-makers' level of importance of the data quality attribute. Here, '1' refers to "no importance," '2' refers to "low importance," '3' refers to "somehow important," '4' refers to "important," and '5' refers to "high importance" [73].

Step 3: Data Analysis

The data were analysed by using the software package SPSS 25. The analysis was carried out in two parts. The first part analysed the data's reliability using Cronbach's alpha test. It was found to be 0.875 at a 5% significance level greater than 0.5. Hence, it confirmed the reliability of the data. The dimensions were ranked according to their mean value to measure the consensus in the experts' opinions. However, when the mean values of two or more dimensions were identical, the dimensions with the lowest standard deviation were placed higher [74]. The ranking of the dimensions based on the data collected through the questionnaire survey is shown in Table 3.

**Table 3.** Ranking of data quality dimensions.

| S. No. | Data Quality Dimensions | Mean | Std. Deviation | Rank |
|---|---|---|---|---|
| 1 | Accuracy | 4.52 | 0.64 | 1 |
| 2 | Accessibility | 4.40 | 0.70 | 2 |
| 3 | Completeness | 4.36 | 0.76 | 3 |
| 4 | Consistency | 4.28 | 0.67 | 4 |
| 5 | Timeliness | 4.27 | 0.68 | 5 |
| 6 | Structure | 3.90 | 0.90 | 6 |
| 7 | Ambiguity | 3.90 | 0.98 | 7 |
| 8 | Integrity | 3.83 | 0.85 | 8 |
| 9 | Value | 3.72 | 0.88 | 9 |
| 10 | Validity | 3.63 | 1.04 | 10 |
| 11 | Reliability | 3.58 | 1.12 | 11 |
| 12 | Appropriateness | 3.58 | 1.12 | 12 |
| 13 | Relevant | 3.58 | 0.85 | 13 |
| 14 | Definition | 3.50 | 0.81 | 14 |
| 15 | Interpretability | 3.38 | 0.96 | 15 |
| 16 | Understandability | 3.38 | 1.10 | 16 |
| 17 | Believability | 3.36 | 1.01 | 17 |
| 18 | Ease of Operation | 3.35 | 0.99 | 18 |
| 19 | Security | 3.35 | 0.90 | 19 |
| 20 | Conciseness | 3.29 | 0.83 | 20 |

### *4.1. Identification of Critical Data Quality Dimensions of Highway Infrastructure Data*

The descriptive statistical analysis did not yield a whole number for the mean value of the responses. Therefore, for the purpose of interpretation, the impact of each dimension on data quality can be considered to lie between the midpoints of two adjacent scales [75]. The importance of the dimensions about the mean value ($\mu$) greater than or equal to 4.5 was deemed to have a very high impact on the important data quality dimension. Similarly, the range of mean values $4.5 > \mu \geq 3.5$ was treated as having high importance; $3.5 > \mu \geq 2.5$ was treated as having moderate importance; $2.5 > \mu \geq 1.5$ was treated as having low importance; and mean values less than 1.5 were treated as having very low importance on data quality. In the study, the key data quality dimensions for assessing highway infrastructure data for effective decision-making dimensions were deemed to be those that were both very high and of high importance.

### *4.2. Importance of Data Quality Dimensions at Respective Decision-Making Levels*

Based on the questionnaire results, consideration was also given to the importance of dimensions. The data quality requirement may not be the same at all levels of decision-making. For instance, the project level focuses on the primary data collection and format. Hence, the dimensions critical at the project level are not critical at the remaining decision-making levels. Hence, the importance of dimensions at all decision-making levels was considered. The significance of data quality dimensions is determined at the strategic, network, program, project selection, and project levels of highway projects. Based on the ratings for the importance of dimensions at the decision-making level, decision-makers believe that all data quality attributes defined under the semiotic model are considered critical in data usage for information generation at all decision-making levels, with a rating of 4 out of 5. The context of data quality differs at each level of decision-making; consequently, data quality dimensions were determined, and the ranking of data quality dimensions was also calculated at each level of the semiotic framework, i.e., at the syntactic, pragmatic, empirical, and semantic levels.

### *4.3. Ranking of Data Quality Dimensions within the Semiotic Framework*

Along with the level of importance, the decision-makers also prioritise data quality dimensions in each category of the semiotic framework. The priority of data quality requirements has changed from stakeholder to stakeholder at each decision-making level. The semiotic framework comprised the syntactic, empiric, semiotic, and pragmatic categories, which deal with the structure, meaning, information, and knowledge of data characteristics [32]. The prioritisation of dimensions was also taken in the questionnaire survey. The responses to dimensions given by the respondents were converted into a rank using Henry Garrett's ranking technique [76]. This technique provides the change of orders of problems into numerical scores. The prime advantage of this technique over simple frequency distribution is that the dimensions are arranged based on their priority from the point of view of decision-makers. Garrett's formula for converting the ranks into the per cent position is shown below as Equation (1):

$$\text{Percent position} = 100 \times (R_{ij} - 0.5)/N_j \quad (1)$$

where $R_{ij}$ = rank given for ith dimension by jth decision-maker

$N_j$ = number of dimensions ranked by the jth individual.

The per cent position of each rank was converted into sources referring to the table given by Garrett and Woodworth [77]. For each factor, the scores of individual stakeholders were added together and divided by the total number of respondents for whom scores were added. These mean scores for all the dimensions were arranged in descending order; the dimensions were accordingly ranked.

## 5. Results and Discussion

This study identified and evaluated the key data quality dimensions for assessing the data quality of highway infrastructure for decision-making effectiveness. For this purpose, the study considered the critical dimensions throughout the highway infrastructure project, as well as the criticality of dimensions at each level of decision-making, as the data quality requirement varies at each level of decision-making. Using the ranking of dimensions shown in Table 3, the overall critical dimensions and preference of the dimensions for the overall project data were determined. Table 4 illustrates the significance of dimensions at each level of decision-making. From the analysis of Table 4, it is clear that the requirements for decision-makers are no longer the same but vary according to the respective hierarchical levels of decision-making.

The data quality dimensions listed in Table 3 are relevant to ensure the overall quality of data for highway infrastructure projects. Effective decision-making relies on the availability of high-quality data, and addressing each of these dimensions can help to ensure that the data used in decision-making are accurate, complete, consistent, and timely. Based on the mean scores in Table 3 and the analysis of Figure 3, the top five data quality dimensions are accuracy, accessibility, completeness, consistency, and timeliness. Ensuring that data are accurate involves verifying that they are correct and error-free. Accessibility involves making the data available and easily retrievable to authorised stakeholders. Completeness ensures that all required data elements are present and accounted for. Consistency involves verifying that the data are consistent with other data elements within the project. Timeliness ensures that the data are available when needed and up to date. Other dimensions listed in Table 3, such as relevance, interpretability, and believability, are also crucial for effective decision-making. Relevant data are essential to the decision-making process because they ensure that the data are related to the project's objectives. Interpretability ensures that data are presented in a way that is easy to understand. At the same time, believability involves ensuring that the data can be trusted and are not biased.

**Table 4.** Decision-maker's level of importance of dimensions.

| S. No. | Data Quality Attributes | Strategic Level | Network Level | Program Level | Project Selection Level | Project Level |
|---|---|---|---|---|---|---|
| 1 | Accuracy | 4.6 | 4.4 | 4.7 | 4.8 | 4.4 |
| 2 | Consistency | 4.8 | 4.6 | 4.1 | 4.3 | 4.2 |
| 3 | Completeness | 4.8 | 4.4 | 4.2 | 4.6 | 4.3 |
| 4 | Structure | 4.8 | 4.3 | 3.6 | 4.2 | 3.7 |
| 5 | Integrity | 4.8 | 4.0 | 3.6 | 4.1 | 3.7 |
| 6 | Conciseness | 3.0 | 3.4 | 3.4 | 3.3 | 3.2 |
| 7 | Ease of Operation | 3.6 | 3.3 | 3.3 | 3.2 | 3.4 |
| 8 | Accessibility | 4.2 | 4.6 | 4.7 | 4.4 | 4.3 |
| 9 | Timeliness | 4.4 | 4.6 | 4.4 | 3.9 | 4.2 |
| 10 | Security | 3.8 | 3.4 | 3.4 | 3.2 | 3.3 |
| 11 | Definition | 4.2 | 3.8 | 3.6 | 3.8 | 3.3 |
| 12 | Ambiguity | 4.8 | 4.4 | 4.2 | 4.1 | 3.5 |
| 13 | Believability | 3.6 | 3.0 | 3.2 | 4.0 | 3.4 |
| 14 | Interpretability | 3.4 | 3.4 | 3.2 | 3.7 | 3.4 |
| 15 | Reliability | 4.2 | 3.4 | 3.5 | 2.9 | 3.7 |
| 16 | Understandability | 2.8 | 3.3 | 3.4 | 3.4 | 3.4 |
| 17 | Validity | 3.0 | 4.0 | 3.6 | 3.0 | 3.7 |
| 18 | Relevant | 4.2 | 3.8 | 3.6 | 3.7 | 3.4 |
| 19 | Value | 4.2 | 3.8 | 3.9 | 4.0 | 3.6 |
| 20 | Appropriateness | 3.8 | 3.4 | 3.6 | 3.9 | 3.6 |

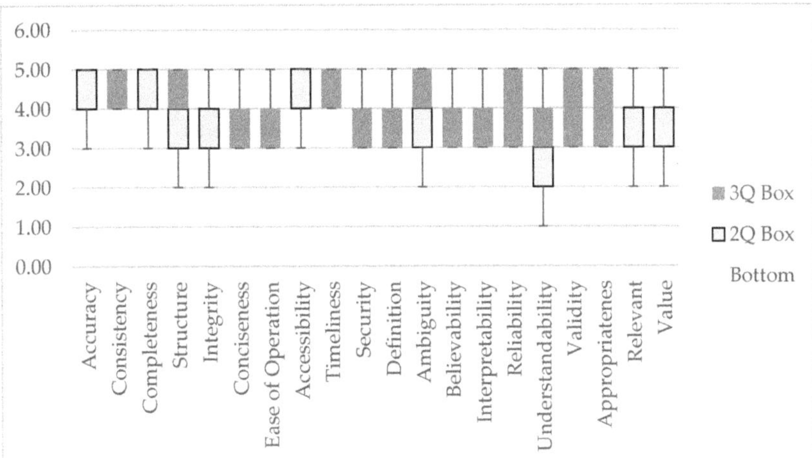


**Figure 3.** Semiotic levels and the data-information-knowledge hierarchy.

At the same time, the stakeholders consider conciseness, ease of operation, and security as the lowest priorities with low mean values. This might be due to the dimension conciseness that implies the compact representation of data, which would create a problem of understanding for all stakeholders for data usage in decision-making. The dimension ease-of-operation implies that data are manipulated and easily customised, which stakeholders feel could cause problems in decision-making if the data are easily manipulated. The dimension security implies keeping data secure and restricting access to the data, and the stakeholders feel the restricting of data would cause issues with the decision-making.

Figure 3 shows the box-and-whiskers plot of the data quality dimensions for assessing overall highway project data quality. It shows that the range of most of the dimensions is between 3 and 5, i.e., the responses from the decision-makers range from somehow important to high importance. Based on the data, it seems that accuracy, completeness, and accessibility are considered to be the most critical dimensions of data quality by more than half of the decision-makers, as they have the highest median value of 5. The dimensions consistency, structure, integrity, timeliness, ambiguity, relevant, and value also have a relatively high median value of 4, indicating that they are still considered essential by many decision-makers. On the other hand, dimensions such as conciseness, ease of operations, security, definition, believability, understandability, validity, and appropriateness have a median value of 3, indicating that they are considered less critical by decision-makers. It is important to note that these findings are based on decision-makers' responses and reflect the objective of data quality measures. Nonetheless, they provide valuable insights into the perceived importance of different dimensions of data quality in the context of highway project data.

### 5.1. Critical Dimensions at Each Level of Decision-Making Hierarchy

The importance of data quality dimensions at each decision-making level, such as strategic, network, program, project selection, and project decision levels of highway infrastructure projects, was also identified, along with key data quality dimensions for assessing the over-project data. Table 4 shows the description of assessment measures and a survey result on the level of importance for semiotic framework data quality attributes obtained from highway decision-makers, respectively. Based on the level of importance, decision-makers think that all data quality dimensions described within the semiotic framework are crucial for generating information at all levels of the decision-making hierarchy for highway infrastructure. However, the results indicate that the conciseness, ease of operational ability, and data understandability dimensions do not significantly influence decision-making processes at all levels of highway infrastructure decision-making, i.e., strategic, network, program, project selection, and project. This may be due to the absence of a system that facilitates the understanding of collected data at these levels. The collected data could be in various formats, including text, images, or numbers, and they could be used as input for decision-making. For data to be used as input, they must be clearly understood according to the judgment of the highway engineers. This may result from the insignificant use of project data at these levels or the continuation of decision-making processes due to limited project scope in the early stages of a project. Therefore, the dimensions with a rating of 4 out of 5 are regarded as crucial for generating information at all levels of the decision-making hierarchy.

#### 5.1.1. Strategic Level

At the strategic level, decision-makers are higher-level authorities, such as the chairman and division heads of NHAI. They deal with policies, guidelines, and the distribution of funds. At the strategic level of the decision hierarchy, the accuracy, consistency, completeness, structure, and integrity dimensions from the syntactic category; the accessibility and timeliness dimensions from the empiric category; the definition, reliability, and ambiguity dimensions from the semantic category; and the relevance and value dimensions from the pragmatic category are crucial decision-making dimensions.

#### 5.1.2. Network Level

At the network level, decision-makers, such as chief general managers, are responsible for determining priorities, developing programs, and determining project objectives. According to the analysis, the critical dimensions at the network level of the decision hierarchy are accuracy, consistency, completeness, structure, and integrity from the syntactic category; accessibility and timeliness from the empirical category; ambiguity from the semantic category; and validity from the pragmatic category. In addition, the network level is subdivided into two decision-making levels, including the program level and project selection level.

#### 5.1.3. Program Level

At the program level of decision-making, the critical dimensions are accuracy, consistency, and completeness from the syntactic category; accessibility and timeliness from the empiric category; and ambiguity from the semantic category. No dimensions from the pragmatic category are crucial for program-level decision-making. Program-level decision-making deals with the programming of the projects. For making decisions at the program level, the decision-makers focus on the form and structure of data, establishing means of communication and data handling and information on the data. The pragmatic category dimensions deal with the knowledge generated from the data, which is not much focused on decision-making at the program level.

#### 5.1.4. Project Selection Level

The project selection level addresses project selection, safety improvement, and traffic control studies at the regional office level. From the syntactic category, the critical dimensions are accuracy, consistency, completeness, structure, and integrity. From the empiric category, the critical dimension is only accessibility. From the semantic category, the critical dimensions are ambiguity and believability. From the pragmatic category, the dimension value is only critical in analysing the questionnaire data.

#### 5.1.5. Project Level

Project-level decisions involve the project director, designers, maintenance engineers, schedulers, and many other engineers responsible for project implementation at the project-implementing unit. For effective decision-making, dimensions such as accuracy, consistency, and completeness from the syntactic level; and accessibility and timeliness from the empiric level are considered critical out of all 20 data quality dimensions. At the project level, its primary concern is data generation, the physical form of data generation, and storage for information generation. Therefore, the dimensions in the semantic and pragmatic categories that deal with information and knowledge generation are not of as high importance as the syntactic and empirics level at the project level.

### *5.2. Ranking of Dimensions within the Semiotic Framework Categories*

Garrett's ranking technique was used to analyse various dimensions for ranking the dimensions within the semiotic framework levels. The decision-makers were asked to rank the dimensions within the framework to understand their preferences for data quality dimensions within the semiotic framework. The semiotic framework comprised syntactic, empiric, semiotic, and pragmatic categories, which deal with the structure, meaning, information, and knowledge of data characteristics [32]. Before ranking the dimensions within the semiotic framework levels, the percentage position for the ranks and their corresponding Garrett value were calculated using Equation (1), as shown in Table 5. The total score was calculated for factors by multiplying the number of stakeholders ranking that dimension (Garrett and Woodworth [76]).

**Table 5.** Percentage position and Garrett value for rank 1 to 7.

| Ranks | Percentage Position | Garret Score |
|---|---|---|
| 1 | 7.14 | 79 |
| 2 | 21.43 | 66 |
| 3 | 35.71 | 57 |
| 4 | 50.00 | 50 |
| 5 | 64.29 | 43 |
| 6 | 78.57 | 34 |
| 7 | 92.86 | 22 |

#### 5.2.1. Syntactics Category

The syntactic category focused on data structures and formats, i.e., the physical form of data rather than its content. In order to understand the data quality requirements in terms of the syntactic category of the data being used by decision-makers, the Garrett ranking technique was used, and the dimensions were ranked as shown in Table 6. Based on the Garrett mean values, stakeholders ranked dimensions as accuracy, consistency, completeness, structure, integrity, conciseness, and ease of operation. This is because the syntactic category is primarily concerned with the physical form rather than the data content; the decision-makers prioritised accuracy over the ease of operation dimension [35]. It is shown that the priority of data quality requirements changed from stakeholder to stakeholder. Hence, we considered most of the responder's ranking as the topmost ranked and followed for other dimensions, as shown in Table 6.

**Table 6.** Ranking of dimensions within the syntactic category of the semiotic framework.

| S. No. | Factors | Rank 1 | 2 | 3 | 4 | 5 | 6 | 7 | Total Number of Stakeholders | Total Score | Total Mean | Rank |
|---|---|---|---|---|---|---|---|---|---|---|---|---|
| 1 | Accuracy | 52 | 16 | 6 | 6 | 11 | 9 | 5 | 105 | 6695 | 63.76 | 1 |
| 2 | Consistency | 12 | 40 | 18 | 19 | 3 | 6 | 7 | 105 | 6051 | 57.63 | 2 |
| 3 | Completeness | 12 | 25 | 42 | 6 | 5 | 5 | 10 | 105 | 5897 | 56.16 | 3 |
| 4 | Structure | 6 | 5 | 15 | 45 | 24 | 6 | 4 | 105 | 5233 | 49.84 | 4 |
| 5 | Integrity | 8 | 6 | 9 | 16 | 49 | 10 | 7 | 105 | 4942 | 47.07 | 5 |
| 6 | Conciseness | 8 | 7 | 5 | 6 | 8 | 44 | 27 | 105 | 4113 | 39.17 | 6 |
| 7 | Ease of Operation | 7 | 6 | 10 | 7 | 5 | 25 | 45 | 105 | 3924 | 37.37 | 7 |

#### 5.2.2. Empiric Category

The empiric category dealt with the issues that arise when data are utilised repeatedly. This category focused on developing means of communication and data handling. Based on the percentage position and Garrett's mean value, the dimensions were ranked as accessibility, timeliness, and security within the empiric category. The dimension accessibility of data was given the highest priority over the security of the data dimension. Table 7 shows the percentage position and Garret score for the ranks as per Equation (1), while Table 8 shows the ranking of the dimensions based on the Garrett mean value. Accessibility refers to how easily users can access data. This includes factors such as the availability of the data, the ease of retrieving them, and the format in which they are presented. Timeliness refers to how up-to-date and relevant the data are. This includes factors such as the frequency of updates and how quickly they are made available. Security refers to data protection from unauthorised access, modification, or disclosure. This includes factors such as the level of encryption used, the strength of access controls, and the measures in place to prevent data breaches. By ranking these dimensions based on their importance, organisations can prioritise their efforts to improve information quality. However, it is essential to note that the relative importance of each dimension may vary depending on the specific context and the users' needs.

**Table 7.** Percentage position and Garrett value for rank 1 to 3.

| Ranks | Percentage Position | Garret Score |
|---|---|---|
| 1 | 16.67 | 69.00 |
| 2 | 50.00 | 50.00 |
| 3 | 83.33 | 31.00 |

#### 5.2.3. Semantic Category

The semantic category deals with the dimensions connected with information rather than data. Information is selected data to which meaning has been assigned in a particular context. It is concerned with meaning. Within the semantic category, the dimensions were ranked as ambiguity, definition, believability, interpretability, reliability, understandability, and data validity, as shown in Table 9. The dimension ambiguity was prioritised over other data dimensions within the category. This might be because the data should be clear for understanding if any ambiguity in data can lead to significant challenges in decision-making for highway projects, potentially resulting in poor design, construction, and long-term consequences for the environment and public safety. The Garrett mean values that were calculated using Equation (1) and the percentage position values are shown in Table 5.

**Table 8.** Ranking of dimensions within the empiric category of the semiotic framework.

| S. No. | Factors | Rank | | | Total Number of Stakeholders | Total Score | Total Mean | Rank |
|---|---|---|---|---|---|---|---|---|
| | | 1 | 2 | 3 | | | | |
| 1 | Accessibility | 46 | 33 | 26 | 105 | 5630 | 53.62 | 1 |
| 2 | Timeliness | 38 | 47 | 20 | 105 | 5592 | 53.26 | 2 |
| 3 | Security | 21 | 25 | 59 | 105 | 4528 | 43.12 | 3 |

**Table 9.** Ranking of dimensions within the semantic category of the semiotic framework.

| S. No. | Factors | Rank | | | | | | | Total Number of Stakeholders | Total Score | Total Mean | Rank |
|---|---|---|---|---|---|---|---|---|---|---|---|---|
| | | 1 | 2 | 3 | 4 | 5 | 6 | 7 | | | | |
| 1 | Ambiguity | 51 | 16 | 7 | 6 | 10 | 9 | 6 | 105 | 6652 | 63.35 | 1 |
| 2 | Definition | 12 | 25 | 41 | 7 | 5 | 5 | 10 | 105 | 5890 | 56.10 | 2 |
| 3 | Believability | 7 | 6 | 10 | 7 | 6 | 24 | 45 | 105 | 3933 | 37.46 | 3 |
| 4 | Interpretability | 8 | 7 | 9 | 16 | 48 | 10 | 7 | 105 | 4965 | 47.29 | 4 |
| 5 | Reliability | 12 | 39 | 17 | 19 | 5 | 6 | 7 | 105 | 6014 | 57.28 | 5 |
| 6 | Understandability | 8 | 7 | 5 | 6 | 8 | 45 | 26 | 105 | 4125 | 39.29 | 6 |
| 7 | Validity | 7 | 5 | 16 | 44 | 23 | 6 | 4 | 105 | 5276 | 50.25 | 7 |

#### 5.2.4. Pragmatic Category

The pragmatic category focused on how individuals use information. It concerns the relationship between data, information, and behaviour in each context. For ranking the dimensions within the pragmatic category, the percentage position of the ranks was calculated using Equation (1), as shown in Table 7. The dimension value of data was given the highest priority over the other dimensions, such as relevant and appropriateness. In the context of highway stakeholders, the dimension value is crucial because it determines the extent to which the data can inform decision-making about highway infrastructure projects, budgeting, and maintenance. Although the dimension appropriateness is critical, it was ranked third in this context because it is a prerequisite for both relevance and value. As per the Garrett ranking technique, the dimensions were ranked as value, relevant, and appropriateness, respectively, as shown in Table 10.

**Table 10.** Ranking of dimensions within the pragmatic category of semiotic framework.

| S. No. | Factors | Rank | | | Total Number of Stakeholders | Total Score | Total Mean | Rank |
|---|---|---|---|---|---|---|---|---|
| | | 1 | 2 | 3 | | | | |
| 1 | Relevant | 33 | 49 | 23 | 105 | 5440 | 51.81 | 2 |
| 2 | Value | 44 | 30 | 31 | 105 | 5497 | 52.35 | 1 |
| 3 | Appropriateness | 28 | 26 | 51 | 105 | 4813 | 45.84 | 3 |

The dimensions were ranked to understand the decision-makers' data quality requirements for decision-making at the individual decision-making levels [30]. As the level of decision-making in the organisation changes, the priority of data quality also changes. At the strategic level, decision-makers focus on policymaking, which could be implemented throughout the organisation. Hence, the data quality requirements at the strategic level differ at the network and project levels. It is important to note that this study utilised semiotic-based quality dimensions to assess data quality at different decision-making levels from the data users' perspective. This proactive assessment of the highway management decision-making hierarchy allows data collectors to determine the level of data quality requirements of highway infrastructure managers and potential decision-makers in a more integrated manner. It allows highway agencies' data management teams to identify the causes behind minimal data usage to improve the quality of generating information and supporting decisions.

## 6. Conclusions

This research was conducted in a multidisciplinary framework that included three primary fields: data quality, big data, and highway infrastructure project data. Even though data quality has been a well-studied topic for the past two decades, the precise terminology for data quality aspects is still lacking. Digitalisation and data management in construction, particularly highway infrastructure, is a developing topic in India, with a scant prior study focusing on data quality. Using data quality dimensions as part of data governance projects is undoubtedly crucial, as it ensures that data users and stakeholders may derive the most significant benefit from data usage. The research discussed in this paper aims to investigate a framework in which data quality dimensions could be more important within the context of highway infrastructure projects in the construction sector. The semiotic framework was adopted from the literature review of various data quality frameworks for this study to establish data quality dimensions for highway infrastructure data. The systematic literature review, semiotic framework, and Garrett ranking were chosen as research methods because of the increasing novelty of vast quantities of data quality and highway infrastructure data, as well as the impracticality of implementing other research methods due to geographical, legal, ethical, and organisational constraints.

Accuracy, accessibility, and consistency are well-discussed data quality dimensions that are supported by the results. Based on this research, the data quality dimensions of completeness and timeliness were added to the three previously mentioned data quality dimensions to produce a list of the five most appropriate data quality dimensions for highway infrastructure data in the construction industry. Considering the results of the semiotic framework of the hierarchical data quality dimensions for the overall highway project data, the contextual category of data quality dimensions was considered to be the most crucial for evaluating data quality. This is easily explained by the breadth of the three domains involved (i.e., data quality, big data, and highway infrastructure data), where thousands of unique data applications used in the highway infrastructure database are possible. Thus, each application's probability of selecting different data quality dimensions increases.

The current research study provides a ranking of the most critical data quality dimensions in the specific context of highway infrastructure projects, as shown in Table 3. This is one of the first studies within this field to use the semiotic framework to achieve this. This research study also considered the level of importance at each decision-making level of the hierarchy, as shown in Table 4. Considering the very contextual nature of data quality, different contexts would be expected to produce a different list of the most critical data quality dimensions. Thus, the study also provided the ranking of the dimensions within the semiotic framework categories using the Garrett raking technique to understand the priorities of the stakeholders.

The comparatively little amount of literature, and more significantly, publications with the perspective of highway infrastructure data, is one of the most significant limitations of this study. Planned are additional research methods that could be applied to the same corpus of literature, with the primary objective of reducing the amount of author bias introduction when evaluating the significance of the other data quality frameworks.

This study serves as a foundation for further research by the authors in highway infrastructure to assess overall data usage in terms of significant data quality using data quality dimensions as features for assessing the current data quality satisfaction levels at decision-making levels from the data users' perspective. There is a need for agencies and data management teams to assess the root cause of the minimal usage of data to improve the quality of generating information and supporting decisions, and they are also required to show the interdependency of various decisions in the final output of a project and address the potential data users' requirements. In ongoing research, the semiotic framework provides a theoretical foundation for developing an instrument, i.e., data quality dimensions, to access the subjective quality of highway project data. The development of quantitative indices for each data quality dimension to quantify the quality would eventually help to develop the decision-making competency of decision-makers. This would help the organisation in the effective execution of projects without delaying the projects and avoid losses due to wrong decisions. By using data quality dimensions as features for machine learning algorithms, further work will distinguish quality data from non-quality data from very large streams of highway datasets. Finally, the ten main data quality dimensions identified serve as a foundation for determining which machine learning algorithms might identify data usage more effectively. Following this, a computationally efficient method for optimum data usage will be designed to use data effectively.

*Practical Engineering and Real-World Applications of Semiotic Framework*

The semiotic framework for assessing data quality is a theoretical framework that analyses data in terms of its essential components: syntactics, pragmatics, empirics, and semantics. This strategy has several real-world and practical engineering applications, such as data integration, business intelligence, data mining, data governance, and data visualisation. In the construction sector context, the semiotic framework of data quality assessment is used in evaluating building designs. Architects and engineers may use this framework to evaluate the accuracy and completeness of their building designs by analysing the signs and symbols used to represent the different design aspects. This may help them uncover design inconsistencies or errors and make the necessary adjustments before construction begins. In engineering applications of the construction industry, the semiotic framework of data quality may be used in several ways, including quality assurance, risk assessment, and compliance. Throughout the project lifecycle of a construction project, a substantial quantity of data must be gathered and evaluated for quality assurance purposes. The semiotic framework may be used to verify that the obtained data are correct and trustworthy, therefore guaranteeing that the project is on track and satisfies all objectives. The semiotic framework may be used to evaluate the risk associated with specific construction activities. Engineers can make informed decisions and reduce the likelihood of accidents or errors by evaluating the data quality used to evaluate risk. The construction industry is highly regulated, and businesses must adhere to various standards and regulations. The semiotic framework can ensure the accuracy and dependability of the data used to demonstrate compliance, thereby reducing the risk of fines.

In conclusion, the semiotic data quality framework has numerous practical engineering applications in the construction industry. Specific to highways, data quality dimensions are indispensable for planning and design, asset management, safety and emergency response, performance measurement, and policy and decision-making. By ensuring the quality of their data, transportation agencies can make more informed decisions, allocate resources more efficiently, and provide more effective transportation systems. Using this framework, architects, engineers, and other construction professionals can guarantee that the data they use is error-free, resulting in improved project outcomes and reduced risk.

**Author Contributions:** Conceptualisation, methodology, validation, formal analysis, investigation, resources, data curation, writing—original draft preparation, C.M.K.; writing—review, K.N.J. and K.R.; writing—editing, C.M.K.; supervision, K.N.J. and K.R. All authors have read and agreed to the published version of the manuscript.

**Funding:** This research received no external funding.

**Data Availability Statement:** The data presented in this study are available on request from the corresponding author. The data are not publicly available due to privacy or ethical.

**Conflicts of Interest:** The authors declare no conflict of interest.

## References


1. Press Information Bureau. NHAI Becomes the First Construction Sector Organisation to Go Fully Digital. 2020. Available online: https://pib.gov.in/indexd.aspx (accessed on 12 June 2020).
2. Snyder, J.; Menard, A.; Spare, N. *Big Data = Big Questions for the Engineering and Construction Industry*; White Paper; First Myanmar Investment (FMI): Yangon, Myanmar, 2019.
3. Thomas, E.; Schott, P.; Bowman, J.; Synder, J.; Spare, N. *Construction Disconnected: Rethinking the Management of Project Data and Mobile Collaboration to Reduce Costs and Improve Schedules*; Plan Grid; First Myanmar Investment (FMI): Yangon, Myanmar, 2018.
4. Deibe, D.; Amor, M.; Doallo, R. Big Data Geospatial Processing for Massive Aerial LiDAR Datasets. *Remote Sens.* **2020**, *12*, 719. [CrossRef]
5. Pierce, L.M.; McGovern, G.; Zimmerman, K.A. *Practical Guide for Quality Management of Pavement Condition Data Collection*; FHWA: Washington, DC, USA, 2013.
6. Oh, E.; Lee, H. An Imbalanced Data Handling Framework for Industrial Big Data Using a Gaussian Process Regression-Based Generative Adversarial Network. *Symmetry* **2020**, *12*, 669. [CrossRef]
7. Zhang, Y.; Kim, C.-W.; Zhang, L.; Bai, Y.; Yang, H.; Xu, X.; Zhang, Z. Long Term Structural Health Monitoring for Old Deteriorated Bridges: A Copula-ARMA Approach. *Smart Struct. Syst. Int. J.* **2020**, *25*, 285–299.
8. Zhang, Z.; Liu, M.; Liu, X.; Wang, X.; Zhang, Y. Model Identification of Durability Degradation Process of Concrete Material and Structure Based on Wiener Process. *Int. J. Damage Mech.* **2021**, *30*, 537–558. [CrossRef]
9. Batini, C.; Rula, A.; Scannapieco, M.; Viscusi, G. From data quality to bid data quality. *J. Database Manag.* **2015**, *26*, 60–82.
10. Lee, I. Big Data: Dimensions, Evolution, Impacts, and Challenges. *Bus. Horiz.* **2017**, *60*, 293–303. [CrossRef]
11. Sadiq, S.; Papotti, P. Big Data Quality-Whose Problem Is It? In Proceedings of the IEEE 32nd International Conference on Data Engineering (ICDE), Helsinki, Finland, 16–20 May 2016; IEEE: Piscataway, NJ, USA, 2016; pp. 1446–1447.
12. Saha, B.; Srivastava, D. Data Quality: The Other Face of Big Data. In Proceedings of the IEEE 30th International Conference on Data Engineering, Chicago, IL, USA, 31 March–4 April 2014; IEEE: Piscataway, NJ, USA, 2014; pp. 1294–1297.
13. Taleb, I.; el Kassabi, H.T.; Serhani, M.A.; Dssouli, R.; Bouhaddioui, C. Big Data Quality: A Quality Dimensions Evaluation. In Proceedings of the 2016 Intl IEEE Conferences on Ubiquitous Intelligence & Computing, Advanced and Trusted Computing, Scalable Computing and Communications, Cloud and Big Data Computing, Internet of People, and Smart World Congress (UIC/ATC/ScalCom/CBDCom/IoP/SmartWorld), Toulouse, France, 18–21 July 2016; IEEE: Piscataway, NJ, USA, 2016; pp. 759–765.
14. Elouataoui, W.; el Alaoui, I.; el Mendili, S.; Gahi, Y. An Advanced Big Data Quality Framework Based on Weighted Metrics. *Big Data Cogn. Comput.* **2022**, *6*, 153. [CrossRef]
15. Cai, L.; Zhu, Y. The Challenges of Data Quality and Data Quality Assessment in the Big Data Era. *Data Sci. J.* **2015**, *14*, 2. [CrossRef]
16. Ghasemaghaei, M.; Calic, G. Can Big Data Improve Firm Decision Quality? The Role of Data Quality and Data Diagnosticity. *Decis. Support Syst.* **2019**, *120*, 38–49. [CrossRef]
17. Haug, A.; Zachariassen, F.; van Liempd, D. The Costs of Poor Data Quality. *J. Ind. Eng. Manag.* **2011**, *4*, 168–193.
18. Laranjeiro, N.; Soydemir, S.N.; Bernardino, J. A Survey on Data Quality: Classifying Poor Data. In Proceedings of the IEEE 21st Pacific Rim International Symposium on Dependable Computing (PRDC), Zhangjiajie, China, 18–20 November 2015; IEEE: Piscataway, NJ, USA, 2015; pp. 179–188.
19. Sadiq, S.; Yeganeh, K.; Indulska, M. Cross-Disciplinary Collaborations in Data Quality Research. *ECIS Proc.* **2011**, *78*, 1–13.
20. Sidi, F.; Ishak, I.; Affendey, L.S.; Jaya, M.I.; Suriani Affendey, L.; Jabar, M.A. A Review of Data Quality Research in Achieving High Data Quality Within Organization. *J. Theor. Appl. Inf. Technol.* **2017**, *30*, 12. [CrossRef]
21. Yonke, C.L.; Walenta, C.; Talburt, J.R. *The Job of the Information/Data Quality Professional*; International Association for Information and data Quality (IAIDQ): Baltimore, MD, USA, 2011.
22. Ballou, D.P.; Pazer, H.L. Modeling Data and Process Quality in Multi-Input, Multi-Output Information Systems. *Manag. Sci.* **1985**, *31*, 150–162. [CrossRef]
23. Ballou, D.; Wang, R.; Pazer, H.; Tayi, G.K. Modeling Information Manufacturing Systems to Determine Information Product Quality. *Manag. Sci.* **1998**, *44*, 462–484. [CrossRef]
24. Wand, Y.; Wang, R.Y. Anchoring Data Quality Dimensions in Ontological Foundations. *Commun. ACM* **1996**, *39*, 86–95. [CrossRef]
25. English, L.P. *Information Quality Applied: Best Practices for Improving Business Information, Processes and Systems*; Wiley Publishing: Hoboken, NJ, USA, 2009; ISBN 047013447X.
26. Redman, T.C. *Data Quality for the Information Age*; Artech House, Inc.: Norwood, MA, USA, 1997; ISBN 0890068836.
27. Coleman, C. *Managing Information Quality: Increasing the Value of Information in Knowledge-Intensive Products and Processes*; Springer: Berlin/Heidelberg, Germany, 2007.

28. Tan, S.G.; Cheng, D. Quality Assurance of Performance Data for Pavement Management Systems. In *Design, Analysis, and Asphalt Material Characterization for Road and Airfield Pavements*; ASCE: Reston, VA, USA, 2014; pp. 163–169.
29. Price, R.; Shanks, G. Chapter 4 Data Quality and Decision Making. In *Handbook on a Decision Support System*; Springer: Berlin/Heidelberg, Germany, 2008; pp. 65–82.
30. Samitsch, C. *Data Quality and Its Impacts on Decision-Making: How Managers Can Benefit from Good Data*; Springer: Berlin/Heidelberg, Germany, 2014; ISBN 3658082003.
31. Krogstie, J. A Semiotic Approach to Data Quality. In Proceedings of the Lecture Notes in Business Information Processing; Springer: Berlin/Heidelberg, Germany, 2013; Volume 147, pp. 395–410.
32. Huang, H. Big Data to Knowledge–Harnessing Semiotic Relationships of Data Quality and Skills in Genome Curation Work. *J. Inf. Sci.* **2018**, *44*, 785–801. [CrossRef]
33. Long, J.A.; Seko, C.E. A New Method for Database Data Quality Evaluation at the Canadian Institute for Health Information (CIHI). In Proceedings of the 7th International Conference on Information Quality (IQ 2002), Tempe, AZ, USA, 24–28 February 2002; pp. 238–250.
34. Lee, Y.W.; Strong, D.M.; Kahn, B.K.; Wang, R.Y. AIMQ: A Methodology for Information Quality Assessment. *Inf. Manag.* **2002**, *40*, 133–146. [CrossRef]
35. Pipino, L.L.; Lee, Y.W.; Wang, R.Y.; Yang, R.Y. Data Quality Assessment. *Commun. ACM* **2002**, *45*, 211–218. [CrossRef]
36. Sukumar, S.R.; Natarajan, R.; Ferrell, R.K. Quality of Big Data in Health Care. *Int. J. Health Care Qual. Assur.* **2015**, *28*, 621–634. [CrossRef]
37. Jankalová, M.; Jankal, R. How to Characterise Business Excellence and Determine the Relation between Business Excellence and Sustainability. *Sustainability* **2020**, *12*, 6198. [CrossRef]
38. Wang, R.Y. A Product Perspective on Total Data Quality Management. *Commun. ACM* **1998**, *41*, 58–65. [CrossRef]
39. Del Pilar Angeles, M.; García-Ugalde, F. A Data Quality Practical Approach. *Int. J. Adv. Softw.* **2009**, *1*, 259–299.
40. Sebastian-Coleman, L. *Measuring Data Quality for Ongoing Improvement: A Data Quality Assessment Framework*; Elsevier: Walthman, MA, USA, 2012; ISBN 0123977541.
41. Vaziri, R.; Mohsenzadeh, M.; Habibi, J. TBDQ: A Pragmatic Task-Based Method to Data Quality Assessment and Improvement. *PLoS ONE* **2016**, *11*, e0154508. [CrossRef] [PubMed]
42. Valverde, C.; Marotta, A.; Panach, J.I.; Vallespir, D. Towards a Model and Methodology for Evaluating Data Quality in Software Engineering Experiments. *Inf. Softw. Technol.* **2022**, *151*, 107029. [CrossRef]
43. Liebenau, J.; Backhouse, J. *Understanding Information: An Introduction*; Palgrave Macmillan: London, UK, 1990; ISBN 0333536800.
44. Azeroual, O.; Jha, M.; Nikiforova, A.; Sha, K.; Alsmirat, M.; Jha, S. A Record Linkage-Based Data Deduplication Framework with DataCleaner Extension. *Multimodal Technol. Interact* **2022**, *6*, 27. [CrossRef]
45. Abedjan, Z.; Chu, X.; Deng, D.; Fernandez, R.C.; Ilyas, I.F.; Ouzzani, M.; Papotti, P.; Stonebraker, M.; Tang, N. Detecting Data Errors: Where Are We and What Needs to Be Done? *Proc. VLDB Endow.* **2016**, *9*, 993–1004. [CrossRef]
46. Wang, R.Y.; Strong, D.M. Beyond Accuracy: What Data Quality Means to Data Consumers. *J. Manag. Inf. Syst.* **1996**, *12*, 5–33. [CrossRef]
47. Crosby, P.B. *Quality Is Free: The Art of Making Quality Certain*; Signet Book: West Bengal, India, 1980; Volume 2247, ISBN 0451622472.
48. Fu, Q.; Easton, J.M. Understanding Data Quality: Ensuring Data Quality by Design in the Rail Industry. In Proceedings of the IEEE International Conference on Big Data (Big Data), Boston, MA, USA, 11–14 December 2017; IEEE: Piscataway, NJ, USA, 2017; pp. 3792–3799.
49. Ramasamy, A.; Chowdhury, S. Big Data Quality Dimensions: A Systematic Literature Review. *J. Inf. Syst. Technol. Manag.* **2020**, *17*. [CrossRef]
50. Madnick, S.; Zhu, H. Improving Data Quality through Effective Use of Data Semantics. *Data Knowl. Eng.* **2006**, *59*, 460–475. [CrossRef]
51. English, L.P. *Improving Data Warehouse and Business Information Quality: Methods for Reducing Costs and Increasing Profits*; John Wiley & Sons, Inc.: Hoboken, NJ, USA, 1999; ISBN 0471253839.
52. Redman, T.C. *Data Quality: The Field Guide*; Digital Press: Oxford, UK, 2001; ISBN 1555582516.
53. Batini, C.; Cappiello, C.; Francalanci, C.; Maurino, A. Methodologies for Data Quality Assessment and Improvement. *ACM Comput. Surv.* **2009**, *41*, 1–52. [CrossRef]
54. Gao, B.; Zhou, Q.; Deng, Y. BIM-AFA: Belief Information Measure-Based Attribute Fusion Approach in Improving the Quality of Uncertain Data. *Inf. Sci.* **2022**, *608*, 950–969. [CrossRef]
55. Madnick, S.; Wang, R.; Dravis, F.; Chen, X. Improving the Quality of Corporate Household Data: Current Practices and Research Directions. *SSRN Electron. J.* **2000**, 365180. [CrossRef]
56. Redman, T.C. Improve Data Quality for Competitive Advantage. *MIT Sloan Manag. Rev.* **1995**, *36*, 99.
57. Hassenstein, M.J.; Vanella, P. Data Quality—Concepts and Problems. *Encyclopedia* **2022**, *2*, 498–510. [CrossRef]
58. Gabr, M.I.; Helmy, Y.M.; Elzanfaly, D.S. Data Quality Dimensions, Metrics, and Improvement Techniques. *Future Comput. Inf. J.* **2021**, *6*, 25–44. [CrossRef]
59. Jesiļevska, S. Data Quality Dimensions to Ensure Optimal Data Quality. *Rom. Econ. J.* **2017**, *20*, 63.
60. Gyulgyulyan, E.; Ravat, F.; Astsatryan, H.; Aligon, J. Data Quality Impact in Business Inteligence. In Proceedings of the 2018 Ivannikov Memorial Workshop, (IVMEM), Yerevan, Armenia, 3–4 May 2018; IEEE: Piscataway, NJ, USA, 2019; pp. 47–51.

61. Loshin, D. *Enterprise Knowledge Management: The Data Quality Approach*; Morgan Kaufmann: London, UK, 2001; ISBN 0124558402.
62. Cappiello, C.; Ficiaro, P.; Pernici, B. HIQM: A Methodology for Information Quality Monitoring, Measurement, and Improvement. In Proceedings of the International Conference on Conceptual Modeling; Springer: Berlin/Heidelberg, Germany, 2006; pp. 339–351.
63. Batini, C.; Cabitza, F.; Cappiello, C.; Francalanci, C.; di Milano, P. A Comprehensive Data Quality Methodology for Web and Structured Data. In Proceedings of the 2006 1st International Conference on Digital Information Management, Bangalore, India, 6–8 December 2006; IEEE: Piscataway, NJ, USA, 2006; pp. 448–456.
64. Moraga, C.; Moraga, M.Á.; Caro, A.; Calero, C. SPDQM: SQuaRE-Aligned Portal Data Quality Model. In Proceedings of the 9th International Conference on Quality Software, QSIC, Jeju, Republic of Korea, 24–25 August 2009.
65. Carlo, B.; Daniele, B.; Federico, C.; Simone, G. A Data Quality Methodology for Heterogeneous Data. *Int. J. Database Manag. Syst.* **2011**, *3*, 60–79. [CrossRef]
66. Falkenberg, E.D. *A Framework of Information System Concepts*; The FRISCO Report (Web Edition); University of Leiden, Department of Computer Science: Leiden, The Netherlands, 1998; ISBN 3901882014.
67. Kahn, M.G.; Raebel, M.A.; Glanz, J.M.; Riedlinger, K.; Steiner, J.F. A Pragmatic Framework for Single-Site and Multisite Data Quality Assessment in Electronic Health Record-Based Clinical Research. *Med. Care* **2012**, *50*, S21–S29. [CrossRef]
68. Knoke, D.; Yang, S. *Social Network Analysis*; SAGE Publication: Thousand Oaks, CA, USA, 2019.
69. Lee, Y.W.; Strong, D.M. Knowing-Why about Data Processes and Data Quality. *J. Manag. Inf. Syst.* **2003**, *20*, 13–39. [CrossRef]
70. Alshikhi, O.A.; Abdullah, B.M. Information Quality: Definitions, Measurement, Dimensions, And Relationship with Decision Making. *Eur. J. Bus. Innov. Res.* **2018**, *6*, 36–42.
71. Jayawardene, V.; Sadiq, S.; Indulska, M. *An Analysis of Data Quality Dimensions*; The University of Queensland: St Lucia, Australia, 2015; pp. 1–31.
72. Tejay, G.; Dhillon, G.; Goyal Chin, A. Data Quality Dimensions for Information Systems Security: A Theoretical Exposition. In *Security Management, Integrity, and Internal Control in Information Systems*; IFIP TC-11 WG 11.1 &WG 11.5 Joint Working Conference 7; Springer: Berlin/Heidelberg, Germany, 2005; pp. 21–39.
73. Tobler, E. A Needs Assessment of Arizona Agricultural Education Equine Science Curriculum. Ph.D. Dissertation, Utah State University, Logan, UT, USA, 2018.
74. Johari, S.; Jha, K. Determinants of Workmanship: Defining Quality in Construction Industry. In *Proceedings of the 35th Annual Conference*; Leeds Beckett University: Leeds, UK, 2019; p. 761.
75. Tripathi, K.K.; Jha, K.N. An Empirical Study on Performance Measurement Factors for Construction Organizations. *KSCE J. Civ. Eng.* **2018**, *22*, 1052–1066. [CrossRef]
76. Assistant Librarian, D.S. Application of Garret Ranking Technique: Practical Approach. *Int. J. Libr. Inf. Stud.* **2016**, *6*, 135–140.
77. Garrett, H.E.; Woodworth, R.S. *Statistics in Psychology and Education*; Vakils, Feffer and Simons Private Ltd.: Bombay, India, 1969; p. 329.

*buildings*

MDPI

*Article*

# Applying Deep Learning and Single Shot Detection in Construction Site Image Recognition

**Li-Wei Lung and Yu-Ren Wang ***

Department of Civil Engineering, National Kaohsiung University of Science and Technology, Kaohsiung 80778, Taiwan
* Correspondence: yrwang@nkust.edu.tw; Tel.: +886-7-381-4526

**Abstract:** A construction site features an open field and complexity and relies mainly on manual labor for construction progress, quality, and field management to facilitate job site coordination and productive results. It has a tremendous impact on the effectiveness and efficiency of job site supervision. However, most job site workers take photos of the construction activities. These photos serve as aids for project management, including construction history records, quality, and schedule management. It often takes a great deal of time to process the many photos taken. Most of the time, the image data are processed passively and used only for reference, which could be better. For this, a construction activity image recognition system is proposed by incorporating image recognition through deep learning, using the powerful image extraction ability of a convolution neural network (CNN) for automatic extraction of contours, edge lines, and local features via filters, and feeding feature data to the network for training in a fully connected way. The system is effective in image recognition, which is in favor of telling minute differences. The parameters and structure of the neural network are adjusted for using a CNN. Objects like construction workers, machines, and materials are selected for a case study. A CNN is used to extract individual features for training, which improves recognizability and helps project managers make decisions regarding construction safety, job site configuration, progress control, and quality management, thus improving the efficiency of construction management.

**Keywords:** construction image; artificial intelligence; deep learning; object detection; single shot multibox detector (SSD)

**Citation:** Lung, L.-W.; Wang, Y.-R. Applying Deep Learning and Single Shot Detection in Construction Site Image Recognition. *Buildings* **2023**, *13*, 1074. https://doi.org/10.3390/buildings13041074

Academic Editors: Maxim A. Dulebenets and Saeed Banihashemi

Received: 20 January 2023
Revised: 28 March 2023
Accepted: 9 April 2023
Published: 19 April 2023



## 1. Introduction

Construction work is tedious and subject to delays, and its quality may be compromised by many factors, such as construction equipment, workers, and materials. Therefore, it is necessary to improve construction quality and progress in today's increasingly competitive market by considering good job site management and meeting construction costs. At a job site currently, a job site manager oversees everything construction-related, including workers, machines, and materials [1–5]. The manager has to take care of virtually everything at the job site [6]. The improvement of management methods using innovative technology helps to not only accelerate the development of the construction industry but also improve a company's competitiveness in the market.

Most general contractors deploy imaging devices, such as photo and video cameras, to document the progress of construction activities throughout the entire process. The image data are collected, in general, by filming with a mobile camera operated by a worker or a video camera set up at a fixed location. Most image data collected are used passively for reference or even just shelved. The others are used to prepare quality documents or demonstrate construction status and progress. Suppose artificial intelligence (AI) is introduced to recognize objects in the images and help job site management identify and tag things in the images. In that case, these image data may serve as an essential basis for decision-making within construction activities, including construction planning and design, job site safety, automated equipment management [7–10], and quality monitoring and maintenance. For example, suppose a specific machine is tagged in video footage of construction activities [11]. In that case, the project team may exploit the captured data for project decisions of route management, machine setup, and site safety [12–15].

When recognizing and classifying objects in many images, a deep learning model may be introduced to accelerate the extraction of high-value digital information crucial for construction management. The mainstream in developing the deep neural network is the convolution neural network (CNN) which extracts critical feature information by including one or more convolution layers and pooling layers through a combination of algorithms and multi-layer computation of convolution neurons as the images are converted into data [16]. The feature information is fed to the neural network for training in a fully connected manner until identical or similar features in the same class of images are identified and documented. The relative locations and features digitally arranged during the recognition of new images are systematically computed and processed to identify the similarities between images for successful image judgment [17,18].

AI is having revolutionary impacts on construction engineering [19]. Thanks to the powerful capability of AI in data processing, analysis, and searching for massive digitization, a model to recognize construction objects at a job site can be built to rapidly and accurately identify workers [20–22], machines [23,24], and materials [25] in job site footage while tagging their relative locations in the images to provide more site-related information for project management, which is a rising topic in the industry in the pursuit of breakthroughs and innovation.

## 2. Literature Review

A construction project has unique complexity. The completion of a project involves an engineering lifecycle consisting of many links, from design and construction to final acceptance. In an era in which the development of technical information evolves at the speed of light, the innovative technologies and management systems used in construction management help not only maintain control over safety and health as the construction work progresses but also facilitate the successful completion of construction projects by reducing uncertainties while focusing on the goal of sustainable development [26].

Artificial intelligence, or AI, is an engineering study focusing on researching and developing intelligent entities. AI includes the use of programs and big data to make computers and machines mimic human thinking and simulate the "intelligent" behaviors of a human being; when AI is the object of study, machine learning (ML) is a model to improve the performance of specific algorithms while learning from experiences, i.e., learning from data collected [27]. However, data learning is based on massive data processed using a multi-layer neural network. A self-learning method is found after linear or nonlinear conversion via multiple processing layers, which automatically extracts features representative of data characteristics in place of the long time taken for traditional feature engineering. Deep learning is a technology that evolved from machine learning [28].

The applications of deep learning in computer vision in recent years are in the following classes [29], as shown in Figure 1: (1) classification: putting an image in one of the established classes by its nature and type; (2) semantic segmentation: identifying pixel blocks by event type instead of classifying into "instances"; (3) classification + localization: tagging a message to a single object with its location and size (w, h); (4) object detection: tagging multiple objects with their locations and sizes; and (5) instance segmentation: tagging "instances"; the objects of the same class are identified by individual locations and sizes, particularly when they are overlapping.

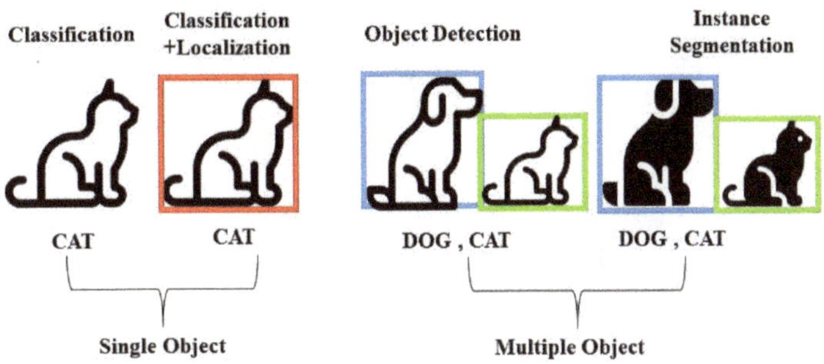


**Figure 1.** Applications in computer vision.

Most recent object detection studies are focused on the use of a CNN for typical model applications in which a matching object is identified before determining in which area a matching thing exists and tagging the location of highest probability with a box, as shown in Figure 2. Two fully connected layers are connected behind the CNN, one for classification and the other for tagging the matching area. There are three algorithms to organize an area: sliding window, region proposal, and grid-based.

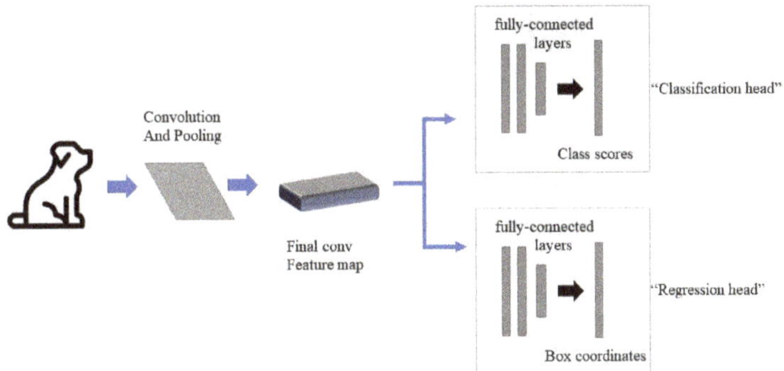


**Figure 2.** Locating algorithm model.

1. Sliding window: a simple but time-consuming method based on the method of exhaustion. It works by establishing windows of various sizes for image scanning and extracting the feature information of every image window. Next, the data is fed to a classifier for object recognition to determine if the probability of the window matching the object to be detected is accurate. This method is the simplest but most time-consuming [30], as presented in Figure 3.

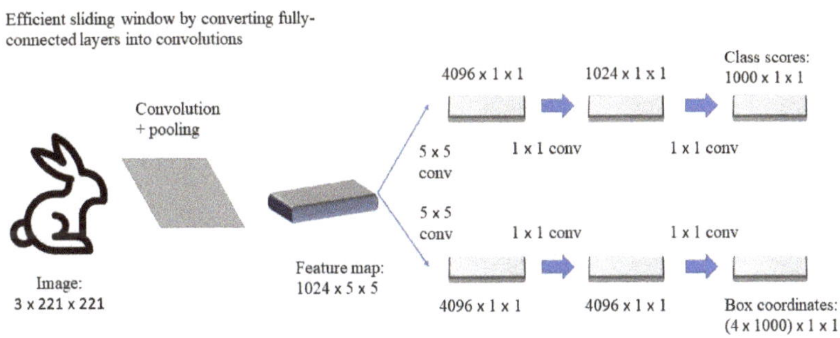


**Figure 3.** Sliding window algorithm.

2. Region proposal: information in the image, such as texture, edges, and color, are used to predetermine the regions of interest (ROI) containing the object and determine the probability of these regions for matching. The high recall is maintained by filtering thousands of regions per second. Similar algorithms are R-CNN, Fast R-CNN, and Faster R-CNN [31–34], as shown in Figure 4.

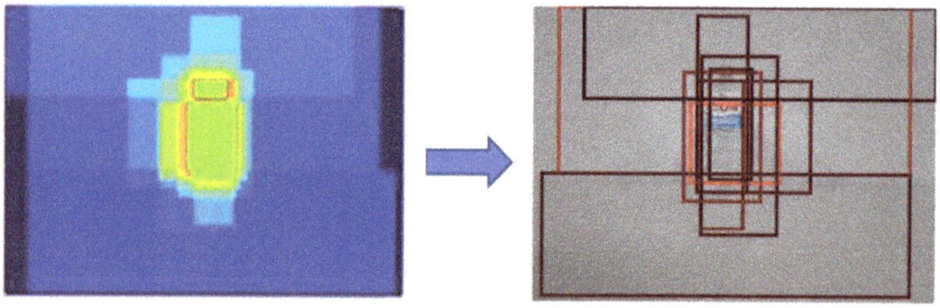


**Figure 4.** Region proposals algorithms.

3. Grid-based regression: a picture is divided into grids, and regions of various sizes are selected with the grids as centers. Regression determines the probability that every bounding box contains the target. This approach is suitable for real-time detection. Similar algorithms are you only look once (YOLO) and single shot multibox detector (SSD) [35], as shown in Figure 5.

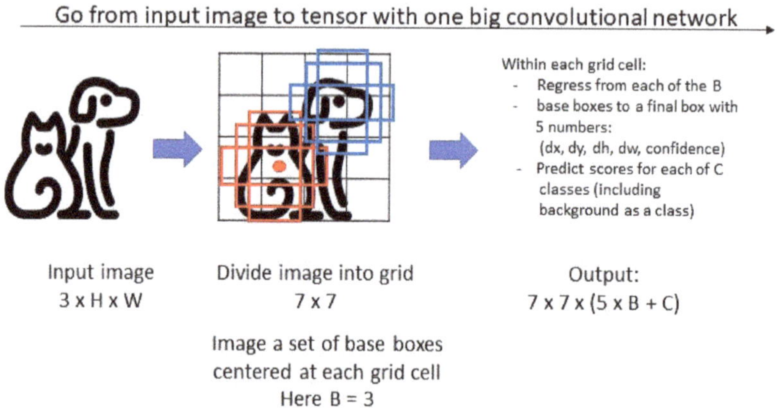


**Figure 5.** Region Proposal algorithms.

You only look once (YOLO) predicts multiple bounding boxes and types of CNNs, realizing end-to-end target detection and identification. This algorithm avoids the weakness that object detection must be trained separately and accelerates the computation dramatically [36], as indicated in Figure 6.

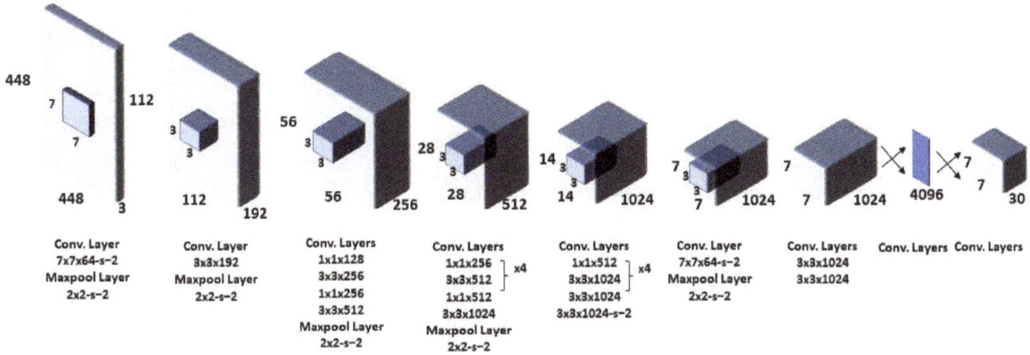


**Figure 6.** Structure of a YOLO model.

The single shot multibox detector (SSD) is based on a feed-forward CNN that generates bounding box sets and scores of different types on the boxes, followed by non-maximum value suppression to complete the final detection process. This explains the incorporation of both the regression concept in YOLO and the anchor mechanism in Faster-CNN in single shot multibox detector (SSD), as regression is performed on the multi-dimensional region features of every location in the entire picture, which retains YOLO's characteristics of being fast while ensuring the window prediction is as accurate as Faster-RCNN [37], as shown in Figure 7.

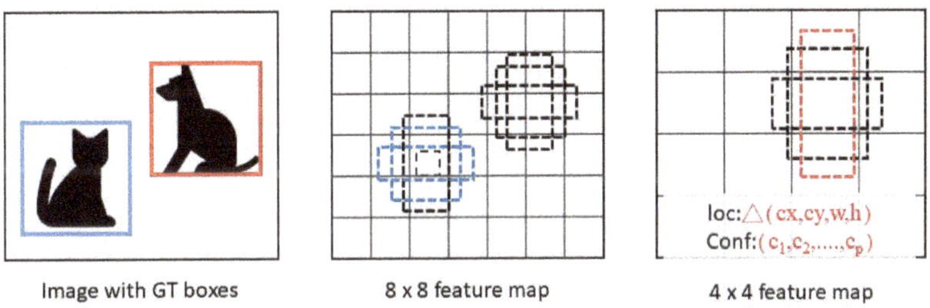


**Figure 7.** Default boxes in the single shot multibox detector model.

Liu et al. (2016) tested the speed and accuracy of different object detection methods. The test results are shown in Table 1:

**Table 1.** Object detection algorithm speed and accuracy comparison.

| Method | FPS | Boxes | mAP |
|---|---|---|---|
| Faster R-CNN | 7 | 6000 | 73.2 |
| Faster YOLO | 155 | 98 | 52.7 |
| SSD300 | 29 | 8732 | 74.3 |

A fast YOLO has faster processing speed but poor mAP. Although Faster R-CNN has a higher accuracy rate (73.2% mAP), it is not significantly more accurate at determining the number of images. In contrast, a single shot multibox detector (SSD) not only has a high accuracy rate but also a fast image detection speed [36].

Single shot multibox detector (SSD) object recognition has been used in many engineering applications. For example, Yudin and Slavioglo [38] used the single shot multi-box detector (SSD) to test how well the model identifies a traffic light, producing good results. Wang et al. [39] proposed an improved single shot multibox detector (SSD) capable of detecting a ship in a noisy background. The results were compared with those from Faster

R-CNN, and it was found that the enhanced single shot multibox detector (SSD) improved detection accuracy.

Much research on image recognition using deep learning has accumulated in recent years. Many people use deep learning technology in artificial intelligence to let computers handle more complex image recognition problems. Table 2 shows the development of deep learning in the construction industry in the past five years of applied research on image recognition.

**Table 2.** Research on the application of deep learning in construction image recognition.

| Author (Year) | Abstract |
|---|---|
| Dorafshan, S., Thomas, R. J., and Maguire, M. (2018) [40] | Compares the performance of deep convolutional neural networks and edge detection algorithms for image-based crack detection in concrete, finding that the neural network approach outperforms traditional edge detection methods. |
| Spencer Jr, B. F., Hoskere, V. and Narazaki, Y. (2019) [41] | Recent advances in computer vision-based civil infrastructure inspection and monitoring techniques, including object detection, semantic segmentation, and deep learning methods, highlight their benefits and challenges. |
| Dung, C. V. (2019) [42] | Proposes an autonomous system for concrete crack detection using a deep, fully convolutional neural network, achieving high accuracy and efficiency compared to traditional manual inspection methods. |
| FANG, Weili, et al. (2020) [43] | A review and discussion of future directions of computer vision for behavior-based safety in construction. |
| Li, Y., Lu, Y. and Chen, J. (2021) [25] | A deep learning approach based on the YOLOv3 detector is proposed for real-time rebar counting on construction sites, which can effectively improve construction efficiency and safety. |
| Chou, J. S. and Liu, C. H. (2021) [24] | An automated system for recognizing trucks in real-time in river dredging areas using computer vision and deep learning. |
| Li, X., Chi, H., Lu, W., Xue, F., Zeng, J., and Li, C. Z. (2021) [44] | An intelligent work packaging system that preserves construction workers' personal image information using federated transfer learning. |
| DEL SAVIO, Alexandre Almeida, et al. (2021) [45] | Artificial intelligence (AI) and computer vision are used to identify objects and equipment on a construction site and how they can improve safety and efficiency. |
| LIN, Chih-Lung, et al. (2022) [22] | Presents a gait-based pedestrian automatic detection and recognition system using a deep learning neural network. |
| Greeshma, A. S. and Edayadiyil, J. B. (2022) [10] | An automated system that uses machine learning and image processing to monitor construction project progress. |
| Del Savio, A., Luna, A., Cárdenas-Salas, D., Vergara, M., and Urday, G. (2022) [11] | A manually classified dataset of construction site images containing 1046 images of eight object classes that can be used to develop computer vision techniques in the engineering and construction fields. |
| Yeşilmen, S. and Tatar, B. (2022) [16] | The efficiency of using convolutional neural networks (CNN) for image classification in monitoring construction-related activities, with a case study on aggregate mining for concrete production. |

Source: This study collated.

Past studies used deep learning algorithms to recognize three postures of construction workers, including standing, bending over, and squatting [20–22]. They provide engineering professionals with comprehensive deep learning solutions for detecting construction vehicles [23,24]. Only single objects, such as people, materials, or engineering vehicles, were seen in the above studies; therefore, the shapes and boundary types recognized were relatively pure. This study uses image automation to simultaneously identify workers, machinery, and materials in the current construction situation, assist the construction site manager in making safety judgments on the location of construction equipment, safety protection measures, and material stacking, and monitor the construction status and maintenance of the construction site to reduce environmental hazards and control progress.

With the continuous evolution of technology, combining big data and artificial intelligence machine learning/deep learning can maximize the value of data. Therefore, this research collects the construction site image data set, imports the object detection system, uses it as artificial intelligence and machine learning training data, and builds AI to automatically identify the personnel, materials, and equipment on the construction site. In the future, continuous learning, modification, and technical improvement can reduce or avoid labor accidents on the construction site, thereby improving construction efficiency and schedule management.

## 3. Methodology

### *3.1. Study Setup*

The CNN in single shot multibox detector (SSD) method of deep learning required massive training images for learning. Firstly, image files of a construction site were collected and converted to matrices by regulating the size of images before data pre-processing, such as optimization. Next, features were extracted using CNN and fed into the fully connected neural network to predict and identify classes. Finally, the trained model was verified by feeding it the test data. The model's learning rate setting would affect the weight adjustment, so this study set the learning rate = 0.00002, epoch = 100, step per epoch = 320, and optimizer type = sgd. The model structure of this proposed method is presented in Figure 8.

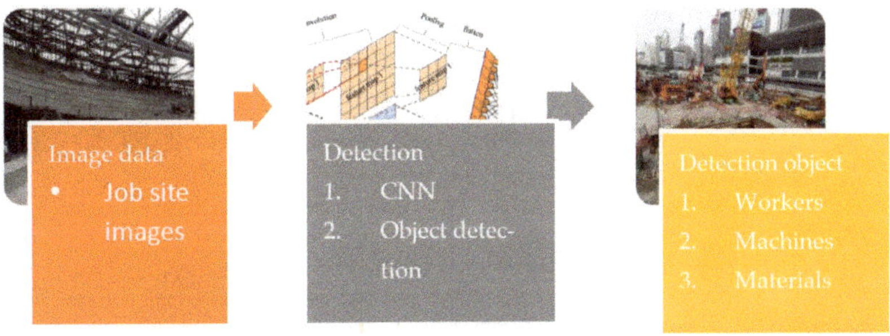


**Figure 8.** Structure of job site image object detection model.

### *3.2. Collection of Job Site Images for a Construction Project*

The multi-class classification in the image classification was selected for the study. Data sets were classified as rebar, worker, and machine. The deep learning model required massive amounts of information for training to improve its recognition accuracy, and the size of the data set was a critical factor for the experiment's success. Data came from three sources, as follows:

1. Legal and free job site pictures obtained from Google under "Creative Commons";
2. Free databases provided by computer vision institutes, such as ImageNet and Labelme of MIT; and
3. Photos of construction job sites taken for the study.

Four hundred sixty-one job site images were collected from the above sources (Figure 9). Feasible data were extracted from the images in the preliminary classification. The job site images collected were manually tagged for workers, machines, and rebar using the image tool provided in "LabelImg." In addition, movements were selected and tagged for classes.

**Figure 9.** Collection of construction site image files.

Two types of files were generated after tagging with LabelImg; one was the image files themselves, and the other was the XML files with image locations tagged. In Figure 10, for example, workers, rebars, and machines are tagged and given specific names in the image. Figure 11 provides an example of the contents of the XML file, including dimensions such as image coordinates. The single shot multibox detector (SSD) deep learning model was established and tested as all images were tagged.

**Figure 10.** LabelImg tagging of a job site photo.

### 3.3. Method of Object Detection (SSD)

Wei Liu [36] devised the single shot multibox detector (SSD), a one-stage method in which a neural network (VGG-16) is used to extract feature maps for classification and regression before the target objects are tested. It incorporates the regression concept in YOLO and identifies the location of the target class in regression. Similar to the anchor mechanism in Faster-RCNN, prior boxes are established and features are extracted from the backbone network. Feature maps of various dimensions are used for prediction, with large feature maps to detect small targets and small maps to detect large targets. Convolution kernel is applied on the feature maps to predict the classes and coordinate offsets of a series of default bounding boxes.

VGG-16 serves as the backbone model for the single shot multibox detector (SSD) structure. The fully connected layer of VGG, fc6, is modified and converted into a 3 × 3 convolution layer, Conv6, and fc7 into a 1 × 1 convolution layer, Conv7, while the pooling layer, pool5, is changed from originally 2 × 2 with stride = 2 to 3 × 3 with stride = 1. 4; convolution layers are added; the test module layer of the 1st feature map is Conv4_3, followed by Conv8_2, Conv9_2, Conv10_2, and Conv11_2 [36,39]. Their sizes are shown in Figure 12.

```
- <object>
    <name>machine</name>
    <pose>Unspecified</pose>
    <truncated>0</truncated>
    <difficult>0</difficult>
  - <bndbox>
      <xmin>1835</xmin>
      <ymin>43</ymin>
      <xmax>2586</xmax>
      <ymax>1055</ymax>
    </bndbox>
  </object>
+ <object>
+ <object>
- <object>
    <name>worker</name>
    <pose>Unspecified</pose>
    <truncated>0</truncated>
    <difficult>0</difficult>
  - <bndbox>
      <xmin>1019</xmin>
      <ymin>761</ymin>
      <xmax>1115</xmax>
      <ymax>898</ymax>
    </bndbox>
  </object>
+ <object>
+ <object>
- <object>
    <name>rebar</name>
    <pose>Unspecified</pose>
    <truncated>1</truncated>
    <difficult>0</difficult>
  + <bndbox>
  </object>
</annotation>
```

**Figure 11.** Contents of the XML file of a tagged job site image.

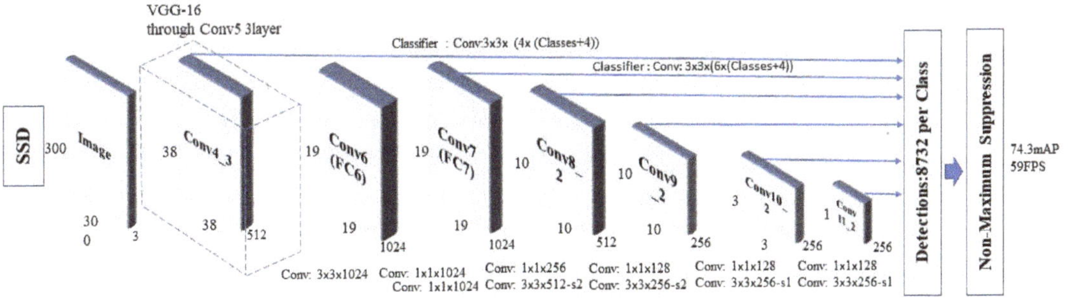


**Figure 12.** Single shot multibox detector model structure.

The size and length–width ratio require consideration for testing the box on a feature map. Every grid on the feature map is scanned to generate corresponding testing boxes (Figure 13). During the training, the ground truth in the picture is checked to match the testing box. The best-fit box is filtered based on intersection over union (IOU). The exact positive and negative sample ratio is close to 1:3. The loss function depends on the weights of location error and confidence error. Data enhancement is carried out via horizontal flipping, random cutting, color twisting, and random sampling of block regions. Top-k prediction boxes with high confidence levels are reserved during the prediction before the object detection algorithm of non-maximum suppression (NMS) is used to filter prediction terms with significant overlapping. The prediction term left at the end is the result [36].

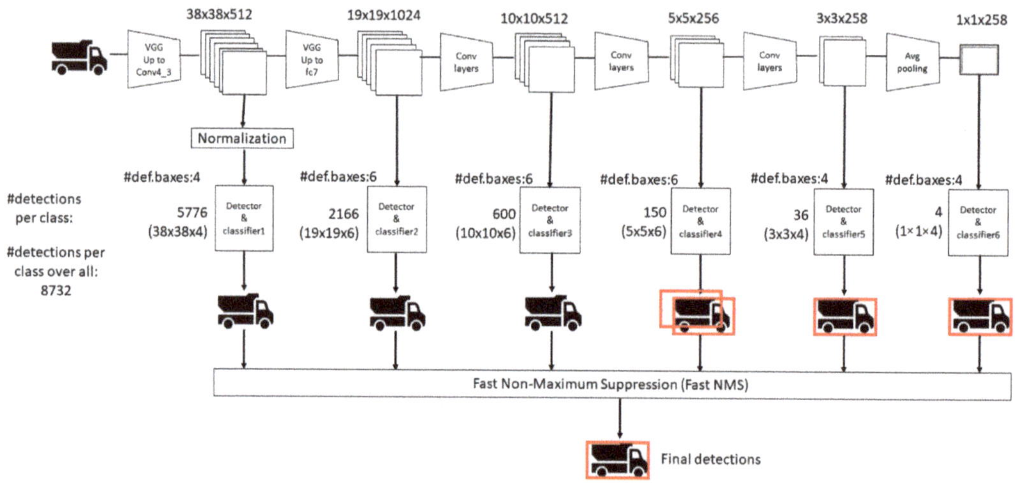


**Figure 13.** Single shot multibox detector target feature detection process.

## 4. Study Contents and Outcomes

### *4.1. Establishment and Testing of Single Shot MultiBox Detector Model*

The main feature extraction program used to establish a single shot multibox detector (SSD) model was vgg.py. Features were extracted using 9 module computati9on feature layers in the sizes of $38 \times 38$, $19 \times 19$, $10 \times 10$, $5 \times 5$, $3 \times 3$, and $1 \times 1$ (Figure 13). At the first convolution computation feature layer, the image fed was $300 \times 300$ in size. Randomly generated $3 \times 3$ filters were used at the convolution layer to extract 64 features, and the activation function of ReLu was adopted to eliminate negative values. Batch normalization was introduced next to improve the stability of data distribution. After two rounds of convolution feature extraction, the pooling layer shrank the image down to $150 \times 150$ in size for the convolution computation of the second set. The filters extracted 128 features at the second set convolution computation feature layer. The same applied to the rest of the computation. Ultimately, the pooling layer reduced the images to $1 \times 1$ in size.

The detect_image feature in the ssd.py program was used for predicting and testing the results. The height and width of the picture were determined after the photo was fed. However, the picture was converted into RGB format to improve detection for the pre-training weight of the image and convenience of color setup in the box. The letterbox_image feature was used to identify the resized image without distortion. The image was normalized based on the batch_size attribute before being fed into the model for regression and type prediction.

Data sets needed to be imported into classes_path while the image training program train.py parameters were established to identify the image classes of rebar, worker, and machine. The pre-training weight, weight_path, was established, and the shape was selected to be $300 \times 300$. The prior box size was defined as anchors_size = [30, 60, 111, 162, 213, 264, 315]. The image training consisted of 2 stages, "freeze" and "unfreeze." The feature extraction network experienced no change during the freezing stage but minor network tuning. Thus, 50 generations were established. The number of data samples captured for one training run was 16. The backbone and feature extraction network experienced changes during the unfreezing stage. Ample memory was used, and, therefore, 100 generations were established. The number of training samples was 8.

The single shot multibox detector (SSD) program selected the pattern to be detected during the establishment test on the training outcome prediction program predict.py. The parameter setting patterns during the detection were single pictures, pre-recorded footage, or images captured directly from the camera. For this study, images were used for the prediction model.

*4.2. Model Training Data Analysis*

In machine learning and deep learning, a loss function is frequently used to evaluate the error between predictions and valid values. The smaller the value, the closer the prediction to the actual value and the more accurate the model. Loss functions commonly used are mean square error (MSE) and cross-entropy; the former is usually used for regression and the latter for classification.

Data outcomes were evaluated based on the performance of the two accuracy indicators, F1 measure and overall accuracy, on the model. Both indicators above were determined using the four factors of the confusion matrix, and they were true positive (TP), true negative (TN), false positive (FP), and false negative (FN). The F1 measure was the harmonized average between accuracy and recall. It was used as an indicator of model performance and expressed as:

$$\text{F1 Measure} = \frac{(2 \times \text{Precision} \times \text{Recall})}{(\text{Precision} + \text{Recall})} \tag{1}$$

The overall accuracy was defined as the ratio of correct prediction of positive and negative samples in the models over all samples and expressed in Equation (2):

$$\text{Overall Accuracy} = \frac{(\text{TP} + \text{TN})}{(\text{TP} + \text{FP} + \text{FN} + \text{TN})} \tag{2}$$

The single shot multibox detector (SSD) was deployed to identify the classes of rebar, worker, and machine in all images collected in the data set. A total of 461 images were collected, including 400 photos of job site activities as machine learning samples, with 80% images for training. In addition, 40 images, accounting for 10% of the data set, served as the test samples during the training; another 40 were used as verification samples, accounting for 10%. In the end, 61 photos the model had not seen were brought in for recognition, and a 1 × 1 confusion matrix was generated, as shown in Table 3.

**Table 3.** Confusion matrix generated by single shot multibox detector model.

| TP | 30 | FN | 18 |
|---|---|---|---|
| FP | 3 | TN | 10 |

A calculation was performed for the two accuracy evaluation indicators based on the four factors generated in the confusion matrix. It was found that the F1 measure was 64%, and the oval accuracy was 66%. The details are provided in Table 4.

**Table 4.** The two accuracy evaluation indicators of the single shot multibox detector model.

| **Indicators** | **Value** |
|---|---|
| F1 Measure | 64% |
| Overall Accuracy | 66% |

The process mentioned above reveals that an SSD-based job site activity image recognition system is built by combining the job site image data collected and deep learning in AI. This system can identify and tag essential objects in a job site image, such as workers, machines, and construction materials. With more job site activity information gained from image recognition, the proposed system may help project managers develop project decisions regarding construction safety, job site configuration, progress control, and quality management, thus improving industrial competitiveness.

### 4.3. Single Shot MultiBox Detector Deep Learning Model Training Outcomes

Three hundred twenty job site activity images, accounting for 80% of the data set, were selected as the training sample for the SS-based job site activity image recognition system proposed herein. In addition, 40 images, or 10% of the data set, were chosen as the test samples during the training. In the end, 61 images the model had not seen were used for recognition; thus, 461 images were collected and used. The visualization outcomes after recognition are presented in Table 5.

**Table 5.** Outcomes of single shot multibox detector image recognition model test.

| Originals1 | Outcomes1 |
| --- | --- |
|  |  |



Image Data Form 1

| Object Name | confidence level | Pixel Coordinates Ymin | Xmin | Ymax | Xmax | Image Number | Time Record |
| --- | --- | --- | --- | --- | --- | --- | --- |
| worker | 0.99 | 733 | 1872 | 1036 | 2223 | img/image_0229.jpg | 2023/3/20 09:15 |
| worker | 0.91 | 750 | 1043 | 992 | 1274 | img/image_0229.jpg | 2023/3/20 09:15 |
| worker | 0.68 | 781 | 566 | 924 | 765 | img/image_0229.jpg | 2023/3/20 09:15 |
| machine | 1.00 | 129 | 1730 | 778 | 2508 | img/image_0229.jpg | 2023/3/20 09:15 |
| machine | 0.95 | 379 | 707 | 852 | 1118 | img/image_0229.jpg | 2023/3/20 09:15 |
| machine | 0.90 | 393 | 1238 | 799 | 1705 | img/image_0229.jpg | 2023/3/20 09:15 |

| Originals 2 | Outcomes 2 |
| --- | --- |
|  |  |



Image Data Form 2

| Object Name | confidence level | Pixel Coordinates Ymin | Xmin | Ymax | Xmax | Image Number | Time Record |
| --- | --- | --- | --- | --- | --- | --- | --- |
| rebar | 0.89 | 992 | 257 | 1772 | 1320 | img/image_0158.jpg | 2023/3/20 09:15 |
| worker | 0.94 | 959 | 431 | 1163 | 626 | img/image_0158.jpg | 2023/3/20 09:15 |
| machine | 0.97 | 0 | 1749 | 1336 | 2505 | img/image_0158.jpg | 2023/3/20 09:15 |
| machine | 0.77 | 1049 | 1448 | 1280 | 1709 | img/image_0158.jpg | 2023/3/20 09:15 |

Automated generation of EXCEL forms for the recognized results included object names, confidence level, pixel coordinates, and time record. The timestamp was based on the computer time when the form was generated, which could be used as the basis for specific management items (Table 5):

1. Monitoring the operation status of construction site personnel and equipment: real-time monitoring of the operation status of construction site personnel and equipment, including entry and exit times, the number of construction personnel, and the number of equipment appearing at that time, thereby effectively improving construction safety and efficiency.
2. Ensuring the supply of construction site materials: effectively monitoring the entry and exit of construction site materials and inventory status, ensuring the timely use of materials, and ensuring the adequate and timely supply of materials on site.
3. Improving the efficiency of construction site management: automatically recording the entry and exit time, location, and other information of construction site personnel and equipment, reducing the cost and risk of manual management, and improving the efficiency and accuracy of site management.
4. Optimizing construction site scheduling: using image recognition technology to record construction logs and monitor the progress of various works at the construction site, adjusting the schedule promptly, improving construction efficiency, and reducing construction delays.

Construction activities at a job site vary widely. The machines subject to image recognition are excavators, loaders, dump trucks, cranes, and concrete mixer trucks, and the recognition accuracy is 69%, on average. The workers are wearing work clothing and reflective vests without a uniform standard, and they are at various locations within the job site performing various tasks, resulting in difficulties in recognition due to the bright side, dark side, and body position, and the recognition accuracy is 53%. The accuracy is 28% for the rebar. The reason for the low recognition accuracy could be that they are similar materials divided into two different classes; also, there are more than civil work activities at the job site; for example, there are plumbing and electrical tasks at a job site, and their materials, such as pipes and cables, may affect the recognition results, as shown in Table 6.

**Table 6.** Model performance indices.

| | mAP | Recall (Threshold = 0.5) | Precision (Threshold = 0.5) | F1-Score (Threshold = 0.5) |
|---|---|---|---|---|
| Rebar | 0.29 | 0.09 | 1.00 | 0.17 |
| Worker | 0.53 | 0.37 | 0.86 | 0.52 |
| Machine | 0.69 | 0.62 | 0.95 | 0.75 |

This study uses automatic identification of construction site workers, material locations, and construction environment conditions of equipment. The resulting photos can identify more than two items simultaneously, providing site supervisors with active warnings of potential occupational safety hazards and increasing construction efficiency through image automation.

## 5. Conclusions and Suggestions

A construction job site covers the building footprint, work area, or material storage. With the simultaneous recognition of objects, such as workers, machines, and materials using a single shot multibox detector (SSD) in this case, it was found that the recognition performed better for large machines, including excavators, cranes, dump trucks, and concrete mixer trucks, with recognition accuracy close to 70%. Recognition accuracy was 53% for workers, and rebar was the least accurately identified of the three.

This study used the single shot multibox detector model with the VGG-16 neural network as its backbone network and VGG-16 is a 16-layer convolutional neural network, including 13 convolutional layers and 3 fully connected layers. A total of 320 construction site construction images (80%) were trained, and the results could mark personnel, machinery, and materials simultaneously. The complexity of each on-site construction image was different; therefore, the time required for each image recognition was also different, but the average single image recognition time was 6 s. The object detection process encountered the following problems:

1. Regarding detection personnel: For construction personnel, posture changes, construction site brightness changes, and object occlusion these problems would lead to false detections.
2. Regarding detection materials: densely packed rebar would produce different degrees of joint and section difficulties; in addition, in the single target detection algorithm, the stacking between the background and the foreground was different, which may have led to a decrease in the sensitivity of the model to the sample. It resulted in false detections.
3. Detection of equipment: Construction equipment detection items included excavators, shovel loaders, dump trucks, cranes, concrete mixer trucks, etc. There were more data sets than construction personnel and materials, and their identification performance was better. But to enhance the training of another project may have led to further overfitting.

Based on the above, this study proposes future research directions regarding technology application, database construction, and algorithm optimization to enhance the accuracy and applicability of detection items:

1. The evolutionary many-objective optimization algorithm with new techniques, such as domain decomposition and multi-objective optimization decomposition can improve the efficiency and accuracy of construction site management and enhance image recognition in construction engineering [46].
2. Optimizing truck scheduling through algorithms can improve the efficiency and accuracy of material transportation and scheduling at construction sites, leading to intelligent and automated material transportation and ultimately enhancing construction efficiency and quality [47].
3. Multi-objective optimization algorithms can significantly enhance the efficiency and accuracy of construction sites management tasks, such as material transportation, equipment scheduling, and personnel management. Integrating image recognition applications with these algorithms enables the intelligent and automated monitoring and control of construction sites, improving construction efficiency and quality [48].
4. Image recognition technology can monitor the construction site in real time, detect potential risk factors, and determine the direction of improvement. At the same time, efficient dock scheduling algorithms can optimize construction materials and equipment logistics, reduce waiting time, and improving overall productivity [49].
5. The direction is to combine image recognition technology to monitor the safety of construction sites in real time, detecting potential safety hazards early, and using NSGA-II and MOPSO algorithms for ambulance routing to improve rescue efficiency and emergency response capabilities [50].
6. Applying the augmented self-adaptive parameter control method to a broader range of construction scenarios can improve construction efficiency and safety. Further research will explore combining the technique with other optimization algorithms to enhance its effectiveness and reduce construction costs [51].
7. To enhance the simultaneous detection of personnel, equipment, and materials, upcoming methods will include feature pyramid, complete intersection over union (Ciou) loss, focal loss, and bag of freebies target detection optimization [52].

Construction engineering is characterized by complexity; therefore, image recognition technology at construction sites enhances the safety and efficiency of construction site management. This technology enables more detailed identification and improvement of production efficiency and quality in the construction industry, thereby providing more significant development opportunities for the future of construction engineering.

**Author Contributions:** Conceptualization, Y.-R.W.; Methodology, L.-W.L.; Investigation, L.-W.L.; Resources, L.-W.L.; Writing—original draft, L.-W.L.; Writing—review & editing, L.-W.L.; Supervision, Y.-R.W.; Project administration, Y.-R.W.; Funding acquisition, Y.-R.W. All authors have read and agreed to the published version of the manuscript.

**Funding:** This research received no external funding.

**Data Availability Statement:** The Image in the research content is all owned by private companies, so they cannot be published.

**Conflicts of Interest:** The authors declare no conflict of interest.

## References


1. Yang, J.; Arif, O.; Vela, P.A.; Teizer, J.; Shi, Z. Tracking multiple workers on construction sites using video cameras. *Adv. Eng. Inform.* **2010**, *24*, 428–434. [CrossRef]
2. Riveiro, B.; Lourenço, P.B.; Oliveira, D.V.; González-Jorge, H.; Arias, P. Automatic morphologic analysis of quasi-periodic masonry walls from LiDAR. *Comput.-Aided Civ. Infrastruct. Eng.* **2016**, *31*, 305–319. [CrossRef]
3. Thakar, V.; Saini, H.; Ahmed, W.; Soltani, M.M.; Aly, A.; Yu, J.Y. Efficient Single-Shot Multi-Box Detector for Construction Site Monitoring. In Proceedings of the 2018 IEEE International Smart Cities Conference (ISC2), Kansas City, MO, USA, 16–19 September 2018; pp. 1–6.
4. Zhu, Z.; Ren, X.; Chen, Z. Visual tracking of construction jobsite workforce and equipment with particle filtering. *J. Comput. Civ. Eng.* **2016**, *30*, 04016023. [CrossRef]
5. Wang, Q.; Cheng, J.C.; Sohn, H. Automated estimation of reinforced precast concrete rebar positions using colored laser scan data. *Comput.-Aided Civ. Infrastruct. Eng.* **2017**, *32*, 787–802. [CrossRef]
6. Nimmo, J.; Green, R. Pedestrian avoidance in construction sites. In Proceedings of the 2017 International Conference on Image and Vision Computing New Zealand (IVCNZ), Christchurch, New Zealand, 4–6 December 2017; pp. 1–6.
7. Alizadehslehi, S.; Yitmen, I. A Concept for Automated Construction Progress Monitoring: Technologies Adoption for Benchmarking Project Performance Control. *Arab. J. Sci. Eng.* **2018**, *44*, 4993–5008. [CrossRef]
8. Fang, W.; Ding, L.; Luo, H.; Love, P.E. Falls from heights: A computer vision-based approach for safety harness detection. *Autom. Constr.* **2018**, *91*, 53–61. [CrossRef]
9. Mahami, H.; Nasirzadeh, F.; Ahmadabadian, A.H.; Esmaeili, F.; Nahavandi, S. Imaging network design to improve the automated construction progress monitoring process. *Constr. Innov.* **2019**, *19*, 386–404. [CrossRef]
10. Greeshma, A.S.; Edayadiyil, J.B. Automated progress monitoring of construction projects using Machine learning and image processing approach. *Mater. Today Proc.* **2022**, *65*, 554–563.
11. Del Savio, A.; Luna, A.; Cárdenas-Salas, D.; Vergara, M.; Urday, G. Dataset of manually classified images obtained from a construction site. *Data Brief* **2022**, *42*, 108042. [CrossRef]
12. Fang, Q.; Li, H.; Luo, X.; Ding, L.; Luo, H.; Rose, T.M.; An, W. Detecting non-hardhat-use by a deep learning method from far-field surveillance videos. *Autom. Constr.* **2018**, *85*, 1–9. [CrossRef]
13. Fang, W.; Ding, L.; Zhong, B.; Love, P.E.; Luo, H. Automated detection of workers and heavy equipment on construction sites: A convolutional neural network approach. *Adv. Eng. Inform. Rmatics* **2018**, *37*, 139–149. [CrossRef]
14. Kim, Y.; Choi, Y. Smart Helmet-Based Proximity Warning System to Improve Occupational Safety on the Road Using Image Sensor and Artificial Intelligence. *Int. J. Environ. Res. Public Health* **2022**, *19*, 16312. [CrossRef] [PubMed]
15. Buniya, M.K.; Othman, I.; Sunindijo, R.Y.; Kashwani, G.; Durdyev, S.; Ismail, S.; Antwi-Afari, M.F.; Li, H. Critical Success Factors of Safety Program Implementation in Construction Projects in Iraq. *Int. J. Environ. Res. Public Health* **2021**, *18*, 8469. [CrossRef] [PubMed]
16. Yeşilmen, S.; Tatar, B. Efficiency of convolutional neural networks (CNN) based image classification for monitoring construction related activities: A case study on aggregate mining for concrete production. *Case Stud. Constr. Mater.* **2022**, *17*, e01372. [CrossRef]
17. Lee, H.; Grosse, R.; Ranganath, R.; Ng, A.Y. Convolutional deep belief networks for calable unsupervised learning of hierarchical representations. In Proceedings of the 26th Annual International Conference on Machine Learning, Montreal, QC, Canada, 14–18 June 2009; pp. 609–616.
18. Makantasis, K.; Protopapadakis, E.; Doulamis, A.; Doulamis, N.; Loupos, C. Deep convolutional neural networks for efficient vision based tunnel inspection. In Proceedings of the 2015 IEEE International Conference on Intelligent Computer Communication and Processing (ICCP), Cluj-Napoca, Romania, 3–5 September 2015; pp. 335–342.
19. Pan, Y.; Zhang, L. Roles of artificial intelligence in construction engineering and management: A critical review and future trends. *Autom. Constr.* **2021**, *122*, 10357. [CrossRef]

20. Yan, X.; Li, H.; Wang, C.; Seo, J.; Zhang, H.; Wang, H. Development of ergonomic posture recognition technique based on 2D ordinary camera for construction hazard prevention through view-invariant features in 2D skeleton motion. *Adv. Eng. Inform.* **2017**, *34*, 152–163. [CrossRef]
21. Sepas-Moghaddam, A.; Etemad, A. Deep Gait Recognition: A Survey. *IEEE Trans. Pattern Anal. Mach. Intell.* **2022**, *45*, 264–284. [CrossRef]
22. Lin, C.-L.; Fan, K.-C.; Lai, C.-R.; Cheng, H.-Y.; Chen, T.-P.; Hung, C.-M. Applying a Deep Learning Neural Network to Gait-Based Pedestrian Automatic Detection and Recognition. *Appl. Sci.* **2022**, *12*, 4326. [CrossRef]
23. Arabi, S.; Haghighat, A.K.; Sharma, A. A deep-learning-based computer vision solution for construction vehicle detection. *Comput.-Aided Civ. Infrastruct. Eng.* **2020**, *35*, 753–767. [CrossRef]
24. Chou, J.-S.; Liu, C.-H. Automated Sensing System for Real-Time Recognition of Trucks in River Dredging Areas Using Computer Vision and Convolutional Deep Learning. *Sensors* **2021**, *21*, 555. [CrossRef]
25. Li, Y.; Lu, Y.J.; Chen, J. A deep learning approach for real-time rebar counting on the construction site based on YOLOv3 detector. *Autom. Constr.* **2021**, *124*, 103602. [CrossRef]
26. Cha, Y.J.; Choi, W.; Büyüköztürk, O. Deep learning-based crack damage detection using convolutional neural networks. *Comput.-Aided Civ. Infrastruct. Eng.* **2017**, *32*, 361–378. [CrossRef]
27. Chang, C.W.; Lin, C.H.; Lien, H.S. Measurement radius of reinforcing steel bar in concrete using digital image GPR. *Constr. Build. Mater.* **2009**, *23*, 1057–1063. [CrossRef]
28. CS231n Convolutional Neural Networks for Visual Recognition, Stanford. 2016. Available online: http://cs231n.stanford.edu/ (accessed on 27 March 2023).
29. Martinez, P.; Al-Hussein, M.; Ahmad, R. A scientrometic analysis and critical review of computer vision applications for construction. *Autom. Constr.* **2019**, *107*, 102947. [CrossRef]
30. Sermanet, P.; Eigen, D.; Zhang, X.; Mathieu, M.; Fergus, R.; LeCun, Y. Overfeat: Integrated recognition, localization and detection using convolutional networks. *arXiv* **2013**, arXiv:1312.6229.
31. Girshick, R.; Donahue, J.; Darrell, T.; Malik, J. Rich feature hierarchies for accurate object detection and semantic segmentation. In Proceedings of the IEEE Conference on Computer Vision and Pattern Recognition, Columbus, OH, USA, 23–28 June 2014; pp. 580–587.
32. Girshick, R.; Donahue, J.; Darrell, T.; Malik, J. Region-based convolutional networks for accurate object detection and segmentation. *IEEE Trans. Pattern Anal. Mach. Intell.* **2015**, *38*, 142–158. [CrossRef]
33. Girshick, R. Fast r-cnn. In Proceedings of the IEEE International Conference on Computer Vision, Santiago, Chile, 11–18 December 2015; pp. 1440–1448.
34. Ren, S.; He, K.; Girshick, R.; Sun, J. Faster r-cnn: Towards real-time object detection with region proposal networks. *Adv. Neural Inf. Process. Syst.* **2015**, *28*, 91–99. [CrossRef]
35. Redmon, J.; Divvala, S.; Girshick, R.; Farhadi, A. You only look once: Unified, real-time object detection. In Proceedings of the IEEE Conference on Computer Vision and Pattern Recognition, Las Vegas, NV, USA, 27–30 June 2016; pp. 779–788.
36. Liu, W.; Anguelov, D.; Erhan, D.; Szegedy, C.; Reed, S.; Fu, C.Y.; Berg, A.C. Ssd: Single shot Multi-box detector. In *European Conference on Computer Vision*; Springer: Cham, Switzerland, 2016; pp. 21–37.
37. Yudin, D.; Slavioglo, D. Usage of fully convolutional network with clustering for traffic light detection. In Proceedings of the 2018 7th Mediterranean Conference on Embedded Computing (MECO), Budva, Montenegro, 10–14 June 2018; pp. 1–6.
38. Wang, Y.; Wang, C.; Zhang, H. Combining a single shot Multi-box detector with transfer learning for ship detection using sentinel-1 SAR images. *Remote Sens. Lett.* **2018**, *9*, 780–788. [CrossRef]
39. Deshpande, A. A Beginner's Guide to Understanding Convolutional Neural Networks. Retrieved March 2017; Volume 31. Available online: https://adeshpande3.github.io/A-Beginner\T1\textquoterights-Guide-To-Understanding-Convolutional-Neural-Networks/ (accessed on 27 March 2023).
40. Dorafshan, S.; Thomas, R.J.; Maguire, M. Comparison of deep convolutional neural networks and edge detectors for image-based crack detection in concrete. *Constr. Build. Mater.* **2018**, *186*, 1031–1045. [CrossRef]
41. Spencer, B.F., Jr.; Hoskere, V.; Narazaki, Y. Advances in computer vision-based civil infrastructure inspection and monitoring. *Engineering* **2019**, *5*, 199–222. [CrossRef]
42. Dung, C.V. Autonomous concrete crack detection using deep fully convolutional neural network. *Autom. Constr.* **2019**, *99*, 52–58. [CrossRef]
43. Fang, W.; Love, P.E.; Luo, H.; Ding, L. Computer vision for behaviour-based safety in construction: A review and future directions. *Adv. Eng. Inform.* **2020**, *43*, 100980. [CrossRef]
44. Li, X.; Chi, H.L.; Lu, W.; Xue, F.; Zeng, J.; Li, C.Z. Federated transfer learning enabled smart work packaging for preserving personal image information of construction worker. *Autom. Constr.* **2021**, *128*, 103738. [CrossRef]
45. Del Savio, A.A.; Luna, A.; Cárdenas-Salas, D.; Vergara Olivera, M.; Urday Ibarra, G. The use of artificial intelligence to identify objects in a construction site. In Proceedings of the International Conference on Artificial Intelligence and Energy System (ICAIES) in Virtual Mode, Jaipur, India, 12–13 June 2021.
46. Zhao, H.; Zhang, C. An online-learning-based evolutionary many-objective algorithm. *Inf. Sci.* **2020**, *509*, 1–21. [CrossRef]
47. Dulebenets, M.A. An Adaptive Polyploid Memetic Algorithm for scheduling trucks at a cross-docking terminal. *Inf. Sci.* **2021**, *565*, 390–421. [CrossRef]

48. Pasha, J.; Nwodu, A.L.; Fathollahi-Fard, A.M.; Tian, G.; Li, Z.; Wang, H.; Dulebenets, M.A. Exact and metaheuristic algorithms for the vehicle routing problem with a factory-in-a-box in multi-objective settings. *Adv. Eng. Inform.* **2022**, *52*, 101623. [CrossRef]
49. Dulebenets, M.A. A novel memetic algorithm with a deterministic parameter control for efficient berth scheduling at marine container terminals. *Marit. Bus. Rev.* **2017**, *2*, 302–330. [CrossRef]
50. Rabbani, M.; Oladzad-Abbasabady, N.; Akbarian-Saravi, N. Ambulance routing in disaster response considering variable patient condition: NSGA-II and MOPSO algorithms. *J. Ind. Manag. Optim.* **2022**, *18*, 1035. [CrossRef]
51. Kavoosi, M.; Dulebenets, M.A.; Abioye, O.F.; Pasha, J.; Wang, H.; Chi, H. An augmented self-adaptive parameter control in evolutionary computation: A case study for the berth scheduling problem. *Adv. Eng. Inform.* **2019**, *42*, 100972. [CrossRef]
52. Tan, M.; Pang, R.; Le, Q.V. Efficientdet: Scalable and efficient object detection. In Proceedings of the IEEE/CVF Conference on Computer Vision and Pattern Recognition, Seattle, WA, USA, 13–19 June 2020; pp. 10781–10790.

MDPI
St. Alban-Anlage 66
4052 Basel
Switzerland
Tel. +41 61 683 77 34
Fax +41 61 302 89 18
www.mdpi.com

*Buildings* Editorial Office
E-mail: buildings@mdpi.com
www.mdpi.com/journal/buildings

www.ingramcontent.com/pod-product-compliance
Lightning Source LLC
Chambersburg PA
CBHW040839250926
51850CB00007BA/1356